Books a

Organic Indoor Air Pollutants

Edited by
Tunga Salthammer

WILEY-VCH

Organic Indoor Air Pollutants

Occurrence – Measurement – Evaluation

Edited by
Tunga Salthammer

WILEY-VCH

Weinheim · Chichester · New York · Toronto · Brisbane · Singapore

Dr. Tunga Salthammer
Wilhelm-Klauditz-Institut (WKI)
Fraunhofer-Institut für Holzforschung
Bienroder Weg 54 E
D-38108 Braunschweig

This book was carefully produced. Nevertheless, authors, editor and publisher do not warrant the information contained therein to be free of errors. Readers are advised to keep in mind that statements, data, illustrations, procedural details or other items may inadvertently be inaccurate.

Libary of Congress Card No: applied for

British Libary Cataloguing-in-Publication Data: A catalogne record for this book is available from the British Library

Die Deutsche Bibliothek – CIP-Einheitsaufnahme

Organic indoor air pollutants : occurence - measurement - evalution / ed. by Tunga Salthammer. - Weinheim ; New York ; Chichester ; Brisbane ; Singapore ; Toronto : Wiley-VCH, 1999
ISBN 3- 527-29622-0

Printed on acid-free and chlorine-free paper.

Composition: Mitterweger & Partner GmbH, D-68723 Plankstadt
Printing: Strauß-Offsetdruck, D-69509 Mörlenbach
Bookbinding: Buchbinderei J. Schäffer, D-67269 Grünstadt
Cover Design: Wolfgang Scheffler, D-55128 Mainz

Printed in the Federal Republic of Germany.

Preface

Air pollution caused by human beings goes hand in hand with civilization. Before industrialization, pollution of the air one breathes was mainly due to the use of open fireplaces, which emitted smoke, CO, CO_2 and gaseous organic combustion products, and the formation of urban structures then gradually led to a significant deterioration in the outdoor air quality in centres of population. The birth of the industrial era led to important sources of emission of polluting substances in the indoor and outdoor air, and reports of diseases caused by air pollution began to increase in number. Today, most people living in modern industrial regions, particularly during the colder months, spend 90 % of their time in closed rooms with only short airing periods.

Heat insulating measures significantly lowered the energy consumption in dwellings, office-blocks and state-maintained buildings, but at the same time the fresh air supply was drastically decreased, leading to an increased accumulation of emitted substances, sometimes exceeding the values measured in outdoor air. Moreover, in the sixties and seventies, materials containing unhealthy substances were used uncritically, creating sources of long-term emission and, as a consequence, a clean-up requirement in many buildings. Well-known examples are formaldehyde (emissions from wood-based materials), pentachlorophenol and its water-soluble sodium salt (wood preservatives) and polychlorinated biphenyls (insulating materials). In addition, many health problems are caused by biological particles such as fungi, dander from pets, and bacteria.

All the above-mentioned factors have led to numerous investigations over approximately 30 years into the deterioration of indoor hygiene caused by foreign substances introduced into interiors. Scientific findings and reports are closely linked to the development of air-analytical measuring methods in trace quantities. Nevertheless, the subject of "indoor air" is a relatively young branch of chemical and medical science. So far, only a few well-spread specialist articles and bookshave appeared, so that the most important sources of information for scientists are the proceedings of the great indoor air conferences such as "Indoor Air" and "Healthy Buildings". In contrast, there is an avalanche of articles and books at the popular-science level.

The principal aim of the present book is to give a summarizing and concentrated presentation of the subject "indoor air" from a chemical-analytical point of view. A treatise on all methods would have been too ambitious and would have been at the expense of the details which make this book of real value to the reader. It therefore deals exclusively and in detail with the detection and determination of volatile organic compounds (VOC), semi-volatile compounds (SVOC) and biological contaminants. Other components such as dust, fibers and inorganic agents are not taken into consideration. It is addressed to chemists, physicists, biologists, doctors and medical students at universities, in industry and in environmental laboratories who deal with the impact of emissions and immissions of organic compounds occurring in research and industrial practice. Students of these sciences would be able to use the book to supplement their lectures and obtain an insight into the problems and methods. During the last five years especially, the subject of VOC and SVOC has made tremendous progress owing to improved methods of analysis, new types of emission chambers and uniform test concepts. Great emphasis has been placed in this book on an up-to-date list of references, drawing the reader's attention to important sources not normally easy to find. Each contribution has been written by an experienced author with profound scientific knowledge acquired through research and teaching.

I would like to thank all authors for their engagement and cooperation. The realization of this book would not have been possible without the support of experienced colleagues, who agreed to prepare high quality contributions in parallel to performing their other daily duties. Special thanks are due to my colleague and friend Dr. Michael Wensing, who was enthusiastic for the project from the first moment. I am equally indebted to Mrs. Emilie Massold, who carefully read all manuscripts and offered valuable comments. Last but not least I am grateful to Dr. Steffen Pauly from Wiley-VCH for his interest in the subject and for his editorial work.

Braunschweig, June 1999 Tunga Salthammer

Contents

1. Measuring organic indoor pollutants

1.1 Application of solid sorbents for the sampling of volatile organic compounds in indoor air 3
E. Uhde, Germany

1.2 Sampling and analysis of aldehydes, phenols and diisocyanates 15
M. Schulz and T. Salthammer, Germany

1.3 Sampling and analysis of wood preservatives in test chambers 31
O. Jann and O. Wilke, Germany

1.4 Sampling and analysis of PCDDs/PCDFs, PAHs and PCBs 45
M. Ball and T. Salthammer, Germany

1.5 Application of diffusive samplers 57
D. Crump, UK

1.6 Real-time monitoring of organic compounds 73
L.E. Ekberg, Sweden

1.7 Assessment methods for bioaerosols 85
P.S. Thorne, USA and D. Heederik, The Netherlands

1.8 Standard test methods for the determination of VOCs and SVOCs in automobile interiors 105
H. Bauhof and M. Wensing, Germany

1.9 Nomenclature and occurrence of glycols and their derivatives in indoor air 117
P. Stolz, N. Weis and J. Krooss, Germany

2. Environmental test chambers and cells

2.1 Environmental test chambers 129
M. Wensing, Germany

2.2 The field and laboratory emission cell – FLEC 143
H. Gustafsson, Sweden

2.3 Mathematical modeling of test chamber kinetics 153
S. Kephalopoulos, Italy

3. Release of organic compounds from indoor materials

3.1 Occurring of volatile organic compounds in indoor air 171
S.K. Brown, Australia
3.2 Emission from floor coverings 185
K. Saarela, Finland
3.3 Indoor air pollution by release of VOCs from wood-based furniture 203
T. Salthammer, Germany
3.4 Volatile organic ingredients of household and consumer products 219
T. Salthammer, Germany
3.5 Occurrence of biocides in the indoor environment 233
W. Butte, Germany
3.6 Secondary emission 251
L. Gunnarsen and U.D. Kjaer, Denmark
3.7 Release of MVOCs from microorganisms 259
J. Bjurman, Sweden
3.8 Indoor bioaerosols – sources and characteristics 275
P.S. Thorne, USA and D. Heederik, The Netherlands

4. Investigation concepts and quality guidelines for organic air pollutants

4.1 Indoor air quality guidelines 291
P. Pluschke, Germany
4.2 The TVOC concept 305
L. Mølhave, Denmark

List of Contributors

Dr. Michael Ball
ERGO Forschungsgesellschaft mbH
Geierstraße 1
D-22305 Hamburg
Germany

Dr. Helmuth Bauhof
TÜV Nord e.V.
Große Bahnstraße 31
D-22525 Hamburg
Germany

Prof. Dr. Jonny Bjurman
Swedish University of Agricultural Sciences
Department of Forest Products
Box 7008
S-75007 Uppsala
Sweden

Stephen K. Brown
Division of Building, Construction and Engineering (CSIRO)
P.O. Box 56
Highett, Victoria 3190
Australia

Prof. Dr. Werner Butte
Carl-von-Ossietzky Universität
Fachbereich Chemie
P.O. Box 2503
D-26111 Oldenburg
Germany

Dr. Derrick Crump
Building Research Establishment Ltd.
Organic Materials Division
Building 4
Garston, Bucknalls Lane
Watford WD2 7JR
U. K.

Dr. Lars E. Ekberg
Department of Building Services Engineering
Chalmers University of Technology
S-41296 Göteborg
Sweden

Dr. Lars Gunnarsen
Danish Building Research Institute (SBI)
Dr. Neergaards Vej 15
DK-2970 Hoersholm
Denmark

Hans Gustafsson
Swedish National Testing and Research Institute, SP
Box 857
S-50115 Borås
Sweden

Dr. Dick Heederik
Department of Environmental Sciences
Environmental and Occupational
Health Unit
Wageningen Agricultural University
Ritzema Bosweg 32A
P.O. Box 238
NL-6700 AE Wageningen
Netherlands

Dr. Oliver Jann
Bundesanstalt für Materialforschung
und -prüfung
Laboratory IV.22
D-12200 Berlin
Germany

Dr. Stelios D. Kephalopoulos
Joint Research Centre
Environment Institute – Air Quality Unit
I-21020 Ispra (VA)
Italy

Dr. Ulla D. Kjaer
Danish Building Research Institute (SBI)
Dr. Neergaards Vej 15
DK-2970 Hoersholm
Denmark

Dipl.-Biol. Jürgen Krooss
Bremer Umwelt Institut (BUI)
Wielandstraße 25
D-28203 Bremen
Germany

Prof. Dr. Lars Mølhave
Department of Environmental &
Occupational Medicine
Aarhus University
Vennelyst Boulevard 6
DK-8000 Århus
Denmark

Dr. Peter Pluschke
Chemisches Untersuchungsamt der
Stadt Nürnberg
Adolf-Braun-Str. 55
D-90429 Nürnberg
Germany

Dr. Kristina Saarela
Technical Research Centre of Finland,
VTT
Chemical Technology
P.O. Box 1403
FIN-02044 Espoo
Finland

Dr. Tunga Salthammer
Wilhelm-Klauditz-Institut (WKI)
Fraunhofer-Institut für Holzforschung
Bienroder Weg 54 E
D-38108 Braunschweig
Germany

Dipl. Chem. Mark Schulz
Wilhelm-Klauditz-Institut (WKI)
Fraunhofer-Institut für Holzforschung
Bienroder Weg 54 E
D-38108 Braunschweig
Germany

Dr. Peter Stolz
Bremer Umwelt Institut (BUI)
Wielandstraße 25
D-28203 Bremen
Germany

Dr. Norbert Weis
Bremer Umwelt Institut (BUI)
Wielandstraße 25
D-28203 Bremen
Germany

Prof. Dr. Peter S. Thorne
Department of Preventive Medicine
and Environmental Health
The University of Iowa
100 Oakdale Campus, 124 IREH
Iowa City, IA 52242–5000
USA

Dr. Erik Uhde
Wilhelm-Klauditz-Institut (WKI)
Fraunhofer-Institut für Holzforschung
Bienroder Weg 54 E
D-38108 Braunschweig
Germany

Dr. Michael Wensing
Gesellschaft für Umweltschutz
TÜV Nord mbH
Große Bahnstraße 31
D-22525 Hamburg
Germany

Dipl. Chem. Olaf Wilke
Bundesanstalt für Materialforschung
und -prüfung
Laboratory IV.22
D-12200 Berlin
Germany

Common Synonyms and Abbreviations in Indoor Air Sciences

A	sample surface [m^2]
A_C	surface area of a test chamber [m^2]
ADI	acceptable daily intake
AM	arithmetic mean
APM	airborne particulate matter
AQG	air quality guidelines
ASHRAE	American Society of Heating, Refrigerating and Air-Conditioning Engineers
ASTM	American Society for Testing and Materials
BA	biological agents
BaP	benzo[a]pyrene
BHT	2,6–di-tert-butyl-4-methyl-phenol
BREC	building related environmental complaints
BRI	building related illness
BRS	building related symptoms
BTV	breakthrough volume [l/g]
C or $C(t)$	concentration in air [$\mu g/m^3$]
C_0	initial concentration [$\mu g/m^3$]
C_s	vapor pressure [mg/m^3] (expressed as concentration)
CEC	Commission of the European Communities
CEN	European Committee for Standardization
CIB	National Council for Building Research, Studies and Documentation
d	distance [mm, cm, m]
δ	boundary layer thickness [mm]
D	molecular diffusity [m^2/h]
DEHP	diethyl-hexyl-phthalate
DOP	dioctylphthalate
ECA	European Collaborative Action
ECD	electron capture detector
EP	emission profile

EPA	US Environmental Protection Agency
ER or *ER*(*t*)	emission rate [μg/h]
ER^0	initial emission rate [μg/h]
ETS	environmental tobacco smoke
FID	flame ionization detector
FLEC	field and laboratory emission cell
FT-IR	Fourier transform infrared spectroscopy
GC	gas chromatography
GM	geometric mean
GSD	geometric standard deviation
h	height [mm, cm, m]
HPLC	high performance liquid chromatography
HVAC	heating ventilating air conditioning system
IAP	indoor air pollution
IAQ	indoor air quality
ISIAQ	International Society of Indoor Air Quality and Climate
ISO	International Organization for Standardization
k_1	source strength [μg/h]
k_2	air exchange (modeling) [h^{-1}]
k_i (i>2)	rate constants [h^{-1}]
L	loading factor [m^2/m^3]
LOAEL	lowest observed adverse effect level
LOEL	lowest observed effect level
LR	leak rate (test chamber) [h^{-1}]
M	mass in source [mg/m^2]
M_0	initial mass in source [mg/m^2]
MCS	multiple chemical sensitivity
MD	median
MS	mass spectrometry
MVOC	microbiological originated VOC
N (or *n*)	air exchange rate [h^{-1}]
NDIR	non dispersive infra red
NOAEL	no observed adverse effect level
NOEL	no observed effect level
OEL	Occupational Exposure Limit
OSHA	Occupational Safety and Health Administration
OT	odour threshold [mg/m^3]
PAD	photoacoustic detector
PAH	polycyclic aromatic hydrocarbons

PAN	peroxyacetyl nitrate
PAS	photo acoustic spectroscopy
PCB	polychlorinated biphenyls
PCDD	polychlorinated dibenzo-p-dioxins
PCDF	polychlorinated dibenzofurans
PCP	pentachlorophenol
PID	photoionization detector
POM	particulate organic matter
PPN	peroxypropionyl nitrate
PUF	polyurethane foam
q	area specific flow rate (N/L) [$m^3/(m^2$ h)]
Q	air flow rate [ml/min, l/h]
QSAR	Quantitative Structure-Activity Relationship
r	correlation coefficient
RH	relative humidity [%]
RI	retention index
RSD	relative standard deviation
RSP	respirable particles
RT	retention time [min]
SBS	sick building syndrome
SD	solvent desorption
SER_A or $SER_A(t)$	area specific emission rate [µg/(m^2 h)]
SER_u or $SER_u(t)$	unit specific emission rate [µg/(unit h)]
SER_V or $SER_V(t)$	volume specific emission rate [µg/(m^3 h)]
SER_l or $SER_l(t)$	length specific emission rate [µg/(m h)]
SIM	single ion mode
ΣVOC	sum of VOC
SVOC	semi volatile organic compounds
t	time [s, h, d]
T	temperature [°C]
TD	thermal desorption
Texanol	2,2,4–trimethyl-1,3-pentanediol-monoisobutyrate
TIC	total ion chromatogram
TLV	threshold limit values
TVOC	total volatile organic compounds
TXIB	2,2,4-trimethyl-1,3-pentanediol-diisobutyrate
v	air velocity [m/s, cm/s]
V	volume [m^3, l, cm^3]
VOC	volatile organic compounds
VVOC	very volatile organic compounds
WAGM	weighted average geometric mean
WHO	World Health Organisation

1. Measuring organic indoor pollutants

1.1 Application of Solid Sorbents for the Sampling of Volatile Organic Compounds in Indoor Air

Erik Uhde

1.1.1 Introduction

The use of building materials, furniture, carpets and household products almost invariably gives rise to the presence of volatile organic compounds (VOCs) in indoor air. Several hundred different compounds with boiling points in the range 50–260 °C have been identified during the last decade. Since most air pollutants occur in low concentrations in the range 1–1000 µg/m^3, highly sensitive detection methods as well as efficient separation methods are needed to analyze air samples.

Continuously working analytical devices such as flame ionization detectors, photoacoustic detectors or ion trap mass spectrometers offer a high time resolution, but often lack the required sensitivity and selectivity. Therefore discontinuous techniques with a sample preconcentration step during or after the sample collection are still preferred, especially in the case of toxic substances where detection limits of less than 10 µg/m^3 are demanded. The sensitivity can easily be increased by a factor of 1000 to 100 000 if an appropriate air sample volume of 1–100 l is drawn through the trap where the organic ingredients are retained. In general there are three possibilities for enriching the components of an air sample:

- absorbing the compounds in a suitable liquid,
- condensing them at low temperatures (cryotrapping) or
- adsorbing them on a porous solid material.

Liquid absorption is a common technique for enriching compounds in reactive liquids such as solutions of DNPH (for aldehydes) or acetylacetone (for formaldehyde). For both solutions the procedure combines trapping and derivatization of the target compound. Another possibility is the use of dissolved alkalis or acids to trap certain substances by the formation of salts in the solution.

Cryotrapping is often used in combination with solid sorbents and therefore is less important as a stand-alone sampling method.

Solid sorbents play an important role for the determination of volatile organic compounds (VOCs) in indoor air. They overcome some serious disadvantages of liquid absorbents:

- The direct analysis of the absorption liquid normally will not give the desired sensitivity, so additional steps of concentration are required. These may lead to a loss of the more volatile compounds present in the sample. Moreover, the possibility of contamination, e.g. with phthalates, cannot be excluded if the samples get in contact with glassware.

- The injection of a solvent into the gas chromatograph makes it difficult to identify and quantify the compounds which are eluted close to the solvent retention time. Furthermore, impurities in the solvent may interfere with the sample compounds and may therefore make the quantification more difficult.
- The use of derivatization agents may increase the sensitivity, but also makes the method specific to a small range of compounds.
- Preconditioning of a solid phase is much easier to achieve than purification of a liquid phase. Therefore problems with blanks are reduced.
- The handling of solid sorbents, often used in prepacked tubes, is more convenient than the use of organic solvents to trap VOCs from the air.

If solid sorbents with a low affinity for water are chosen, they even overcome the main disadvantage of cryotrapping: water may get trapped in the device and may lead to serious analytical problems as well as mechanical problems due to the formation of ice in the trap.

Of the variety of solid sorbents presently available, Tenax, activated charcoal and Carbotrap are the most widely used ones. This is mostly due to the versatility they offer, especially for sampling of VOCs typically found in indoor air (C6 to C16).

1.1.2 Solid Sorbents – a Brief Overview

Three general types of solid sorbents are mainly used for trapping volatile organic compounds in air: inorganic sorbents, porous materials based on carbon and organic polymers. Table 1.1-1 shows some properties of common sorbents.

The surface area of a sorbent has an impact on the amount of a given substance that can be withheld by the medium, and the surface polarity determines the type of compounds a sorbent can be used for. That way the sorbents offer a different suitability for VOC analysis depending on the type and amount of substance to be sampled. The inorganic sorbents are used to trap hydrocarbons and PCBs.

Carbon-based sorbents with a large surface area are useful to trap very low-boiling compounds, whereas it gets increasingly difficult to desorb substances with higher boiling points. Reactive compounds may even decompose at active sites on the sorbent surface.

Porous polymers with a comparatively small surface area allow the adsorption and desorption of high-boiling compounds such as glycols and phthalates and also the trapping of reactive substances such as aldehydes or acrylates. On the other hand it is difficult to sample low-boiling compounds like C2-C5 alkanes.

A special case of porous polymer sorbent is polyurethane foam, which can be used to collect large air samples up to 100 m^3 (Ligocki and Pankow, 1985). A major application of PU foams is the trapping of high boiling organopesticides such as lindane or permethrine.

The different characteristics of the presented sorbents show the need to carefully choose the right adsorption medium for a given VOC mixture.

Table 1.1-1. Properties of some solid sorbents according to Quintana et al. (1992), ECA (1994) and Figge et al. (1987).

Type	Structure	Surface area (m^2/g)	Products	Desorption	Compounds tested (starting at b.p.)	Polarity	Thermal stability	Water affinity
Inorganic	Silica gel	1–30	Volasphere	Solvent	PCBs, pesticides	High	~ 400 °C	
	Molecular sieve			Solvent	Permanent gases	High	< 400 °C	
	Aluminum oxide	~300	Alumina F1	Solvent	Hydrocarbons	High	300 °C	
Carbon based	Activated charcoal	800–1200		Solvent	Non-polar and slightly polar VOCs (> 50 °C)	Medium	> 400 °C	High
	Carbon molecular sieves	400–1000	Carbosieve, Ambersorb	Solvent/ Thermal	Non-polar and slightly polar VOCs (>−80 °C)	Low	> 400 °C	Low
	Graphitized carbon blacks	12–100	Carbotrap	Thermal	Non-polar VOCs (> 60 °C)	Low	> 400 °C	Low
Porous polymers	Styrene polymers	300–800	Porapak, XAD, Chromosorb,	Thermal/ Solvent	Non-polar and moderately polar VOCs (> 40 °C)	Variable	< 250 °C	Low
	Phenyl-phenylene oxide polymers	20–35	Tenax	Thermal	Non-polar VOCs (> 60 °C)	Low	< 375 °C	Low
	PU foams			Solvent	Pesticides	Low	–	Low

Several workers examined the suitability of the available sorbents for a general VOC analysis. Figge et al. (1987) tested the retention volumes of 26 different sorbents using a VOC mixture of 29 compounds with boiling points from 21 °C to 361 °C. They rated the sorbents in four groups with decreasing sorption strength. Some of the carbon-based sorbents such as Carbosieve SII showed the overall highest ability to retain organic compounds. Adsorbents with good retention properties for higher boiling compounds regardless of their polarity were e.g. Porapak Q, Chromosorb 106 and XAD-4. The third group included Porapak S, Tenax GC and Carbopack B, which showed good retention properties only for higher boiling, non-polar VOCs. For PTFE, Chromosorb T and other weak sorbents only poor retention volumes were found.

Brown (1996) reported the results of a multilaboratory study concerning the suitability of different sorbents for the measurement of VOCs in the workplace environment. In this study 20 test compounds were used and the sorbents had to fulfill a number of test criteria. Chromosorb 106 was found to be the most versatile sorbent and especially useful for the sampling of very volatile and polar compounds. Other sorbents which satisfied the acceptance criteria were Carbotrap, Tenax TA, Tenax GR and Carbopack B.

Rothweiler et al. (1991) and De Bortoli et al. (1992) compared two widely used adsorbents, Tenax TA and Carbotrap, regarding their performance for VOC sampling. Both authors found the two sorbents to be convenient for the sampling of nonpolar organic compounds regarding background emission and analyte recovery. They both agree in reporting significant analyte losses when using Carbotrap with more reactive compound, such as aldehydes or acrylates.

Measurements of the thermal desorption efficiencies of different sorbents were done by Cao and Hewitt (1993). They noticed a loss of terpenes on Carbotrap.

A very comprehensive review article of Matisová and Skrabáková (1995) discusses the suitability of various carbon-based sorbents.

1.1.3 Active vs Passive Sampling

Two different sampling strategies for collecting air samples exist: actively drawing air through a sorbent-filled cartridge or tube or passively letting the compounds penetrate a well-defined sorbent bed simply through gradient-driven diffusion.

The active sampling process is obviously a faster way to take an air sample and is, especially in cases where a high sensitivity and therefore a large sample volume is needed (e.g. sampling of pesticides), the recommended method. The accuracy of the sample is clearly determined by the sample volume, and measuring the correct air volume is certainly one of the most important aspects when using active sampling. Calibrated pumps, mass flow controllers or pumps in combination with appropriate air meters are commonly used devices to control the sample volume. If the collected sample volume is small and sampling time short, active sampling allows the measuring of dynamic processes with a higher sensitivity than continuously working analytical devices (FIDs, PADs). For this reason it is up to now still the preferred method for studying kinetic experiments in emission test chambers.

The use of a passive sampler is characterized by long sampling times, which are needed to allow the airborne compounds to enter the sampler and get trapped on the sorbents surface. Sampling times often exceed several days. The long sampling times show a pitfall of this sampling technique: due to the long time the sampler is in contact with the air to be sampled, very volatile compounds may not only enter the sampler, but may also have the chance to leave it on the same path because of low interaction with the sorbent. In this case an underestimation of the very volatile substances can be expected.

Passive samplers are not useful to monitor peak concentrations; on the other hand they offer a convenient way to assess e.g. long term exposure of persons in an indoor environment. A more detailed description of the passive sampling technique is given in Sec. 1.5.

The decision whether active or passive sampling should be used strongly depends on the type of experiment to be carried out: for measuring the mean concentration over a long period of time in a given environment, passive sampling often is the easier way. Lewis et al. (1985) even managed to collect an analytically sufficient sample with a special Tenax-based passive sampler within only 1 h. For the determination of indoor air concentrations during fast, dynamic processes active sampling normally is preferred.

1.1.4 Thermal Desorption vs Solvent Extraction

After collection on a sorbent, the VOCs from an air sample have to be transfered into the analytic device, which often is a gas chromatograph or an HPLC system.

Depending on the activity of the sorbent and the characteristics of the analyte there are two general possibilities of removing the sample from the sorbent. One of these is solvent extraction, which is commonly used for inorganic sorbents as well as for PU foams and the carbon-based sorbents with a high surface activity. For the extraction a suitable organic solvent is selected. Dimethylformamide, carbon disulfide (preferred because of its low response on a flame ionization detector), dichloromethane or other polar solvents, often combined with a desorbing agent such as methanol or water, are used to recover the trapped substances from the adsorbent. The solvent as well as the desorbing agent need to be chosen carefully; ideally they should not interfere with any of the sample peaks (therefore their purity should be high and their peak on the used separation column should be narrow), they should not react with the sample or the sorbent, their affinity to the sorbent should be high enough to remove the sample compound completely and their response on the used detector should be as low as possible.

Thermal desorption, on the other hand, makes use of the fact that the ability of a sorbent to retain compounds dramatically decreases at elevated temperatures. Therefore heating of the sorbent under a continuous stream of an inert carrier gas can be used to transfer the adsorbed compounds into the GC system. In general the desorption temperature should be at least 20 °C above the boiling point of the adsorbed compounds, so the range of compounds which can be analyzed with thermal desorption is limited by the thermal stability of the sorbent. In order to avoid a loss of analyte due to decomposition on the sorbents surface at elevated temperatures the surface activity needs to be carefully selected.

The main advantages of thermal desorption are the higher sensitivity and the absence of a solvent peak, which could interfere with a detector (e.g. MSD) or mask potential analyte peaks.

Modern thermal desorption units allow the use of multibed tubes, where different sorbents of increasing surface area and activity are combined. They can be used to trap substances within a wide boiling range. High-boiling compounds get trapped in the first zone of the adsorbent bed (often a porous polymer), and the more volatile substances pass through this zone and are collected on a sorbent with greater activity. Desorption must occur in the reverse direction (backflush) to prevent the high-boiling compounds from getting in contact with the more active sorbent.

1.1.5 Breakthrough Volumes

Every sorbent has a limited capacity for a given analyte, and this depends on the characteristics of the sorbent, on the type of compound to be trapped and on certain sampling parameters such as temperature and humidity of the air (Bertoni et al.).

Breakthrough of a substance can occur if

- the sampling speed is too high and the compounds are flushed through the tube without enough time to interact sufficiently with the sorbent surface
- the concentration is far to high, the sorbent surface gets saturated with the compound and the excess passes the sorbent without adequate retention
- the retention ability for the given amount of substance is not sufficient, so that the compound is retained but eluted again in the ongoing sampling procedure.

While the first two cases can easily be avoided by either using a low sampling rate (100 ml/min for Tenax tubes and 1–2 l/min for activated charcoal tubes is normally a safe sampling speed) or by using small enough sample volumes (Tenax: 1–8 l, activated charcoal: 5–100 l). The third case can only be overcome by selecting another sorbent or carefully choosing the right sampling temperature and a small sampling volume.

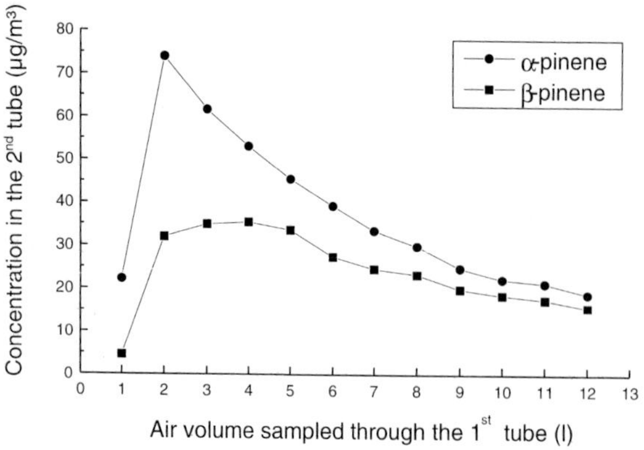

Figure 1.1-1. Breakthrough of volatile substances on Tenax TA. Terpene concentrations in the backup tubes during sampling of 12 l of air through a Tenax tube (Uhde, 1999).

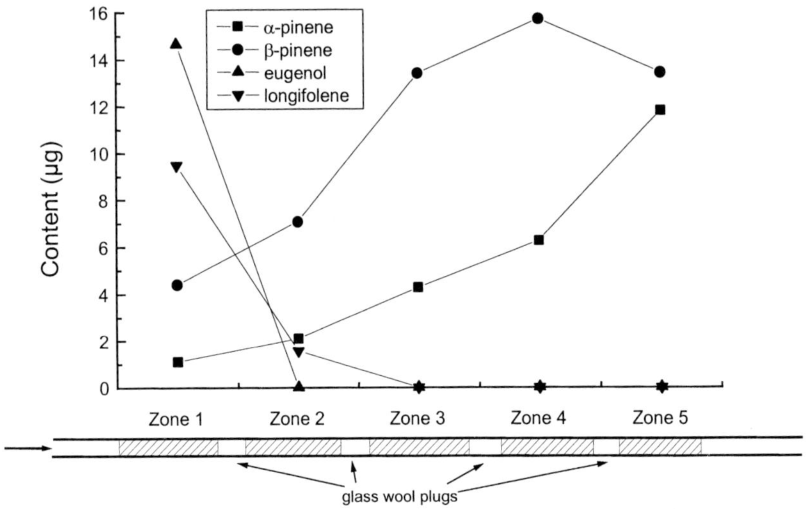

Figure 1.1-2. Distribution of volatile compounds on an adsorbent tube after spiking and sampling 2 l of air. The sampling direction is marked with an *arrow* (Uhde, 1999).

Breakthrough volumes can be determined by a direct method (monitoring the effluent during sampling air with known concentration of a test compound) or by an indirect method in which the sorbent tube is used as a gas-chromatographic column and the retention time of an injected compound is used to calculate the breakthrough volume (Brown and Purnell, 1979).

Figures 1.1-1 and 1.1-2 show the results of two experiments illustrating the breakthrough of volatile terpenes on Tenax TA. In the first case a Tenax-filled tube was spiked with a terpene mixture in methanol. A second tube was mounted behind it as a backup tube and 1 l of air was drawn through the assembly. The procedure was repeated until 12 l were drawn through the first tube, each litre on a separate backup tube. The concentrations of two low-boiling terpenes in the backup tubes is shown in Fig. 1.1-1. A significant breakthrough of both compounds is detectable after sampling 2 l. The overall breakthrough of α-pinene and β-pinene was 57 % and 39 %, respectively.

The second experiment visualizes the distribution of analytes in a special adsorbent tube filled with five equal amounts of Tenax TA, each separated by a silanized glass wool plug. After spiking with high concentrations of different terpenes and sampling 2 l of air the content of each Tenax zone was analyzed separately. Figure 1.1-2 shows the content in the different zones. The effect of chromatography-in-the-tube is clearly visible: the highest-boiling compound (longifolene) is almost completely retained in the first zone, the low-boiling compounds (α-pinene and β-pinene) have already partially left the tube. While the β-pinene "peak" is still located inside the tube (zone 4), the α-pinene concentration increases towards the end of the tube, which indicates that major amounts of the substance have already left the tube.

1.1.6 Water Affinity – a Chromatographic Problem

A high water content of a given sample is often a gas chromatographic problem. Ice can lead to a clogging in purge-and-trap samplers and other cryofocussing units. Water gives a high background noise level in GC/MS chromatograms and can thoroughly influence the signal of other detectors. It is also known to cause damage on fused silica columns due to a hydrolysis of silicon film material at elevated temperatures.

For these reasons the ideal solid sorbent should have a low affinity to water, a requirement easily met by the porous organic polymers. The carbon-based sorbents, especially activated charcoal, show a comparatively high uptake of water.

1.1.7 Degradation Products and Sorbent Background

Almost all of the above-described solid sorbents show a kind of background emission after a certain period of time. This is caused by thermal or photochemical degradation of the sorbent material. In addition to this, many of the sorbents react with adsorbed compounds or reactive gases such as O_2, O_3 and NO_x (Hanson et al., 1981; Neher and Jones, 1977; Clausen and Wolkoff, 1997) to form further products which appear in the sorbent background.

As long as the degradation products of a used sorbent are known and do not interfere with the collected sample compounds, the sorbent degradation itself is not an analytical problem. But often, at least if the compounds to be sampled are not known in advance, it is difficult to determine whether e.g. benzene is a blank of the used sorbent Tenax TA or a compound trapped from the air sample. Figure 1.1-3 shows a chromatogram of a glass sample tube filled with Tenax TA and stored in daylight for 2 weeks. It presents the important substances to be expected as Tenax blanks. Helmig (1996) discusses methods to control and avoid sorbent background peaks.

Other adsorbents show a specific background as well: Tirkkonen et al. (1995) present a compilation of compounds that were found under thermal desorption conditions for seven carbon-based and porous-polymer sorbents.

Figure 1.1-4 gives an example of the formation of a reaction product on the surface of a sorbent. The result of a solvent extraction of activated charcoal with CS_2 and CS_2/methanol, respectively, shows quite different artifact formation after a 48 h storage period. Four new compounds could be found in the sample containing methanol as a desorption agent. All of them could be identified as substances formed by the reaction of CS_2 and methanol in the presence of activated charcoal.

The blank problem becomes more severe when volatile organic compounds are sampled in the presence of reactive gases. Pellizzari et al. (1984) and Zielinska et al. (1986) showed the influence of different reactive inorganic gases on the decomposition of Tenax GC and showed possibilities to protect the adsorbent with mild reduction agents. Clausen and Wolkoff (1997) tried to utilize the amount of degradation products formed on Tenax as an indicator for the presence of reactive species in indoor air.

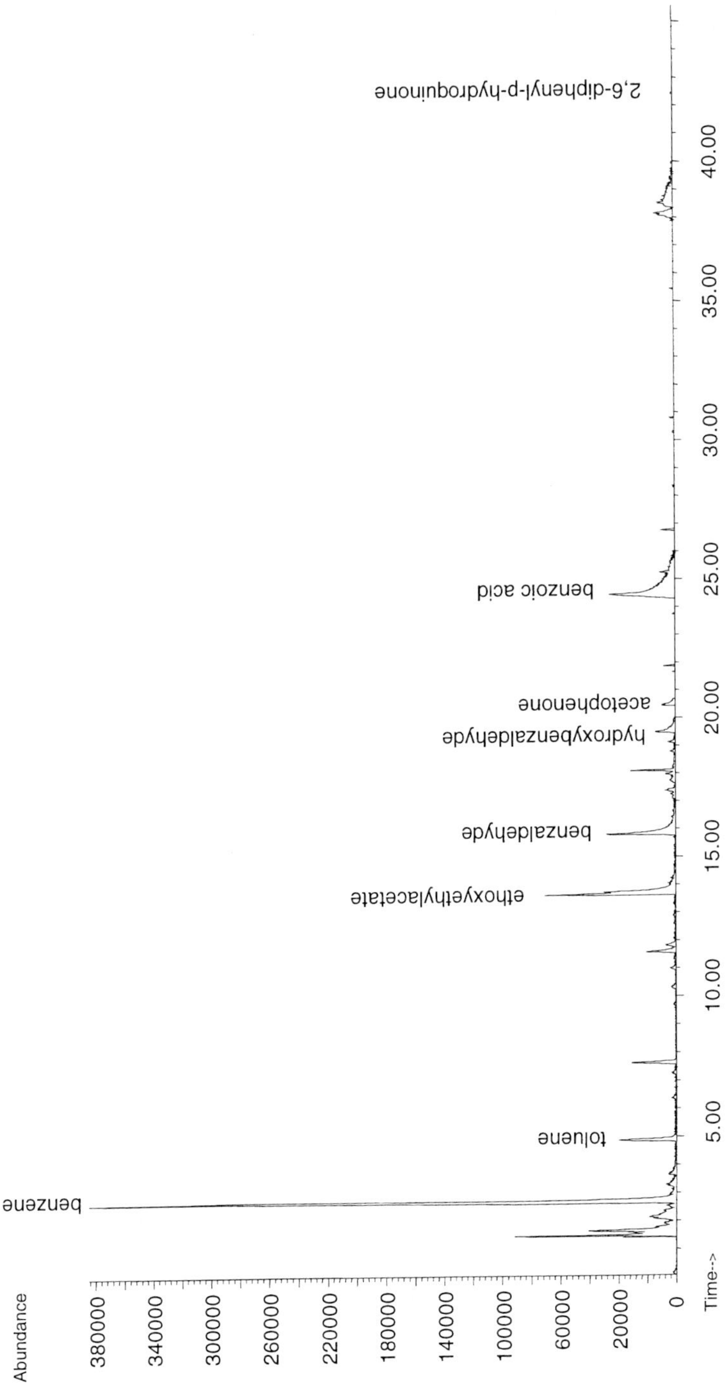

Figure 1.1-3. Background of Tenax TA after cleaning and a 2-week storage period. Ethoxyethyl acetate is emitted from the tube seals.

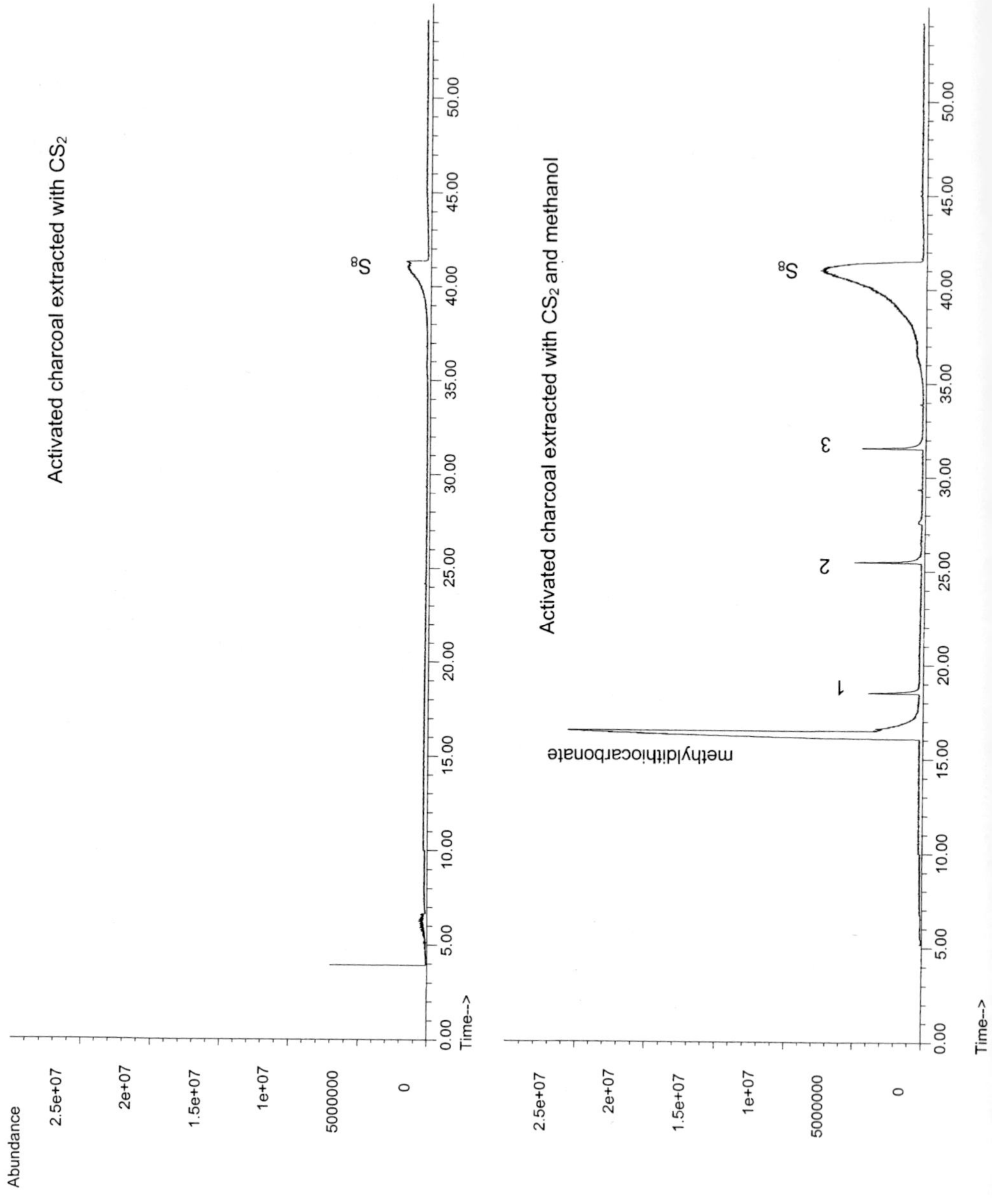

Figure 1.1-4. Background of activated charcoal after extraction with CS_2 and CS_2/methanol. 1,2, and 3 are condensation products of CS_2 and CH_3OH.

1.1.8 Conclusions

Solid sorbents offer a convenient means to enrich the contents of an air sample. The variety of sorbents currently available allows sampling of gaseous compounds from VVOC to SVOC. However, none of the existing sorbents is capable of retaining all compounds, so either a combination of adsorbents with different characteristics (multibed tubes) or a sorbent specially chosen for the actual analytical problem has to be used.

Although new techniques of analyzing VOCs such as PADs and sensor systems have been introduced in the last decade, the use of solid sorbents in combination with gas-chromatographic separation still seems to be the preferred method because of its sensitivity, its selectivity and, at least if thermal desorption is used to transfer the sample into the GC, its convenience.

References

Bertoni G., Bruner F., Liberti A. and Perrino C. (1981): Some critical parameters in collection, recovery and gas chromatographic analysis of organic pollutants in ambient air using light adsorbents. J. Chromatogr., *203*, 263–270.

Brown R. H. (1996): What is the best sorbent for pumped sampling-thermal desorption of volatile organic compounds? Experiences with the EC Sorbents Project. Analyst, *212*, 1171–1175.

Brown R. H. and Purnell C. J. (1979): Collection and analysis of trace organic vapour pollutants in ambient atmospheres. J. Chromatogr., *178*, 79–80.

Cao X.-L. and Hewitt C. N. (1993): Thermal desorption efficiencies for different adsorbate/adsorbent systems typically used in air monitoring programs. Chemosphere, *27*, No. 5, 695–705.

Clausen P. A., Wolkoff P. (1997): Degradation products of Tenax TA formed during sampling and thermal desorption analysis: indicators of reactive species indoors. Atmos. Environ., *31*, No. 5, 715–725.

De Bortoli M., Knöppel H., Pecchio E., Schauenburg H. and Vissers H. (1992): Comparison of Tenax and Carbotrap for VOC sampling in indoor air. Indoor Air, *2*, 216–224.

ECA (1994): Sampling strategies for volatile organic compounds (VOCs) in indoor air. Report No. 14, EUR 16051 EN Luxembourg: Office for Official Publications of the European Community.

Figge K., Rabel W. and Wieck A. (1987): Adsorptionsmittel zur Anreicherung von organischen Luftinhaltsstoffen. Fresenius Z. Anal. Chem., *327*, 261–278.

Hanson R. L., Clark C. R., Carpenter R. L. and Hobbs C. H. (1981): Evaluation of Tenax-GC and XAD-2 as polymer adsorbents for sampling fossil fuel combustion products containing nitrogen oxides. Envir. Sci. Tech., *15*, 6, 701–705.

Helmig D. (1996): Artefact-free preparation, storage and analysis of solid adsorbent sampling cartridges used in the analysis of volatile organic compounds in air. J. Chrom. A. *732*, 414–417.

Lewis R. G., Mulik J. D., Coutant R. W., Wooten G. W. and McMillin C. R. (1985): Thermally desorbable passive sampling device for volatile organic chemicals in ambient air. Anal. Chem., *57*, 214–219.

Ligocki M. P. and Pankow J. F. (1985): Assessment of adsorption/solvent extraction with polyurethane foam and adsorption/thermal desorption with Tenax GC for the collection and analysis of ambient organic vapors. Anal. Chem., *57*, 1138–1144.

Matisová E. and Skrabáková S. (1995): Carbon sorbents and their utilization for the preconcentration of organic pollutants in environmental samples. J. Chromatogr., *A707*, 145–179.

Neher M. B. and Jones P. W. (1977): In situ decomposition product isolated from Tenax GC while sampling stack gases. Anal. Chem., *49*, 3, 512–513.

Pellizzari E., Demian B. and Krost K. (1984): Sampling of organic compounds in the presence of reactive inorganic gases with Tenax GC. Anal. Chem., *56*, 793–798.

Quintana M., Uribe B. and Lopez Arbeloa J. (1992): Sorbents for active sampling. In: Brown R., Curtis M., Saunders, K. and Vanderschiesche S. (eds) Clean air at work – new trends in assessment and measurement for the 1990s. The Royal Society of Chemistry, Cambridge, 124–134.

Rothweiler H., Wäger P. A. and Schlatter C. (1990): Comparison of Tenax TA and Carbotrap for sampling and analysis of volatile organic compounds in air. Atmos. Environ., *25B*, 2, 231–235.
Tirkkonen T., Mroueh U.-M. and Orko I. (1995): Tenax as a collection medium for volatile organic compounds. NKB Committee and Work Reports 1995 : 6 E.
Uhde E. (1999): Sampling of monoterpenes in indoor air. In preparation.
Zielinska B., Arey J., Ramdahl T., Atkinson R. and Winer A.M. (1986): Potential for artifact formation during Tenax sampling of polycyclic aromatic hydrocarbon. J. Chromatogr., *363*, 382–386.

1.2 Sampling and Analysis of Aldehydes, Phenols and Diisocyanates

Mark Schulz and Tunga Salthammer

1.2.1 Introduction

Given the fact that nowadays most people spend more than 20 out of 24 hours indoors, the significance of a healthy indoor climate becomes obvious. Emissions of volatile organic compounds from new types of building materials – in combination with sealing of windows and doors for better insulation properties and energy saving – have often promoted a remarkable deterioration of indoor air quality. In Germany, complaints about odor and health effects first became evident in the 1970s. These were triggered by formaldehyde concentrations of up to 1 ppm as found by measurement. The main reason for these high indoor air concentrations was emissions from ceiling boards as well as from furniture made of particle board. Insufficient ventilation exacerbated the accumulation of formaldehyde in the indoor environment.

Table 1.2-1. Various methods for the determination of aldehydes, ketones, phenols and diisocyanates.

Method	Detection	Compounds
Pararosaniline [1)]	Colorimetry	Formaldehyde
Chromotropic acid [1)]	Colorimetry	Formaldehyde
MBTH [1)]	Colorimetry	Σ(aldehydes, C_1–C_3)
AHMT [1)]	Colorimetry	Σ(aldehydes)
Acetylacetone [1)]	Colorimetry/fluorimetry	Formaldehyde
DNPH [1)]	HPLC/UV	Aldehydes, ketones
Tenax [2)]	GC/FID	Aldehydes, ketones
Dimedone [1)]	HPLC/fluorescence	Aldehydes
Iodometric [1)]	Titrimetry	Formaldehyde
p-Nitroaniline [3)]	Colorimetry	(phenols)
Solvent absorption [4)]	HPLC/fluorescence	Phenols
2PP [5)]	HPLC/fluorescence	HDI, MDI,
MPP [6)]	HPLC/ELCD	2,4-TDI, 2,6-TDI
Dibutylamine [7)]	HPLC/MS	

1) Vairavamurthy et al. (1992); Roffael (1993)
2) Uhde and Borgschulte (1999)
3) VDI 3485, Part 1 (1988)
4) Ogan and Katz (1981)
5) Schulz and Salthammer (1998)
6) Schmidtke and Seifert (1990)
7) Tinnerberg et al. (1996)

As a consequence, great efforts were made to reduce formaldehyde emissions from wood-based materials. This was accomplished by optimizing urea-formaldehyde resins with respect to reduction of free formaldehyde as well as replacement of formaldehyde by other substances such as phenol and diisocyanates as reactants. Other emission sources of formaldehyde are coating materials, as for example acid-curing lacquers. With the inclusion of formaldehyde into the „Gefahrstoffverordnung" (German regulation on hazardous materials) in 1988 and into the „Chemikalienverbotsverordnung" (German regulation on the prohibition of chemicals) in 1993, formaldehyde emissions from building products were limited by law. In the following years guidelines for other products and substances were issued (Pluschke, 1996). The German RAL-UZ 76 (1995) gives regulations for wood and wood-based materials which are labelled with the so-called "Blue Angel". Limit values for aldehydes/ketones, formaldehyde, phenols, 4,4'-diphenylmethane-diisocyanate (MDI) and breakthrough are now fixed.

The concentrations of aldehydes and phenols in ambient air or in test chambers are generally well below 100 $\mu g/m^3$. The concentrations of diisocyanate monomers are in the magnitude of ng/m^3. This requires suitable measurement strategies and sensitive techniques for sampling and analysis (see Table 1.2-1).

1.2.2 Aldehydes and Ketones

Formaldehyde is known to be a ubiquitously occurring chemical pollutant and is now the most abundant carbonyl compound in our atmospheric environment (Finlayson-Pitts and Pitts, 1986). Furthermore, formaldehyde exhibits a special role among aldehydes because of its widespread use in building products. For this reason, various methods for the analytical determination of formaldehyde in indoor air have been developed (Kleindienst et al. 1988). Due to the low concentration in urban air of 0.0001 – 0.2 ppm formaldehyde and its photophysical properties (Okabe, 1978), its direct spectroscopic determination is highly sophisticated. More convenient and therefore very popular are sampling methods in which the compound is trapped in liquid absorbers or on impregnated filters and cartridges. Here, formaldehyde is derivatized and the resulting chromophore can be determined by spectroscopic or chromatographic techniques. However, a common problem is the specificity for formaldehyde, because nearly all reagents also react with other aldehydes and ketones. Today, titrimetric methods are no longer used for the analysis of formaldehyde in indoor air. Nevertheless, iodometry is still of importance for the preparation of calibration standards. Here, formaldehyde is oxidized by iodine in alkaline solution, and the excess iodine is then back-titrated with sodium thiosulfate (Salthammer, 1992). Many methods for the determination of formaldehyde were reviewed by Menzel et al. (1981), Roffael (1993) and Vairavamurthy et al. (1992).

In recent years, other saturated and unsaturated aldehydes also attracted considerable attention, because many of them are highly odorous and suspected to be irritative to the eyes and mucous membranes. Such aldehydes are typical oxidative degradation products of unsaturated fatty acids, which are components of many building products for indoor use. In recent years, a variety of analytical techniques have been developed or modified for the simultaneous determination of formaldehyde and higher aldehydes in ambient air.

1.2.2.1 Pararosaniline Method

A purple dye is formed from formaldehyde and pararosaniline in the presence of sodium sulfite (see Eq. 1). Miksch et al. (1981) postulated the formation of a colorless Schiff base from acidified pararosaniline and formaldehyde in the first step. Under acidic conditions this intermediate reacts with SO_2 to form the chromophore. The strong absorbance at 570 nm is used for colorimetric detection. Other aldehydes such as acetaldehyde, acrolein and propanal interfere, but at pH $\leq$ 1.0 the reaction is specific for formaldehyde. In the presence of atmospheric SO_2 a toxic Hg(II)-reagent is necessary for the elimination of sulfite formed from SO_2. Although the sensitivity of the modified pararosaniline method is limited and it is susceptible to interference, is has been one of the most widely used techniques for the determination of formaldehyde (VDI 3484, Part 1; Roffael, 1993).

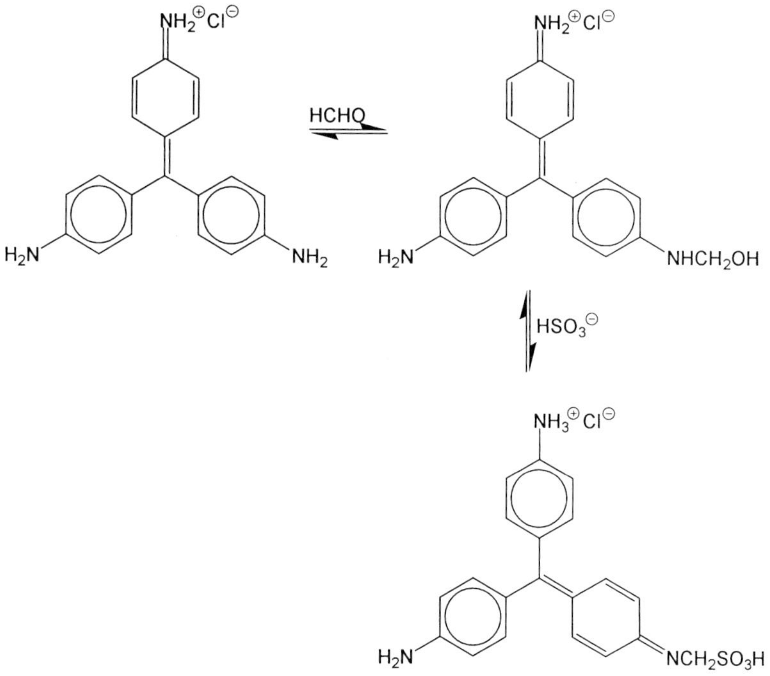

(1)

1.2.2.2 Chromotropic Acid Method

In the presence of concentrated sulfuric acid, chromotropic acid (1,8-dihydroxynaphthalene-3,6-disulfonic acid) reacts with formaldehyde to give a red-violet hydroxydiphenylmethane derivative, as shown in Eq. (2) (Eegriwe, 1937), which is soluble in acid. In the second step of the reaction, a violet quinoid oxidation product is formed with atmospheric oxygen. The concentrated sulfuric acid behaves as a catalyst for dehydration and oxidation. The absorption maximum at 580 nm is used for colorimetric detection.

The reaction is specific for formaldehyde if the pH value is < 1.0. Interference can be caused by phenols, some other organic substances and strong oxidizers (Menzel et al., 1981). The main disadvantages of this method are the low stability of chromotropic acid in solution and the necessity for using at least 80 % sulfuric acid.

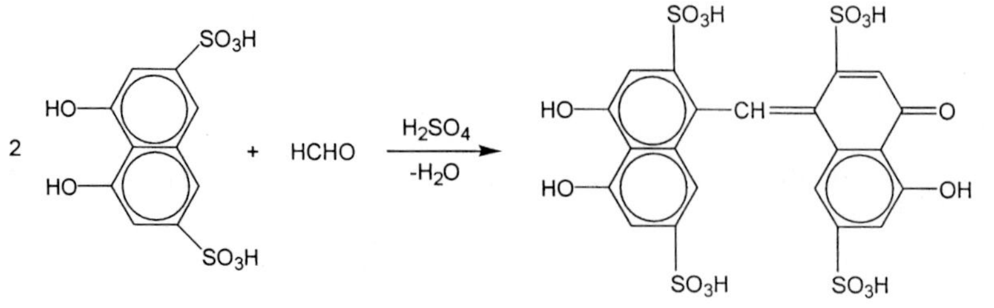

(2)

1.2.2.3 MBTH Method

The MBTH method (3-methyl-2-benzothiazolinone-hydrazone) is a nonselective colorimetric method for the determination of aliphatic aldehydes of low molecular weight (VDI 3862, Part 1, 1990). MBTH reacts on aldehydes to give an azide. In parallel, a reactive cation is formed by oxidation of MBTH with Fe(III). In a further step a blue ionic dye is formed (see Eq. 3). The absorbance is monitored at 628 nm.

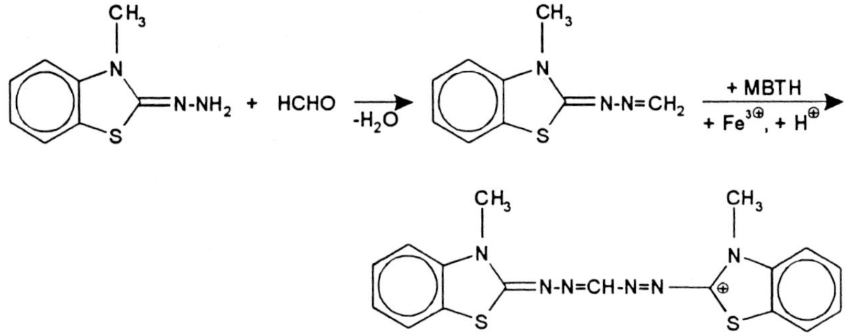

(3)

This reaction is less sensitive than that of the pararosaniline method. Other aldehydes undergo an analogous reaction, but the yield is generally lower. The MBTH method quantifies total aldehydes in ambient air in terms of their formaldehyde equivalents. Strong reducing agents can interfere with the determination of aldehydes.

1.2.2.4 AHMT Method (Purpald Method)

AHMT (4-amino-3-hydrazino-5-mercapto-4H-1,2,4-triazole) reacts on aldehydes in strongly alkaline media to form a colorless intermediate product. This product is oxidized by atmospheric oxygen to give a purple-colored dye. The resulting chromophore has four conjugated double bonds; the molecular formula shown in Eq. (4) is one of corresponding formaldehyde chromophore by colorimetry at a wavelength of 550 nm (VDI 3862, Part 4, 1991). The sensitivity of the AHMT-reagent varies in its dependence on the type of reacting aldehyde. The highest molar absorption coefficent is obtained for formaldehyde with $\varepsilon(550\text{nm}) = 2.41 \times 10^4\ \text{l mol}^{-1}\ \text{cm}^{-1}$ (König, 1985). The other aldehydes show different absorption maxima and significantly lower molar absorbance coefficients.

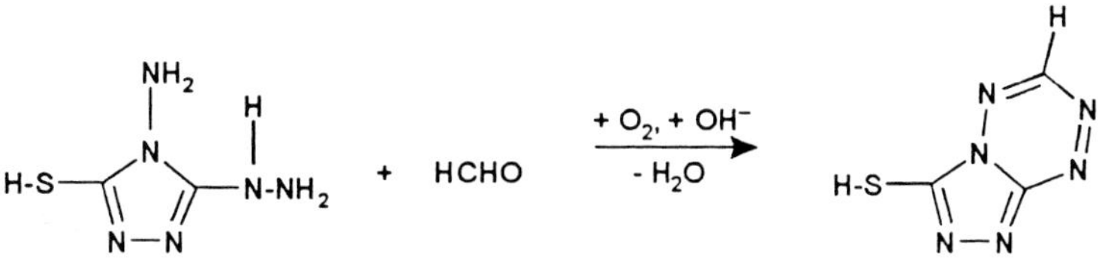

(4)

1.2.2.5 Acetylacetone Method

Among the derivatization methods, the procedure described by Nash (1953) is now widely applied for a number of different compounds. The reaction, which is based on the Hantzsch synthesis, involves the cyclization of 2,4-pentanedione (acetylacetone), ammonium acetate and formaldehyde, forming dihydropyridine-3,5-diacetyl-1,4-dihydrolutidine (DDL) as shown in Eq. (5). An analogous method was developed by Sawicki and Carnes (1968), who used 5,5-dimethyl-1,3-cyclohexanedione (dimedone) instead of 2,4-pentanedione (Compton and Purdy, 1980). The quantification of DDL can be performed by colorimetry at 412 nm [$\varepsilon(412\ \text{nm}) = 7450\ \text{l mol}^{-1}\ \text{cm}^{-1}$]. DDL also exhibits a broad and unstructured fluorescence with a maximum at 510 nm which offers the possibility of a very sensitive and selective fluorimetric determination of formaldehyde (Belman, 1963). Other aldehydes give analogous reactions, but the resulting lutidine derivatives only show very low fluorescence responses. The photophysical properties of DDL have been investigated as a function of solvent, temperature and light intensity. It was demonstrated that the absorption and fluorescence behavior is highly sensitive to the influence of heat or irradiation. Nevertheless, the quantitative determination of formaldehyde via colorimetry or fluorimetry of DDL can easily be performed if these effects are taken into account (Salthammer, 1993).

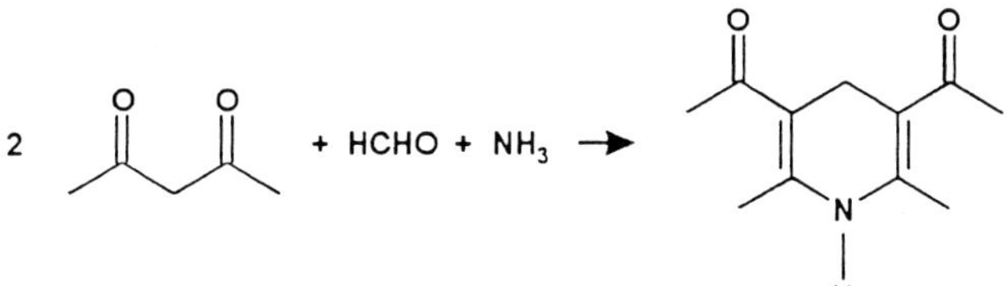

(5)

The acetylacetone method was proposed as a standard technique for the determination of formaldehyde (VDI 3484, Part 2, 1991) because of its simple handling and high reliability (DIN EN 717-1, 1996). A typical concentration profile of a formaldehyde measurement using the acetylacetone method in a private living room is shown in Fig. 1.2-1. First, the windows in the room were opened for 10–15 min to increase the air exchange rate and to decrease the concentration of pollutants. Sampling started immediately after

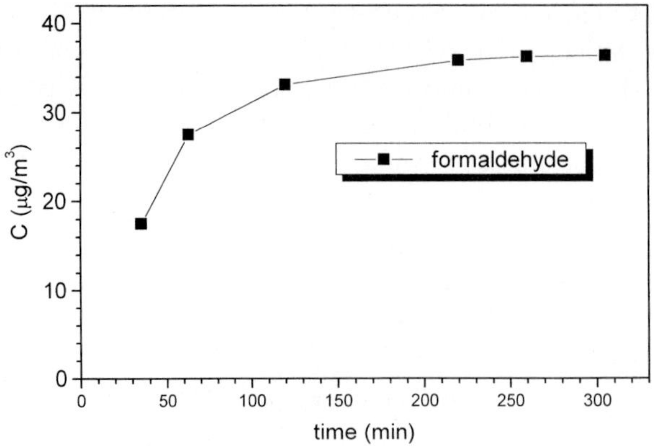

Figure 1.2-1. Concentration vs time function for formaldehyde in a living room. Sampling started immediately after ventilation.

ventilation and was performed by absorption of formaldehyde in distilled water by use of two Muencke absorbers connected in series (see Fig. 1.2-2). The air flow was 2 l/min and the total sampling volume was 80 l. Under the assumption of a constantly and continuously emitting source, the steady-state level with $C(\infty) = 0.029$ ppm was established after 4–5 h. For a volatile compound such as formaldehyde the progress towards the steady state can be estimated from the simple equation $C(t) = C(\infty)[1-\exp(-Nt)]$. In this case, with a measured air exchange rate of $N = 0.58\ \mathrm{h}^{-1}$, 95 % of the steady state is reached in ca. 5 h. The VDI guidelines 4300 Part 1 (1995) and Part 3 (1997) recommend strategies for sampling of formaldehyde in indoor air; the technical procedure using the acetylacetone method is described by Salthammer (1992).

1.2.2.6 DNPH Method

There are many HPLC methods for the determination of aldehydes in ambient air. They differ in sampling technique, elution and chromatographic separation (Tuss et al., 1982). 2,4-Dinitrophenylhydrazine DNPH (LIS, 1989; VDI 3862, Part 2, 1990; Selim, 1977; Druzik et al., 1990; Lipari and Swarin, 1982; Slemr, 1991) is now widely used for derivatization because this reagent is also suitable for the simultaneous analysis of long-chain aldehydes, aromatic aldehydes and ketones. In acidic solution, DNP-hydrazone derivatives are formed from DNPH and aldehydes or ketones as shown in Eq. (6). The reaction proceeds with nucleophilic addition to the carbonyl group followed by elimination of water.

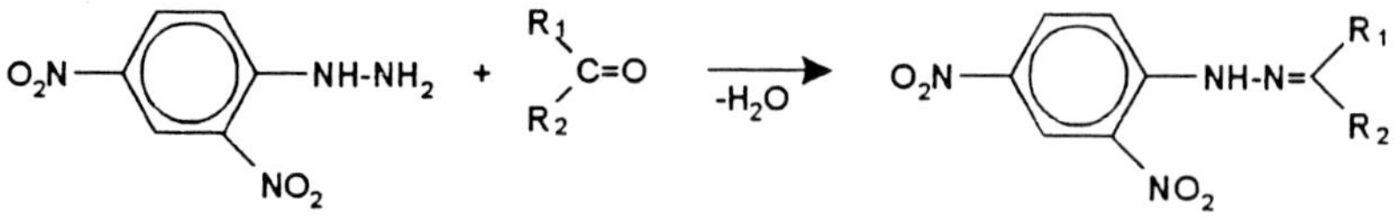

(6)

For sampling, air is sucked through cartridges coated with a DNPH solution (see Fig. 1.2-2). Ready-to-use cartridges are now commercially available or can be prepared as described by Zhang et al. (1994a; 1994b). Air samples are typically collected for 1–3 h

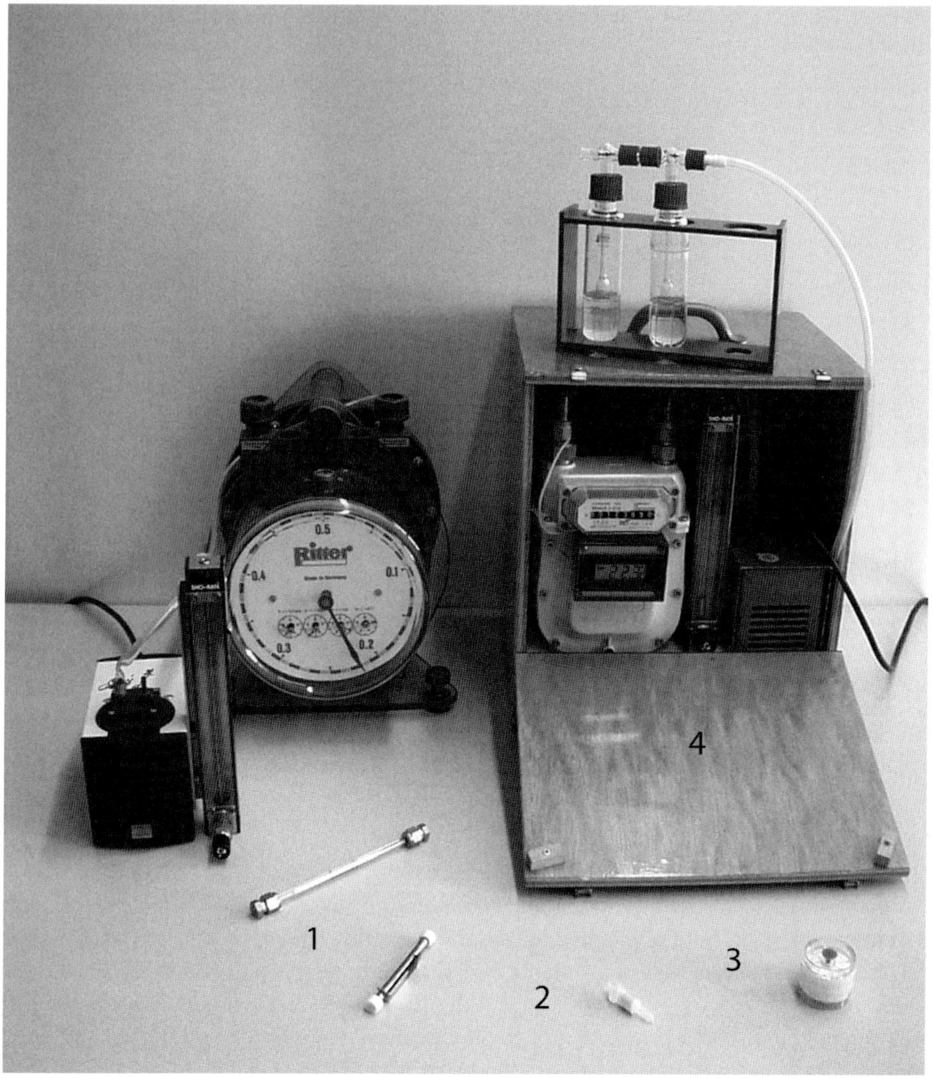

Figure 1.2-2. Different types of sampling devices for aldehydes, phenols and MDI. 1 = Tenax tube with gas meter and pump; 2 = DNPH cartridge; 3 = MDI cartridge; 4 = Muencke absorber with gas meter and pump.

at a flow rate of 0.5–1.0 l/h. The sampled cartridge is slowly eluted with acetonitrile. This eluate is then directly used for HPLC analysis. The separation of the carbonyl-DNP-hydrazones is achieved by use of a C_{18} column and water/acetonitrile solvent combinations with binary or ternary gradients. The separation can be optimized for specific aldehydes (Crump and Gardiner, 1987). A UV detector is commonly used; absorption maxima of different hydrazones range from 345 nm to 427 nm and were summarized by Vairavamurthy et al. (1992). Many modern types of HPLC equipment enable multi-

Table 1.2-2. HPLC-parameters for the separation and detection of aldehydes and ketones by use of the DNPH-method.

Parameter	Analytical conditions			
Derivatizing agent	Dinitropenylhydrazine (DNPH)			
Column	Knauer B6 Y535, 250 × 4.0 mm Eurospher 100-C_{18} with precolumn			
Oven temperature	30 °C			
Tray temperature	10 °C			
Solvent A	60 %/40 % acetonitrile/H_2O			
Solvent B	100 % acetonitrile			
Flow	1.0 ml/min			
Solvent gradient	100 % A → 50 %/50 % A/B in 10 min 50 %/50 % A/B → 100 % B in 6 min 100 % B → 50 %/50 % A/B in 4.5 min 50 %/50 % A/B → 100 % B in 0.5 min			
Retention time	Formaldehyde	5.08 min	Acetaldehyde	6.09 min
	Acetone	7.31 min	Butanal	9.08 min
	Benzaldehyde	9.91 min	Cyclohexanone	10.72 min
	Pentanal	11.88 min	Hexanal	12.46 min
	Heptanal	14.11 min	Octanal	15.53 min
	Nonanal	16.73 min	Decanal	17.77 min
UV detection	$\lambda_{ex(1)} = 355$ nm ± 10 nm $\lambda_{ex(2)} = 365$ nm ± 10 nm			

wavelength detection to be carried out or the recording of full spectra, which significantly improve the sensitivity of the method and the possiblities for identification of carbonyls. The analytical parameters of a suitable HPLC method are given in Table 1.2-2. NO_2 also reacts with DNPH with formation of 2,4-dinitrophenyl azide. This product has similar chromatographic behavior to that of the hydrazones and may cause interference (Karst et al., 1993).

1.2.2.7 GC methods for the Analysis of Aldehydes

Methods based on derivatization and GC analysis have been developed (Hoshika and Muto, 1978) but are of low popularity. A rapid analytical method for aldehydes is based on sampling on a solid adsorbent followed by thermal desorption and GC/FID detection. The required collection time is < 1 h per sample and no further sample preparation steps are needed. Uhde and Borgschulte (1999) have tested several combinations of adsorbents for sampling aldehydes from acetic aldehyde up to decanal. Tenax TA was found to be suitable for retaining the aliphatic aldehydes from butanal to decanal without breakthrough. Figure 1.2-3 shows the results of a static experiment in a 1-m^3 stainless steel

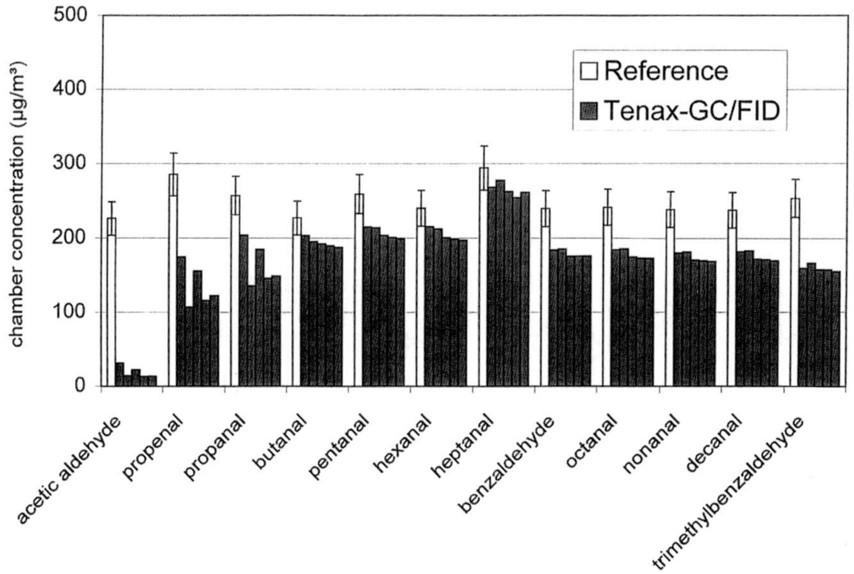

Figure 1.2-3. Tenax-GC/FID (■) analysis on different aldehydes. The experiment was perfomed under static conditions (no air exchange) in a 1-m^3 stainless steel chamber. The (□) bars represent the estimated concentrations in the chamber.

test chamber at 23 °C. The (□) bars demonstrate the estimated concentrations in the test chamber with estimated errors. A test mixture was injected by use of a microsyringe. The resulting concentrations were calculated for a chamber volume of 1 m^3 neglecting possible sinks, which certainly leads to overestimation. The (■) bars represent the results of 5 consecutive measurements using the Tenax-GC/FID method. The sampling volume was 1 l. It becomes obvious that the results are not reliable for acetaldehyde, propanal and propenal. However, the suitability of Tenax sampling for higher aldehydes is evident. The increase of sensitivity achieved with the Tenax-GC/FID method makes it possible to analyze aldehydes from butanal with air volumes of 2 l or less, which is useful for measuring emission processes with a high time resolution. A standard device for Tenax sampling is shown in Figure 1.2-2.

1.2.3 Analysis of Phenols

There are many chemical processes leading to emissions of phenol and its derivatives into urban air. Phenolic compounds are present in exhaust gases of motor vehicles and in waste gases of other combustion processes. In the indoor environment phenols are released from phenol-formaldehyde bonded wood-based products.

1.2.3.1 *p*-Nitroaniline Method

The determination of phenols according to VDI 3485, Part 1 (1988) is a sum method. The phenolic compounds react on diazotized *p*-nitroaniline to produce an azo dye (see Eq. 7). In the first step of the reaction, a diazonium ion is formed from *p*-nitroaniline and sodium nitrite in acidic solution. This diazonium ion is a strong electrophilic reagent and attacks phenol in the *para* position with formation of an azo coupling by substitution.

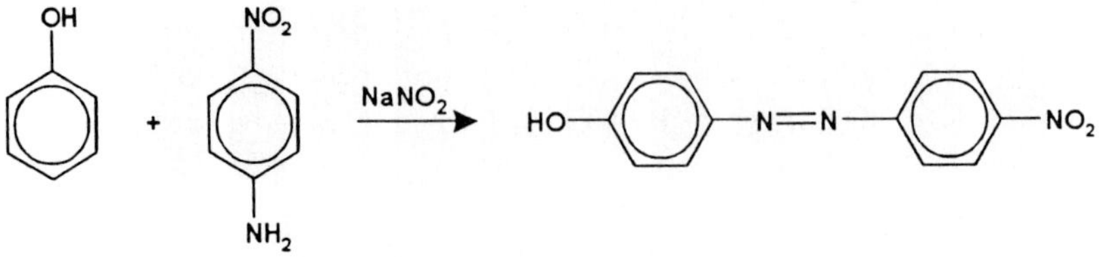

(7)

For sampling, air is sucked through a solution of 0.1 M NaOH by use of Muencke absorbers with a flow rate of 2 l/min. After neutralization with hydrochloric acid, the *p*-nitroaniline reagent is added to form 4-nitro-4'-hydroxy-azobenzene. After 45 min the reaction is complete and the absorbance is measured against water at 490 nm. The derivatization with *p*-nitroaniline is not specific for phenol. Therefore other phenolic compounds are also monitored. However, in comparison to 4-nitro-4'-hydroxy-azobenzene, the molecular absorption coefficients of the resulting azo dyes are significantly lower and the UV spectra are strongly shifted. Interference is mainly caused by aromatic amines, hydrogen sulfide and SO_2. With a total sampling volume of 100 l, the detection limit of this method is about 0.8 $\mu g/m^3$.

1.2.3.2 Direct Analysis of Phenols by HPLC

HPLC in combination with UV and/or fluorescence detection enables a sensitive and specific determination of phenol without derivatization to be performed. Air sampling is done in alkaline solution by the use of Muencke absorbers. This solution is directly analyzed by HPLC after neutralization. Different HPLC techniques using isocratic conditions or solvent gradients have been proposed (Kaschani, 1991; Engelsma et al., 1990). Suitable wavelengths for UV detection are 254 nm and 280 nm. In aqueous solution phenol exhibits a broad fluorescence in the 280–350 nm region (Berlman, 1971), which enables a highly sensitive fluorimetric determination to be performed. Ogan and Katz (1981) achieved a better signal-to-noise ratio and significantly lower detection limits using fluorescence with stopped-flow scanning for phenol detection instead of a UV detector at 274 nm when investigating alkylphenols by HPLC.

1.2.4 Diisocyanates

Diisocyanates are an important class of chemicals of commercial interest, which are frequently used in the manufacture of indoor materials such as adhesives, coatings, foams and rubbers (Ulrich, 1989; Oertel, 1985). In some types of particle board, the diisocyanates have substituted formaldehyde. Isocyanates are characterized by the electrophilic –N=C=O group, which may easily react with molecules containing "active" hydrogen, such as water or alcohols. On hydrolysis with water, primary amines are formed, while a reaction with alcohols leads to carbamates (urethanes). Polyurethane (PUR) products are then obtained from a polyaddition of diisocyanate and diol components. Common industrially applied compounds are 2,6-toluene diisocyanate (2,6-TDI), 2,4-toluene diisocyanate (2,4-TDI), 4,4'-diphenyl-methane diisocyante (MDI) and hexamethylene diisocyanate (HDI). The diisocyanate monomers are known as respiratory sensitizers and cause irritation of eyes, skin and mucous membrane. Therefore it is desired to measure even trace quantities in the atmosphere of workplace environments or private dwellings.

On sampling, an immediate derivatization for the formation of suitable compounds is necessary and a number of different procedures have been published (Andersson et al., 1982; Renman et al., 1986). A standard colorimetric method was introduced by Marcali (1957) and later modified (Pilz et al., 1970; Rando et al., 1985). More sensitive techniques use gas chromatography (Bishop et al., 1983) or HPLC with UV, fluorescence or electrochemical detection (Hakes et al., 1986; Sangö, 1979; Andersson et al., 1983; Audunsson et al., 1983; Sangö, et al., 1980; Dalene et al., 1988). Sensitive techniques using dibutylamine for derivatization and LC-UV/LC-MS for detection of diisocyanates and amines have been reported in a series of papers (Spanne et al., 1996; Tinnerberg et al., 1996; Tinnerberg et al., 1997).

The current state of the art involves the use of piperazine compounds which react with diisocyanates to form electrochemically detectable or fluorescent urea derivatives (Eickeler, 1990; Warwick et al., 1981; Schmidtke and Seifert, 1990; Brown et al., 1987; Bagon et al., 1984; Ellwood et al., 1981). Two OSHA methods (nos. 42 and 47) (OSHA, 1989a; OSHA, 1989b) describe a derivatization with 1-(2-pyridyl)-piperazine (2PP) to 1,6-bis-[4-(2-pyridyl)-1-piperazine-carbamyl]-hexane (HDI-2PP), 2,6-bis-[4-(2-pyridyl)-1-piperazine-carbamyl]-toluene (2,6-TDI-2PP), 2,4-bis-[4-(2-pyridyl)-1-piperazine-carbamyl]-toluene (2,4-TDI-2PP) and N,N'-(methylenediphenylene)-bis-4-(2-pyridinyl)-1-piperazine-carboxamide (MDI-2PP) on a coated filter, using the combination HPLC/fluorescence for separation and detection, respectively.

The most sensitive HPLC-method has been described by Schmidtke and Seifert (1990). When using 1-(2-methoxy-phenyl)-piperazine (MPP) for derivatization and a column switching technique, detection limits of 1 ng/m^3 and 30 ng/m^3 could be achieved by electrochemical (ELCD) and UV monitoring, respectively. However, the MPP-ELCD method is very susceptible to trouble and therefore hardly suitable for routine analysis.

1.2.4.1 Analysis of MDI

The OSHA method no. 47 (OSHA, 1989b) has been developed for workplace areas with concentrations > 1 μg/m^3 and describes the derivatization of MDI to form MDI-2PP on a coated filter, using the combination HPLC/fluorescence for separation and detection, respectively (see Eq. 8). Some relevant properties of MDI-2PP have been investigated with regard to the analysis of diisocyanates by HPLC/fluorescence (Salthammer et al., 1997). High molecular absorbance coefficients (Hardy and Walker, 1979) and sufficient fluorescence response were found. Hence it was concluded that derivatization using 2PP is also very suitable for determination of airborne diisocyanate in living spaces where concentrations of MDI are generally well below 100 ng/m^3. In order to establish a highly sensitive and easy-to-handle routine method, the OSHA-method no. 47 could be distinctly improved (Schulz and Salthammer, 1998).

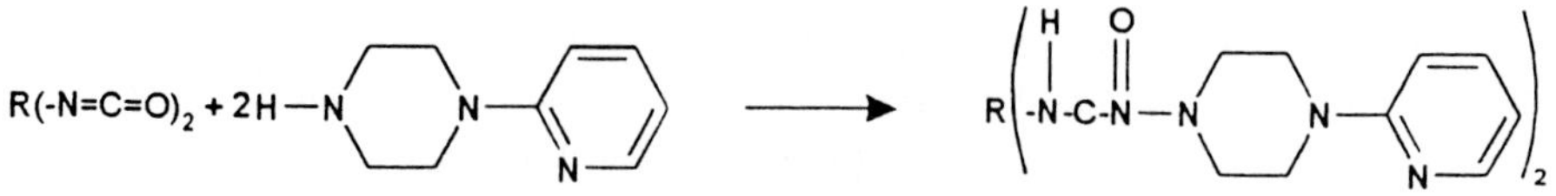

(8)

Sampling cassettes with 2PP-coated filters can be purchased commercially (see Fig. 1.2-2). Air is sucked through the cassette with a flow of 60 l/h. MDI-2PP is extracted from the filter in a 4 ml vial by use of an acetonitrile/DMSO mixture in a ratio of 90 : 10. This extract is directly used for HPLC analysis. A suitable column for separation is a 250 × 4.6 mm Nucleosil 100 C_8. Oven and tray temperatures are kept at 30 °C and 10 °C, respectively. A solvent gradient using the components acetonitrile, H_2O and aqueous buffer solution is used. The flow is 1.0 ml/min. The MDI-2PP chromophore is excited at 240 ± 20 nm. The fluorescence response is monitored at 370 ± 20 nm. The selected HPLC conditions enable a shift of the MDI-2PP signal to a higher retention time of 22.3 min, resulting in a better separation from underivatized 2PP and possible interference.

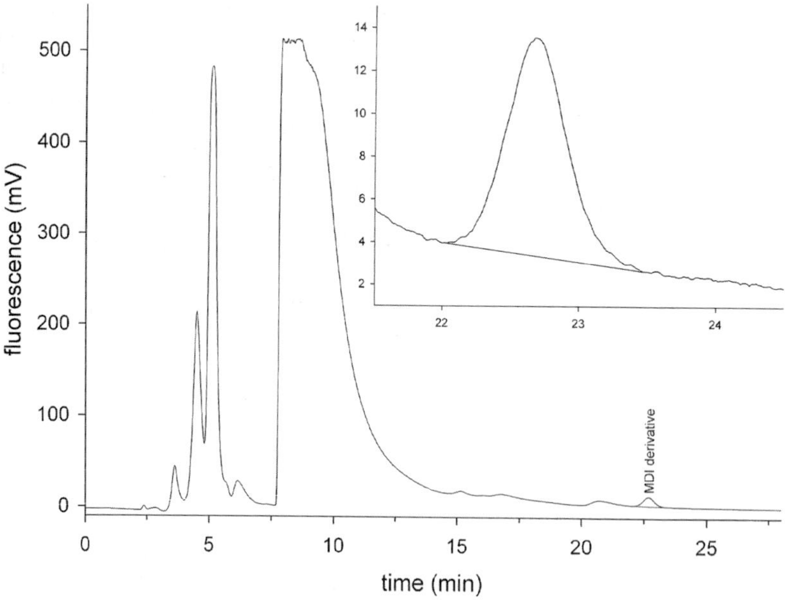

Figure 1.2-4. HPLC/UV-signal of the MDI-2PP derivative. The retention time is 22.3 min, the concentration of the MDI solution was 7.8 ng/l (Schulz and Salthammer, 1998).

Table 1.2-3. HPLC-parameters for the separation and detection of MDI by use of the 2PP-method (Schulz and Salthammer, 1998).

Parameter	Analytical conditions
Derivatizing agent	1-(2-Pyridyl)-piperazine (2PP)
Column	Nucleosil 100 C_8, 250 × 4.6 mm I.D.
Oven temperature	30 °C
Tray temperature	10 °C
Solvent A	37.5%/62.5% acetonitrile/H_2O; 0.01 M NH_4Ac pH 6.2, acetic acid
Solvent B	50%/50% acetonitrile/H_2O; 0.01 M NH_4Ac pH 6.0, acetic acid
Flow	1.0 ml/min
Solvent gradient	100 % A for 10 min 100 % A → 100 % B in 2 min 100 % B for 13 min 100 % B → 100 % A in 1 min
Retention time	MDI 22.3 min
UV detection	$\lambda_{ex} = 254$ nm ± 20 nm
Fluorescence detection	$\lambda_{ex} = 240$ nm ± 20 nm $\lambda_{em} = 370$ nm ± 20 nm

The limits of detection and determination can be calculated from calibration curves as recommended by DIN 32645 (1994) (see also Einax et al., 1997). A chromatogram for the MDI calibration concentration of 7.8 ng/ml is shown in Fig. 1.2-4. All relevant parameters for HPLC-analysis are summarized in Table 1.2-3.

The analytical method described here is based on standard equipment and provides sufficient sensitivity to investigate traces of MDI monomers in test chambers, living spaces and workplace areas. To achieve a detection limit well below 100 ng/m^3, a total sampling volume of about 100 l is necessary. A total sampling volume of 1000 l gives a detection limit of 9.0 ng/m^3 (Schulz and Salthammer, 1998). The complete procedure including sampling, elution and analysis is easy to handle, not very time-consuming and therefore highly suitable as a routine method for monitoring airborne MDI monomers.

References

Andersson K., Gudéhn A., Levin J.-O. and Nilsson C.-A. (1982): Analysis of gaseous diisocyanates in air using chemosorption sampling. Chemosphere, *11*, 3–10.

Andersson K., Gudéhn A., Levin J.-O. and Nilsson C.-A. (1983): A comparative study of solvent and solvent-free sampling methods for airborne 4,4'-diphenylmethane diisocyanate (MDI) generated in polyurethane production. Am. Ind. Hyg. Assoc. J., *44*, 802–808.

Audunsson G. and Mattiasson L. (1983): Simultanous determination of amines and isocyanates in working atmospheres by gas-liquid chromatography, J. Chromatogr., *261*, 253–264.

Bagon D.A., Warwick C.J. and Brown R.H. (1984): Evaluation of total isocyanate-in-air method using 1-(2-methoxyphenyl)piperazine and HPLC. Am. Ind. Hyg. Assoc. J., *45*, 39–43.

Belman S. (1963): The fluorimetric determination of formaldehyde. Anal. Chim. Acta, *29*, 120–126.

Berlman I.B. (1971): Handbook of Fluorescence Spectra of Aromatic Molecules, Academic Press, New York.

Bishop R.W., Ayers T.A. and Esposito G.G. (1983): A gas chromatographic procedure for the determination of airborne MDI and TDI. Am. Ind. Hyg. Assoc. J., *44*, 151–155.

Brown R.H., Ellwood P.A., Groves J.A., Robertson S. M. (1987): New method for the determination of airborne isocyanates. Cellular Polymers, *6*, 1–8.

Compton B.J. and Purdy W.C. (1980): The mechanism of the reaction of the Nash and the Sawicki aldehyde reagent. Can. J. Chem., *58*, 2207–2211.

Crump D.R. and Gardiner D. (1987): Sources and concentrations of aldehydes and ketones in indoor environments in the UK. Proceedings of Indoor Air 87, Berlin, Vol. 2, 666–670.

Dalene M., Mathiasson L., Skarping G., Sangö C. and Sandström F. (1988): Trace analysis of airborne aromatic isocyanates and related aminoisocyanates and diamines using high-performance liquid chromatography with ultraviolet and electrochemical detection. J. Chromatogr., *435*, 469–481.

DIN 32645 (1994): Chemical analysis; decision limit, detection limit and determination limit; estimation in case of repeatability, term, methods, evaluation. Beuth, Berlin.

DIN EN 717–1 (1996): Wood based panels–Determination of formaldehyde release. Part 1: Formaldehyde emission by the chamber method, Beuth, Berlin.

Druzik C.M., Grosjean D., Van Neste A. and Parmar. S.S. (1990): Sampling of atmospheric carbonyls with small DNPH-coated C18 cartriges and liquid chromatrography analysis with diode array detection. Intern. J. Environ. Anal. Chem., *38*, 495–512.

Eegriwe E. (1937): Reaktionen und Reagenzien zum Nachweis organischer Verbindungen IV. Z. Anal. Chem., *110*, 22–25.

Eickeler E. (1990): Aspects of quality assurance in the determination of 2,4-toluene diisocyanate by HPLC. Fresenius J. Anal. Chem., *336*, 129–131.

Einax J.W., Zwanziger H.W. and Geiss S. (1997): Chemometrics in environmental analysis, Wiley-VCH, Weinheim.

Ellwood P.A., Hardy H.L. and Walker R.F. (1981): Determination of atmospheric isocyanate concentrations by high-performance thin-layer chromatography using 1-(2-pyridyl)-piperazine reagent. Analyst, *106*, 85–93.

Engelsma M., Kok W.Th. and Smit H.C. (1990): Selective determination of trace levels of phenol in river water using electrochemical concentration modulation correlation chromatography. J. Chromatogr., *506*, 201–210.

Finlayson-Pitts B.J. and Pitts J.N. (1986): Atmospheric Chemistry. Wiley, New York.

Hakes D.C., Johnson G.D. and Marhevka J.S. (1986): An improved high pressure liquid chromatographic method for the determination of isocyanates using "nitro reagent". Am. Ind. Hyg. Assoc. J., *47*, 181–184.

Hardy H.L. and Walker R.F. (1979): Novel reagent for the determination of atmospheric isocyanate monomer concentrations. Analyst, *104*, 890–891.

Hoshika Y. and Muto G. (1978): Rapid separation of lower carbonyl compounds by gas-liquid-solid chromatography. J. Chromatogr., *152*, 533–537.

Karst U., Binding N., Cammann K. and Witting U. (1993): Interferences of nitrogen dioxide in the determination of aldehydes and ketones by sampling on 2,4-dinitrophenylhydrazine-coated sorbent. Fresenius J. Anal. Chem., *345*, 48–52.

Kaschani T. (1991): Bestimmung von Phenolen. Deutsche BP, Wedel.

Kleindienst T.E., Shepson P.B., Nero C.M., Arnts R.R., Tejada S.B., Mackay G.I., Mayne L.K., Schiff H.I., Lind J.A., Kok G.L., Lazrus A.L., Dasgupta P.K. and Dong S. (1988): An intercomparison of formaldehyde measurement techniques at ambient concentration. Atmos. Environ., *22*, 1931–1939.

König H. (1985): Photometrische Aldehydbestimmung. Staub–Reinhalt. Luft, *45*, 423–425.

Lipari F. and Swarin S. (1982): Determination of formaldehyde and other aldehydes in automobile exhaust with an improved 2,4-dinitrophenylhydrazine method. J. Chromatogr., *247*, 297–306.

LIS–Landesamt für Immissionsschutz (1989): LIS Bericht Nr. 92, Essen.

Marcali K. (1957): Microdetermination of toluenediisocyanates in atmosphere. Anal. Chem., *29*, 552–558.

Menzel W., Marutzky R. and Mehlhorn L. (1981): Formaldehyd-Messmethoden. WKI-Bericht Nr. 13, Braunschweig.

Miksch R.R., Anthon D.W., Douglas W., Fanning L.Z., Hollowell C.D., Revzan K. and Glanville J. (1981): Modified pararosaniline method for the determination of formaldehyde. Anal. Chem., *53*, 2118–2123.

Nash T. (1953): The colorimetric estimation of formaldehyde by means of the Hantzsch reaction. Biochem. J., *55*, 416–421.

Oertel G. (ed) (1985): Polyurethane Handbook. Hanser, Munich.

Ogan K. and Katz E. (1981): Liquid chromatographic separation of alkylphenols with fluorescence and ultraviolet detection. Anal. Chem., *53*, 160–163.

Okabe H. (1978): Photochemistry of small molecules. Wiley, New York.

OSHA Analytical Laboratory (1989a): Diisocyanates: 1,6-hexamethylene diisocyanate (HDI), toluene-2,6-diisocyanate (2,6-TDI), toluene-2,4-diisocyanate (2,4-TDI), OSHA Method 42, Salt Lake Technical Center, Salt Lake City, Utah.

OSHA Analytical Laboratory (1989b): Methylene bisphenyl isocyanate (MDI), OSHA Method 47, Salt Lake Technical Center, Salt Lake City, Utah.

Pilz W., Johann I. (1970): Spezielle analytische Methoden für die Arbeitsmedizin und Industriehygiene. IV. Mikrochimica Acta, 351–358.

Pluschke P. (1996): Luftschadstoffe in Innenräumen. Springer, Berlin.

RAL–Deutsches Institut für Gütesicherung und Kennzeichnung e. V. (1995): Emissionsarme Holzwerkstoffplatten (RAL-UZ-76). Beuth, Berlin.

Rando R.R., Hammad Y.Y. (1985): Modified Marcali method for the determination of total toluenediisocyanate in air. Am. Ind. Hyg. J., *46*, 206–210.

Renman L., Sangö C., Skarping G. (1986): Determination of isocyanate and aromatic amine emission from thermally degraded polyurethanes in foundries. Am. Ind. Hyg. Assoc. J., *47*, 621–628.

Roffael E. (1993): Formaldehyde release from particleboard and other wood based panels. Forest Research Institute, Malayan Forest Records No. 37, Maskha Sdn. Bhd., Kuala Lumpur.

Salthammer T. (1992): Quantitative Bestimmung von Formaldehyd in der Innenraumluft. Praxis der Naturwissenschaft–Chemie., 3/41, 24–26.

Salthammer T. (1993): Photophysical properties of 3,5-diacetyl-1,4-dihydrolutidine in solution: Application to the analysis of formaldehyde. J. Photochem. Photobiol. A: *74*, 195–201.

Salthammer T., Wismach C. and Miertzsch H. (1997): Absorption and fluorescence of 1-(2-pyridyl)-piperazine and four diisocyanate derivatives in solution. J. Photochem. Photobio. A, *107*, 159–164.

Sangö C. (1979): Improved method for determination of traces of isocyanates in working atmospheres by high-performance liquid chromatography. J. Liq. Chromatogr., *2*, 763–774.

Sangö C. and Zimerson E. (1980): A new reagent for determination of isocyanates in working atmospheres by HPLC using UV or fluorescence detection. J. Liq. Chromatogr., *3*, 971–990.

Sawicki E. and Carnes R. A. (1968): Spectrophotofluorimetric determination of aldehydes with dimedone and other reagents. Mikrochim. Acta, 148–159.

Schmidtke F. and Seifert B. (1990): A highly sensitive high-performance liquid chromatographic procedure for the determination of isocyanates in air. Fresenius J. Anal. Chem., *336*, 647–654.

Schulz M. and Salthammer T. (1998): Sensitive determination of airborne diisocyanates by HPLC: 4,4'-Diphenylmethane-diisocyanate (MDI), Fresenius J. Anal. Chem., 362, 289–293.

Selim S. (1977): Separation and quantitative determination of traces of carbonyl compounds as their 2,4-dinitrophenylhydrazones by high-pressure liquid chromatography. J. Chromatrogr., *137*, 271–277.

Slemr J. (1991): Determination of volatile carbonyl compounds in clean air. Fresenius J. Anal. Chem., *340*, 672–677.

Spanne M., Tinnerberg H., Dalene M., Skarping G. (1996): Determination of complex mixtures of airborne isocyanates and amines. Part 1. Liquid chromatography with ultraviolet detection of monomeric and polymeric isocyanates as their dibutylamine derivatives. Analyst, *121*, 1095–1099.

Tinnerberg H., Spanne M., Dalene M., Skarping G. (1996): Determination of complex mixtures of airborne isocyanates and amines. Part 2. Toluene diisocyanate and aminoisocyanate and toluenediamine after thermal degradation of a toluene diisocyanate-polyurethane. Analyst, *121*, 1101–1106.

Tinnerberg H., Spanne M., Dalene M., Skarping G. (1997): Determination of complex mixtures of airborne isocyanates and amines. Part 3. Methylenediphenyl diisocyanate, methylenediphenylamino isocyanate and methylenediphenyldiamine and structural analogues after thermal degradation of polyurethane. Analyst, *122*, 275–278.

Tuss H., Neizert V., Seiler W. and Neeb R. (1982): Method for determination of formaldehyde in air in the pptv-range by HPLC after extraction as 2,4-dinitrophenyl-hydrazone. Fresenius Z. Anal. Chem., *312*, 613–617.

Uhde E. and Borgschulte A. (1999): Thermal desorption GC/FID analysis of aliphatic aldehydes in indoor air. Procedings of the 8th International Conference on Indoor Air and Climate, Edinburgh, in press.

Ulrich H. (1989): Isocyanates, organic, In: Elvers B., Hawkins S., Ravenscroft M., Schulz G. (eds). Ullmann's Encyclopedia of Industrial Chemistry, 5th edn., Vol. A14, VCH, Weinheim.

Vairavamurthy A., Roberts J.M. and Newman L. (1992): Methods for determination of low molecular weight carbonyl compounds in the atmosphere: a review. Atmos. Environ., *26A*, 1965–1993.

VDI 3484 Part 1 (1979): Messen gasförmiger Immissionen–Messen von Aldehyden. Bestimmen der Formaldehyd-Konzentration nach dem Sulfit-Pararosanilin-Verfahren. Beuth, Berlin.

VDI 3484 Part 2 (1991): Messen gasförmiger Immissionen, Messen von Aldehyden. Bestimmung der Formaldehyd-Konzentration nach dem Acetylaceton-Verfahren. Beuth, Berlin, in preparation

VDI 3485 Part 1 (1988): Ambient air measurement–measurement of gaseous phenolic compounds, *p*-nitroaniline method. Beuth, Berlin.

VDI 3862, Part 1 (1990): Gaseous emission measurement–measurement of aliphatic aldehydes (C_1 to C_3) MBTH method. Beuth, Berlin.

VDI 3862, Part 2 (1990): Messen gasförmiger Emissionen–Messen aliphatischer und aromatischer Aldehyde und Ketone nach dem DNPH-Verfahren. Acetonitril-Verfahren, Beuth, Berlin, in preparation.

VDI 3862, Part 4 (1991): Messen gasförmiger Emissionen–Messen aliphatischer Aldehyde nach dem AHMT-Verfahren. In preparation.

VDI 4300, Part 1 (1995): Indoor air pollution measurement–general aspects of measurement strategy. Beuth, Berlin.

VDI 4300, Part 3 (1997): Measurement of indoor air pollutants–measurement strategy for formaldehyde. Beuth, Berlin.

Warwick C.J., Bagon D.A. and Purnell C.J. (1981): Applicaton of electrochemical detection to the measurement of free monomeric aromatic and aliphatic isocyanates in air by high-performance liquid chromatography. Analyst, *106*, 676–685.

Zhang J., He Q. and Lioy P.J. (1994a): Characteristics of aldehydes: Concentrations, sources, and exposures for indoor and outdoor residential microenvironments. Environ. Sci. Technol., *28*, 146–152.

Zhang J., Wilson W.E. and Lioy P.J. (1994b): Indoor air chemistry: Formation of organic acids and aldehydes. Environ. Sci. Technol., *28*, 1975–1982.

1.3 Sampling and Analysis of Wood Preservatives in Test Chambers

Oliver Jann and Olaf Wilke

1.3.1 Introduction

Wood preservative is the term for a mixture of wood preservative agents and other ingredients in a solvent. This chapter focusses on organic wood preservative agents (WPA) such as lindane, PCP, dichlofluanid, tebuconazole and permethrine.

Because wood preservative agents are biocides they might be dangerous to the environment if they volatilize from the wood into the atmosphere or indoor air.

In the past, PCP, lindane and DDT have been used in several instances in Germany to protect wood indoors. In recent years modern biocides such as pyrethroids (e.g. permethrine) or triazoles (e.g. tebuconazole) have been used instead of the classical organic chloro-pesticides to protect wood.

The latest wood preservatives contain a mixture of different biocides. Normally these are combinations of fungicides and insecticides e.g. dichlofluanid, tebuconazole and permethrine in the concentration range 0.03 – 1.0 % in solution together with other ingredients such as plasticizers, pigments and others.

These relatively new biocides have a very low vapor pressure (tebuconazole: 1.3×10^{-6} to 7.2×10^{-7} Pa, permethrine: 1×10^{-4} to 4.5×10^{-5} Pa) compared to former biocides (lindane: 1.2×10^{-3} to 9.4×10^{-4} Pa). Substances such as PCP, lindane or DDT are already found indoors in gaseous form or adsorbed to dust; modern biocides such as pyrethroids are only found adsorbed to dust.

These substances can be classified as semi-volatile organic compounds (SVOC) or particulate organic matter (POM).

Biocides for wood protection are stable against degradation to ensure long-term protection but that also means a potential for long-term volatilization if they do not stay in the wood.

Latest research shows that even the modern substances with very low vapor pressure volatilize into the environment (Jann et al., 1997).

Due to the fact that it is very difficult to determine the exact vapor pressure for substances with very low vapor pressure, the calculation of saturation concentration also includes high uncertainty. In the literature different vapor pressures for the same biocide can be found. Therefore calculations of saturation concentration also give different results and in the case of a wrong vapor pressure wrong maximum air concentration.

1.3.2 Emission Measurements in Test Chambers

A way of estimating real air concentrations of wood preservative agents (WPA) is to study their volatilization under controlled conditions. This can be done in test chambers where different environmental conditions such as temperature, humidity, air velocity, air exchange and surface size can be adjusted. All these parameters influence the emission rate from a treated wood. Some chamber experiments for simulation of biocide volatility have been carried out in the past. Zimmerli (1982) described model experiments and gave equations for the calculation of equilibrium concentrations for some WPA.

He used simple glass cylinders with a diameter of 24 mm and a length between 2 cm and 10 cm. A small piece of treated wood was placed in the bottom and paraffin oil-treated filter paper as sorption medium for passive sampling was placed at the top. The containers were sealed airtight and were convection-free diffusion systems.

Landsiedel and Plum (1981) made investigations of the volatility of tributyltin compounds. They used a plastic chamber of 200 l (58 cm × 58 cm × 58 cm). The surface of treated wood (1.25 m^2) was treated with 200 g/m^2 (3–6 % TBTO in solution). The temperature was around 22 °C and the air exchange was about 0.4 h^{-1} generated by sucking air through the chamber. They used activated carbon for sampling.

Petrowitz (1986) investigated the release of the active substances of wood preservatives from chemically treated building timbers. He used glass containers with 1.5 l volume and an air exchange of 4 h^{-1} at 20 °C. Pine and spruce wood (50 cm^2) were treated with 250 ml/m^2 (lindane and endosulfane 1 %, dichlofluanid 0.5 %, PCP 5 % in Shellsol AB). Sampling was also done with activated carbon (4 g) followed by elution with toluene and acetone. Analysis was done by GC-ECD using aldrine as internal standard.

Prueger and Pfeiffer (1994) made preliminary tests of a laboratory chamber technique intended to simulate pesticide volatility in the field. They also used a 1.5-l bell-shaped glass jar but used glass slides (2.5 cm × 7.5 cm) for treatment with 60 µl of a solution of metolachlor in methanol. The air exchange was 40 h^{-1}, air speed 2.4 mm/s, temperature 34 °C and relative humidity 20 %. For sampling, 2.75 g of XAD-8 sorbent resin were used followed by elution with 35 ml ethyl acetate and GC-NPD analysis.

Kubiak, Maurer and Eichhorn (1993) presented a new laboratory chamber constructed for studying the volatilization of ^{14}C-labeled pesticides from plant and soil surfaces. An air-conditioning system provides for temperature and relative humidity conditions between 10 °C and 30 °C and 35 % to 90 %, respectively. Wind speed simulation between 0.3 m/s and 2.5 m/s is possible.

Horn (1993) investigated dichlofluanid, furmecyclox, IPBC, tebuconazole and permethrine in a 1-m^3 glass chamber. The loading was 1 m^2/m^3 and air exchange was 1 h^{-1}. Temperature and relative humidity were 23 °C and 45 %, respectively. After testing suitability of silica gel, active charcoal, polyurethane foam (PUF), XAD-4 and Tenax he chose PUF for sampling followed by Soxhlet extraction for 6 h with 5 % diethyl ether in *n*-hexane. Analysis was done by GC-ECD and GC-FID.

A method for measuring losses of active ingredients and other preservative ingredients from treated timber is described as a European standard in DIN V ENV 1250-1. A cylindrical chamber (40 cm long and 10.0 cm in diameter) of volume 3 l with an air exchange of 1 h^{-1} is prescribed. Sixteen wood panels sized 50 mm × 25 mm × 15 mm with a total treated surface of 0.06 m^2 have to be placed inside giving a loading of 21.33 m^2/m^3. A description of a sampling and analysis procedure is not given.

All the experiments and methods mentioned above are based on different conditions with different chambers and different aims. It is difficult to compare results for the same biocide if climate, chamber loading (ratio of treated surface to chamber volume) and/or air exchange are different and no knowledge of possible influences of sink effects and time dependence exists.

Furthermore, for all these experiments sampling was done in different ways.

For general sampling of biocides in air, different sorbents and comparisons between them are described in the literature (Figge, Rabel, Wieck 1987) and guidelines exist (VDI 1993 and 1997; ASTM 1989).

The adsorbent must have good trapping efficiency for the analyte (no breakthrough) and the recovery from the adsorbent must also be good enough.

For high boiling substances such as the WPA, low air concentrations are to be expected. Therefore high enrichment is needed to get a sufficient amount of analyte for analysis. This means long sampling time or high sample volume.

The most popular adsorber for biocids in air seems to be polyurethane foam (PUF), which allows high sample volume rates (up to 250 l/min, total volume 300 m^3) (Patton et al. 1992; Zaranski et al. 1991; Hawthorne et al. 1989; Lewis and Jackson 1982; Turner and Glotfelty 1977).

For taking a high sample volume out of chambers the air flow through the chamber has to be sufficient. The air volume for sampling has to be lower than or equal to the air flow needed for the air exchange in the chamber. A higher sample volume would change the air exchange and thus the concentration.

Sampling can also be done with Tenax fixed by glass wool stoppers or glass wool alone with sample volumes of 0.5–40 l with a flow of 6–27 l/h. This procedure was followed by thermal desorption and GC-MS or GC-AED analysis (Jann et al., 1997).

1.3.3 Experimental

1.3.3.1 Test Chambers

The authors of this chapter used different types of test chambers (Fig. 1.3-1) for comparison reasons: 1-m^3 stainless steel chambers, 0.02-m^3 glass chambers and 0.625-l chambers constructed to hold the treated wood as chamber walls to prevent sink effects.

The temperature was 23 °C and the relative humidity 45 %. Loading was 0.2 m^2/m^3 in the 1-m^3 chambers (8 wood panels sized 0.25 m $\times$ 0.05 m treated on both sides with 250 ml/m^2). The air exchange was 1 h^{-1} in the 1-m^3 chambers giving an aera specific air flow rate (ratio of air exchange to loading) of 5 m^3/m^2h. The same rate was adjusted in the 20-l chambers. Loading here was 1.25 m^2/m^3 (1 wood panel as above), air exchange 6.25 h^{-1} (125 l/h, 20 l volume). Investigated were lindane, furmecyclox, dichlofluanid, propiconazole, tebuconazole, permethrine and deltamethrine over a period of more than 125 days.

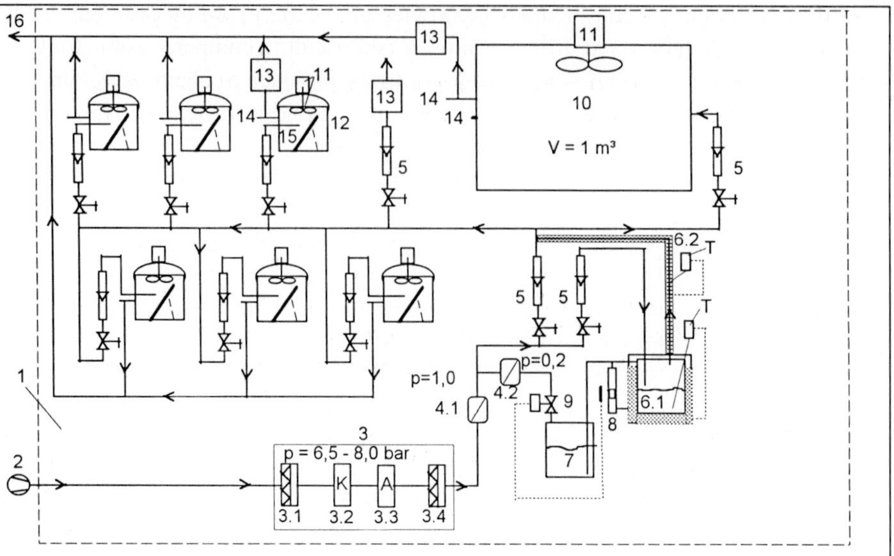

1 Climate chamber
2 Oilfree compressor
3 Clean air unit consisting of
3.1 Filter (Dust + Aerosol)
3.2 Heatless dryer ($\tau \approx$ -40 °C)
3.3 Activated charcoal
3.4 Submicrofilter (Dust)
4.1 Reducing valve 6,5 - 8,0 bar → 1,0 bar
4.2 Reducing valve 1,0 bar → 0,2 bar
5 Gas flow meter and gas flow control
6.1 Temperature controlled water bath
6.2 Temperature controlled air pipe
7 Water storage bottle
8 Niveau control contact free: magnet/reedcontact
9 Electromagnetic valve
10 1-m³ chamber (electropolished stainless steel)
11 Ventilator
12 0,02 m³-glass chamber based on a desiccator in accordance with DIN 12491 (6x)
13 Climate sensor
14 Air sampling port
15 Test piece
16 Air outlet

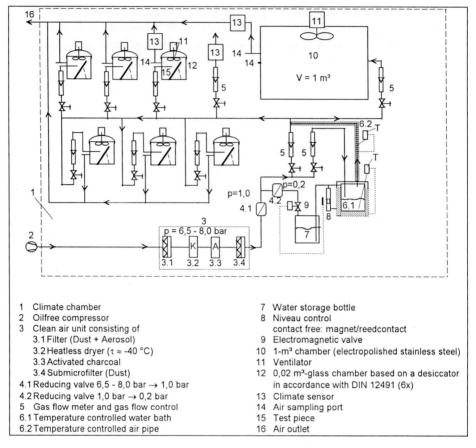

Figure 1.3-1. Test chamber system consisting of 0.02-m^3 and 1-m^3 chambers.

1.3.3.2 Sampling

Sampling was done with polyurethane foam in two dimensions. At the 1-m^{3}chambers with an air exchange of 1 h^{-1} a volume of 750 l/h was sampled for 1–4 days giving a total sample volume of 15–70 m^3. An experienced sampling system (VDI guideline 3498, part 1, 1993) was adapted to the chamber conditions. The PUF was reduced in size to 22 mm in diameter and 50 mm in length (PUF22). This was large enough for a maximum flow of 750 l/h and improved and shortened the sample extraction. The PUF plug (two for breakthrough measurements) was held in a stainless steel container during sampling. This container was 160 mm long and 20 mm wide and had the option of a filter inset (50 mm in diameter) in front of the PUF plugs.

For the 0.02-m^3 chambers with an air exchange of 6.25 h^{-1} the maximal sample volume was 125 l/h. The PUF plug was 65 mm long and 13 mm in diameter (PUF13) and was held in a glass tube of 11 mm diameter and 250 mm length.

At the 1-m^3 chambers both sampling systems were used in parallel at the same time and gave the same results.

The PUF plugs were home-made by cutting them out of a PUF mat which was frozen in liquid nitrogen. The cleaning was first done by repeated water rinsing to remove visible dirt and then for 8 h by Soxhlet extraction with acetone.

1.3.3.3 Sample Treatment

After adsorption of the biocides, these have to be desorbed and made accessible to analysis. For desorption from PUF the Soxhlet extraction is commonly used. As alternatives supercritical fluid extraction (SFE) with CO_2 and elution with different solvents were tested.

For the determination of the recovery of the WPA, all three methods were used to extract PUF which was spiked with 20 µl or 100 µl of a solution of a WPA mixture in *n*-hexane(Fig. 1.3-2). Total amounts were 0.8–1.0 µg and 4–5 µg, respectively. The desorption was started at least 30 min after spiking to evaporate the solvent and simulate adsorption. For some determinations, sampling was simulated by sucking chamber air through the spiked PUF.

Extraction of the spiked PUF was done simulating as much as possible interaction of WPA and PUF; this means the WPA had to cross the complete PUF (the spiked side of PUF was aligned to the desorption solvent inlet).

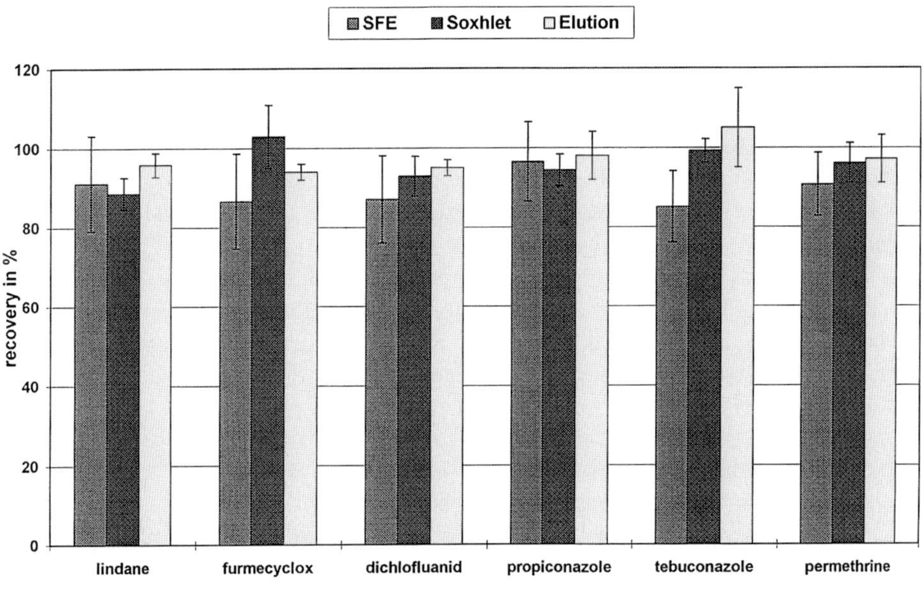

Figure 1.3-2. Comparison of the recoveries for Soxhlet extraction, SFE and elution of WPA from PUF.

Desorption conditions

a) SFE

A SFE-703 from Dionex was used for SFE. Different extraction cells with volumes from 2 ml to 32 ml were tested. For the PUF used in the VDI guideline the biggest cell had to be used because of the PUF size (diameter 56 mm, length 50 mm). In this case SFE gave poor recovery (between 5 % and 25 %) for the WPA. SFE of PUF22 which could be placed in 10-ml extraction cells gave better recovery (between 40 % and 60 %). Compressing the PUF22 in the 10-ml cell to 2 ml by a stainless steel insert gave recoveries between 85 % and 97 % for the WPA.

The SFE was done at temperature 80 °C, pressure 450 bar, extraction time 30 min and restrictor temperature 150 °C; the trapping solvent was dichloromethane.

b) Soxhlet extraction

Soxhlet extraction was performed in a 100-ml extractor with 150 ml of a mixture of 5 % diethyl ether in *n*-hexane. The extraction time was 24 h in the beginning but was decreased to 4 h (25 cycles), which gave the same results.

c) Elution

Elution was first tried for PUF13 in the glass tubes because it could be done without more equipment. Only a pear-shaped flask was needed to collect the eluate. 50 ml (5 times 10 ml) of solvent was used but it was shown that 20 ml was also enough. The best solvent for all investigated WPA turned out to be acetone, but dichloromethane and mixtures of ethyl acetate/*n*-hexane (1 : 1) and *n*-hexane/ether (5 : 1) also gave good recoveries.

After elution, PUF13 still fixed in the glass tube was dried at 50 °C for 1 h and was again ready for sampling.

PUF22 could also be eluted with recoveries of over 90 %. It had to be transferred to a 20-mm diameter glass tube which was tapered to 5 mm at the end. Five 10-ml portions of acetone were used for elution and the PUF22 was afterwards squeezed with a glass stick.

All the extracts from a), b) and c) were concentrated to 1 ml using a rotary evaporator at 30 °C and 2.1×10^4 Pa after adding 1 ml hexane as keeper (not for b) and the internal standard.

All three methods give about the same good recovery. Soxhlet extraction requires most time and solvent. SFE only needs 15 ml solvent for trapping the analytes and takes about 1.5 h but is the most expensive method. Elution is simple to carry out in a very short time. This made it the method of choice for the chamber measurements.

1.3.3.4 Breakthrough Volumes

Breakthrough was tested under sampling conditions with a second PUF plug. For PUF22 there was no breakthrough for 52 m^3 of 18 $\mu g/m^3$ dichlofluanid at a flow of 760 l/h. For lindane and furmecyclox no breakthrough was observed through PUF13 at air concentrations of 134 $\mu g/m^3$ and 129 $\mu g/m^3$ respectively at a flow of 67 l/h and a sample volume of 1.3 m^3. A sample volume of 13 m^3 with 95 l/h was also not high enough to carry out furmecyclox, lindane, dichlofluanid, propiconazole, tebuconazole or permethrine spiked onto PUF13 in concentrations between 4 µg and 5 µg. PUF22 spiked with the same amount of the above-mentioned substances and passed with 55 m^3 air at 750 l/h gave the same recovery as spiked PUF22 without simulation of sampling.

1.3.3.5 Adsorption onto a Dust Filter

With the use of a quartz filter in front of the PUF22 the possible presence of dust in the test chambers was investigated. 50 m^3 chamber air were sucked through the filter. The filter was weighed before and after under the same climate conditions and no dust was found gravimetrically. Even on a special gold-plated filter (nucleopore) which is normally used for determination of asbestos in air there was no detection of dust particles with an electron microscope.

In spite of this observation, extraction of the filters showed nearly the same concentrations for tebuconazole and permethrine as found for sampling with PUF alone (see Table 1.3-1). These results show that for the sampling of biocides with very low vapor pressures (e.g. permethrine and tebuconazole) a quartz filter retains not only particle- or dust-bonded permethrine but also gaseous permethrine. Sampling with a quartz filter in front of the PUF does not allow differentiation between the two.

Table 1.3-1. Comparison of the concentrations of WPA using PUF alone (PUF13) and with filters (nucleopore filter, quartz filter, PUF22) using parallel sampling. Results in ng/m^3 (n.u.: not used n.d.: not detectable).

	PUF22	Quartz filter	Nucleopore filter	Sum	PUF13 without filters
Dichlofluanid	3600	1400	n.u.	5000	5400
Tebuconazole	n.d.	27.2	n.u.	27.2	25.2
Permethrine	n.d.	14.1	n.u.	14.1	14.6
Dichlofluanid	3770	540	n.d.	4310	4900
Tebuconazole	n.d.	6.4	20.3	26.7	31.0
Permethrine	n.d.	11.3	5.4	16.7	15.5
Dichlofluanid	3700	650	n.u.	4350	5600
Tebuconazole	n.d.	26.9	n.u.	26.9	34.9
Permethrine	n.d.	16.8	n.u.	16.8	17.6
Dichlofluanid	3560	430	160	4150	4700
Tebuconazole	n.d.	6.5	20.0	26.5	35.5
Permethrine	n.d.	7.4	9.2	16.6	18.3

1.3.3.6 Analysis

For analysis of the WPA, gas chromatography was used in conjunction with FID, NPD and MSD (Fig. 1.3-3).

Although the concentrations of permethrine were quite low (10–30 ng/m^3) identification by retention time and quantification was possible with GC-FID because the chamber air was very clean and the PUF blank did not interfere with the peaks of the two permethrine isomers. A cold injection system (Gerstel, Mühlheim) did improve the detection limit by better transfer and allowing the injection of 2–10 μl of the extract.

The NPD was employed parallel to the FID (split ratio FID/NPD was 3 : 1) as a selective detector for the nitrogen-containing WPA, e.g. tebuconazole.

MSD gives even better detection limits (1–3 ng/m^3) by selected ion monitoring. It was also possible to identify unknown peaks as decomposition products of dichlofluanid and propiconazole which came out with less than 2 % of the original WPA air concentration.

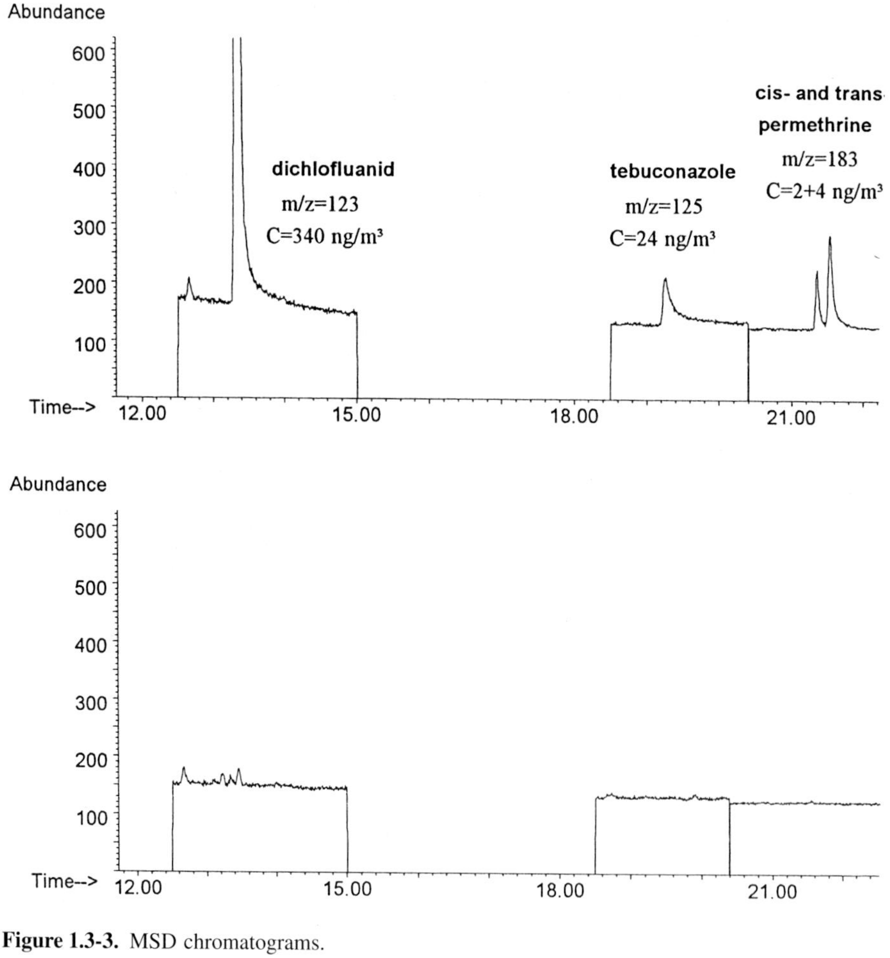

Figure 1.3-3. MSD chromatograms.

For quantification 4-(4'-chlorobenzoyl)pyridine as internal standard was added to the extract. Concentrations of WPA were calculated with determined relative response factors.

Selected ion monitoring (SIM) was done with the following masses (target ion underlined):

Lindane	180.9	182.8	218.85
Chlorobenzoylpyridin	111.0	139.0	217.0
Dichlofluanide	122.9	223.3	225.9
Propiconazole	172.9	174.9	258.9
Tebuconazole	125.0	250.0	252.0
Permethrine	162.9	164.9	183.0

GC-FID/NPD was done with an HP GC 5890 with CIS 3. Helium flow was split 3 : 1 (FID:NPD ratio). 2 µl of the extract was injected. The column was an Rtx-200, 30 m, 0.32 mm, 0.25 µm from Restek. The temperature program was started at 80 °C, held for 1.5 min, increased at 25 °C/min to 170 °C, then 5 °C/min to 260 °C, then 20 °C/min to 280 °C and held for 5 min. CIS had a starting temperature of 20 °C for 15 s solvent venting in split mode, then splitless for 90 s during heating to 280 °C at 10 °C/s.

Two GC-MS systems were employed: an HP GC 5890 II plus with MSD 5972 and an HP GC 5890 with MSD 5971a both with CIS. Columns were Rtx-200 from Restek, 30 m, 0.25 mm, 0.25 µm and HT5 from SGE, 25 m, 0.22 mm, 0.1 µm.

MSD parameters were: EI mode, interface 300 °C, MS temperature 200 °C, selected ion monitoring with 3,06 cycles/s, dwell time 90 ms, mass peak width 0,5 amu.

Temperature program: 60 °C for 1 min, 20 °C/min to 200 °C, 5 °C/min to 250 °C, 20 °C/min to 290 °C for 5 min.

1.3.4 Experimental Results

The concentrations found in the air of a 0.02 – m^3 chamber for a mixture of dichlofluanid (0.5 %), tebuconazole (1.0 %) and permethrine (0.1 %) over a time of more than 800 days are shown in Fig. 1.3-4. More experiments are described by Jann et al. (1997).

The concentration for dichlofluanid decreases with time from about 20 µg/m^3 to less then 1 µg/m^3. This was the expected lapse, although starting concentration was above the saturation concentration if calculated by vapor pressure (1.3×10^{-4} Pa to 1.4×10^{-5} Pa at 20 °C). In contrast to this normal decrease are the curves for permethrine and tebuconazole. At the beginning of the measurements the concentrations were just detectable. During the first weeks there was a slight increase up to 10 ng/m^3. Then there was a further increase to about 30 ng/m^3. After 150 days a slight decrease in the air concentration started. This course of the concentration increase might be explained by sink effects. Because of their low vapor pressure tebuconazole and permethrine are more likely to be adsorbed onto the inside surface of the chamber. Before the air concentration can rise to the steady state concentration the chamber walls must be saturated with the biocides.

Figure 1.3-5 shows the increase in the permethrine concentration in an empty chamber which was connected to the air outlet of the chamber with steady state concentration. It

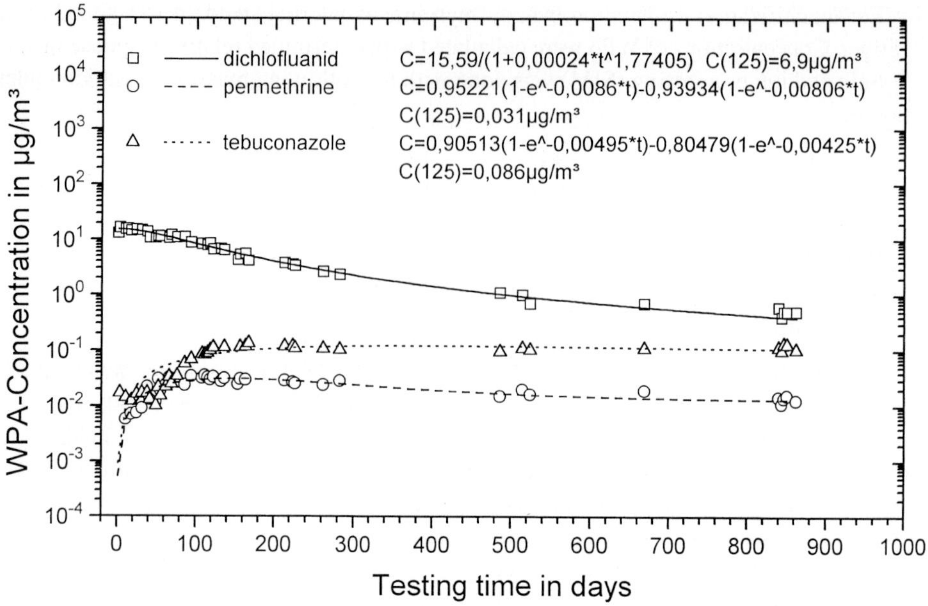

Figure 1.3-4. WPA concentrations: long-term experiment.

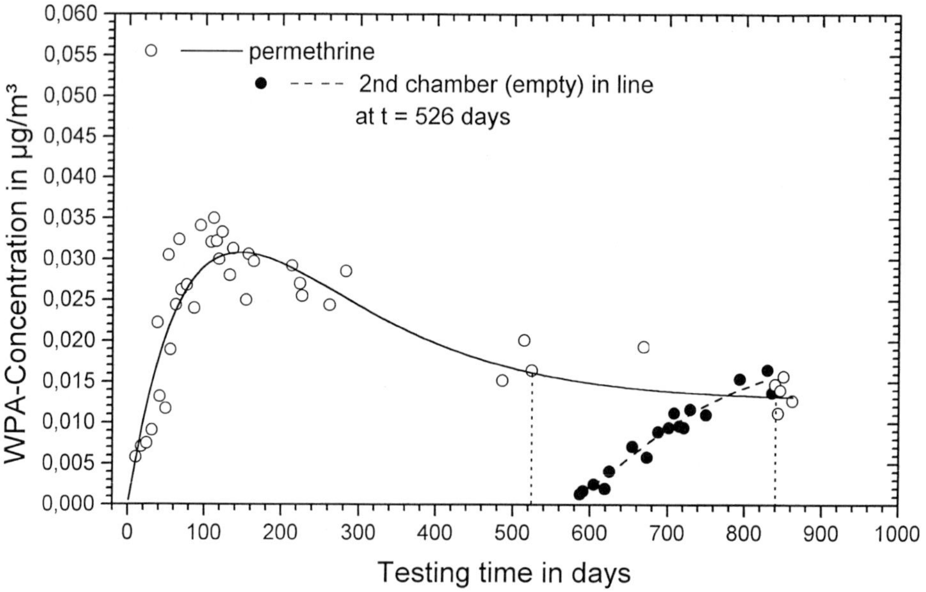

Figure 1.3-5. Empty chamber connected to the test chamber: permethrine.

Table 1.3-2. Permethrine and tebuconazole concentrations on adsorption-surfaces after 23 days in a 0.02-m^3 test chamber.

		Permethrine		Tebuconazole	
		μg	μg/m^2	μg	μg/m^2
Watch glasses	1. elution	2.95	22.7	10.6	81.5
5 pieces 0.026 m^2	2. elution	0	0	0.295	2.3
A = 0,13 m^2	sum	2.95	22.7	10.9	83.8
Stainless steel plates	1. elution	0.5	8	5.19	83.0
5 pieces 0,0125 m^2	2. elution	0	0	1.25	20.0
A = 0,0625 m^2	sum	0.5	8	6.44	103.0
total	sum	3.45	17.9	17.34	90.1

took nearly 300 days to get the same permethrine air concentration in the second chamber. Dichlofluanid concentration was the same after 10 days only. Tebuconazole concentration did not rise to the concentration of the first chamber, because it condensed in the connecting glass tube.

Further evidence for sink effects was that the same air concentration was obtained after removing the treated wood from inside the chambers. Obviously the loaded walls started to emit the adsorbed WPA. Eleven days afterwards tebuconazole and permethrin were still found at about 70 % of the value before the removal. Dichlofluanid concentration decreased faster and fell to 10 % after 11 days.

A characterization of the sink effects was done by inserting freshly cleaned (rinsed with acetone) glass and metal plates inside the chambers. After 23 days the plates were taken out and rinsed twice with about 50 ml acetone. The concentrations found are shown in Table 1.3-2.

In Table 1.3-3 the calculated emission rates and emission factors are given for dichlofluanid, tebuconazole and permethrine calculated from the concentration after 125 days (equilibrium). An input/output calculation over the time of 862 days is also given.

From Table 1.3-3 it can be seen that even over the time period of 862 days (nearly 3 years) less than 1 % of the applied amount of the very low volatile biocides such as tebuconazole and permethrine has been released into the air. In contrast to this the release for dichlofluanid amounts to about 25 %.

Table 1.3-3. Emission characteristics for dichlofluanid, tebuconazole and permethrine in a 0.02-m^3 test chamber.

WPA	C	$SER_A(t)$	ER(t)		WPA	WPA	WPA
	[μg/m^3]	(C × 5 m/h) [μg/hm^2]	(C × 3 m^3/d) [μg/d]	(C × 0,125 m^3/h) [μg/h]	output [μg]	input [mg]	out : in [%]
		after 125 days			over 862 days		
Dichlofluanid	6.9	34.5	20.7	0.8625	6834.0	28.0	24.41
Permethrine	0.031	0.155	0.093	0.003875	50.7	5.5	0.92
Tebuconazole	0.086	0.43	0.258	0.01075	244.0	55.0	0.44

Comparison between the three types of test chambers showed good agreement only for the concentration of dichlofluanid in the 1-m^3 and 0.02-m^3 chambers. For the 0.625-l chambers problems occurred due to leaks because of unsufficient tightness. Permethrine and tebuconazole concentrations in the 1-m^3 chamber did not rise to the same level as in the 0.02-m^3 chamber within the testing time of 125 days. Probably the ratio of chamber inside surface to the surface of the test specimen was too large in the 1-m^3 chamber to get an air/surface equilibrium during that time.

1.3.5 Conclusions

By means of chamber tests in combination with an appropriate sampling and desorption method it is possible to get information about the emission behavior of wood preservative agents out of treated wood or generally of biocides emitted from other materials. In spite of the sink effects determined it is possible to calculate emission rates if an equilibrium between air and wall concentration is reached. Furthermore, it is possible to get information about sink effects if certain materials are introduced into the chamber and analyzed afterwards or the biocide concentration on the chamber walls is determined. From this data, in combination with the knowledge of the amount of biocides used for treatment, relative losses by emissions can be calculated.

1.3.6 Summary

Organic wood preservative agents can be emitted from treated wood. The emission can be measured by means of test chambers in combination with sampling on polyurethane foam and solvent desorption. Especially the simple method of elution has great advantages with respect to reduced costs and time. The emission values and the emission characteristics depend on the volatility of the compounds. For biocides with a wide range of volatility from dichlofluanid to permethrine the described method is applicable. Other biocides such as lindane and furmecyclox have also been investigated by the sampling method, and the method can also be used for these higher-volatility compounds.

The calculation of emission rates is possible after a sufficient time. For the substances investigated the 0.02 m^3 chamber showed the best results. In this chamber after a maximum test duration of 125 days no further increase in the concentrations of the lower-volatility substances occurred. Comparisons with other chambers in some cases showed lower concentrations due to sink effects affecting the chamber air concentration.

Acknowledgement

The authors would like to thank the German Federal Environmental Office (Umweltbundesamt, UBA) for the financial contribution to the investigations (UBA project-no. 12606003).

References

American Society for Testing and Materials (1989): Standard practice for sampling and analysis of pesticides and polychlorinated biphenyls in indoor atmospheres ASTM-Designation: D 4861–88 (1989), 388–401.

DIN V ENV 1250-1 (1995, Vornorm): Holzschutzmittel, Verfahren zur Bestimmung der Abgabe von Wirkstoffen und anderen Schutzmittelbestandteilen aus behandeltem Holz–Teil 1: Laboratoriumsverfahren zur Gewinnung von Analysenproben zur Bestimmung der Abgabe durch Verdunstung.

Figge K., Rabel W. and Wieck A. (1987): Adsorptionsmittel zur Anreicherung von organischen Luftinhaltsstoffen. Fresenius Z. Anal. Chem., *327*, 261–278.

Hawthorne S.B., Krieger M.S. and Miller D.J. (1989): Supercritical carbon dioxide extraction of PCB, PAH, heteroatom-containing PAH and *n*-alkanes from polyurethane foam sorbents. Anal. Chem, *61*, 736–740.

Horn W. and Marutzky R. (1993): Measuring of organic wood preservatives in indoor air–sampling and 1-m^3 chamber tests. Proceedings of Indoor Air, Vol. 2, 513–518.

Horn W. and Marutzky R. (1993): Untersuchungen zur Bestimmung von organischen Holzschutzmittelwirkstoffen bzw. deren Abbauprodukten in der Luft von Wohngebäuden Forschungsbericht Wilhelm-Klauditz-Institut (WKI) Fraunhofer-Arbeitsgruppe für Holzforschung. Abschlussbericht E 90/25–WKI 105988 Braunschweig.

Jann O. (1995): Bestimmung der Emission aus Materialien Neuere Methoden der Umweltanalytik 21. UTECH-Seminar Berlin 93–113.

Jann O., Wilke O., Walther W. and Ullrich D. (1997): Entwicklung eines standardisierbaren Prüfverfahrens zur Bestimmung des Eintrages von Holzschutzmittel-Wirkstoffen aus behandeltem Holz, Altholz und daraus hergestellten Holzwerkstoffen in die Luft. UBA-Forschungsvorhaben 126 06 003, Bundesanstalt für Materialforschung und -prüfung (BAM), Berlin.

Kubiak R., Maurer T. and Eichhorn K.W. (1993): A new laboratory model for studying the volatilization of pesticides under controlled conditions. Science Total Environ., *132*, 115–123.

Landsiedel H. and Plum H. (1981): Untersuchungen zur Flüchtigkeit von Tributylzinn-Verbindungen. Holz als Roh- und Werkstoff, *39*, 261–264.

Lewis R.G. and Jackson M.D. (1982): Modification and evaluation of a high-volume air sampler for pesticides and semivolatile industrial organic chemicals. Anal. Chem., *54*, 592–594.

Patton G.W., McConnell L.L., Zaranski M.T. and Bidleman T.F. (1992): Laboratory evaluation of polyurethane foam-granular adsorbent sandwich cartridges for collecting chlorophenols from air. Anal. Chem., *64*, 2858–2861.

Petrowitz H.-J. (1986): Zur Abgabe von Holzschutz-Wirkstoffen aus behandeltem Holz an die Raumluft. Holz als Roh- und Werkstoff, *44*, 341–346.

Prueger J.H. and Pfeiffer R.L. (1994): Preliminary tests of a laboratory chamber technique intended to simulate pesticide volatility in the field. J. Environ. Qual., *23*, 1089–1093.

Turner B.C. and Glotfelty D.E. (1977): Field air sampling of pesticide vapors with polyurethane foam. Anal. Chem., *49*, 7–10.

van Loy M.D. and Nazaroff W.W. (1997): Dynamic behavior of semivolatile organic compounds in indoor air, 1. Nicotine in a stainless steel chamber. Environ. Sci. Technol., *31*, 2554–2561.

VDI-Richtlinie 3498 Blatt 1 (Entwurf 1993): Messen von Innenraumluft; Messen von polychlorierten Dibenzo-*p*-dioxinen und Dibenzofuranen. VDI-Handbuch Reinhaltung der Luft, Band 5, Beuth, Berlin.

VDI-Richtlinie 4300 Blatt 4 (1997): Messen von Innenraumluftverunreinigungen–Messstrategie für PCP und Lindan in der Innenraumluft. Beuth, Berlin.

Wilke O. and Ullrich D. (1995): Luftprobenahme und Analyse von Holzschutzmittel-Wirkstoffen aus Prüfkammerluft. Neuere Methoden der Umweltanalytik, 21. UTECH-Seminar, Berlin, 115–129.

Zaranski M.T., Patton G.W., McConnell L.L. and Bidleman T.F. (1991): Collection of nonpolar organic compounds from ambient air using polyurethane foam-granular adsorbent sandwich cartridges. Anal. Chem., *63*, 1228–1232.

Zimmerli B. (1982): Modellversuche zum Übergang von Schadstoffen aus Anstrichen in die Luft. In: Aurand K, Seifert B. and Wegner J. (eds) Luftqualität in Innenräumen. Gustav Fischer Verlag, Stuttgart, 235–267.

References

[illegible]

1.4 Sampling and Analysis of PCDDs/PCDFs, PAHs and PCBs

Michael Ball and Tunga Salthammer

1.4.1 Introduction

Polychlorinated dibenzodioxins and polychlorinated dibenzofurans (PCDDs/PCDFs), polycyclic aromatic hydrocarbons (PAHs), and polychlorinated biphenyls (PCBs) are now ubiquitous in the environment. Their occurrence in indoor air mainly results from the frequent use of chemical products such as paints, glue, sealants, fire retardants and wood protection agents up to the mid-1980s. Furthermore, PCDDs/PCDFs and PAHs are always formed to some extent during incomplete combustion processes. A list of possible sources is given in Table 1.4-1.

Except for naphthalene, which belongs to the group of volatile organic compounds (VOCs), PCDDs/PCDFs, PAHs and PCBs are classified as semivolatile organic compounds (SVOCs) or as organic compounds associated with particulate organic matter (POM). SVOCs and POM have low vapor pressures and deposit on surfaces of indoor equipment, dust and other particles. As a consequence of critical toxicological properties and human health risk on exposure, guidelines for evaluation, exposure, sanitation and waste disposal have been recommended by the WHO and several other authorities (see Maroni et al., 1995; Schulz, 1993; Ballschmiter and Bacher, 1996; Anderson and Albert, 1999).

Table 1.4-1. Possible sources of PCDDs/PCDFs, PAHs and PCBs in the indoor environment.

Substance group	Sources
PCDDs/PCDFs	Pentachlorophenol and other biocides in wood preservatives, textiles, leather
	Products containing PCBs
	Open fires in the presence of halogenated materials
PAHs	Tobacco smoke
	Combustion processes (fire-places, exhaust of diesel engines, chimneys, candles, barbecue)
	Paints and glues containing tar oil or bitumen
	Wood protection agents
PCBs	Defective electrical capacitors, e.g. in fluorescent light tubes
	Defective electrical transformers
	Paintings and varnishes containing flame retardants
	Plasticizers in sealing materials
	Wood protection agents

1.4.2 Classification and Occurrence

1.4.2.1 PCDDs/PCDFs

The collective term „dioxins" classifies two groups of chlorinated aromatic ethers: the 75 isomers of chlorinated dibenzodioxin (PCDDs) and the 135 isomers of chlorinated dibenzofuran (PCDFs), a total of 210 congeners. The 49 tetra-octa-CDDs and the 87 tetra-octa-CDFs are of toxicological relevance (Rotard, 1990; Schuster and Dürkop, 1993). Main indoor sources of PCDDs/PCDFs are large wood surfaces treated with wood preservatives in the 1970s. The application of biocides such as pentachlorophenol (PCP) necessarily caused contamination with dioxins and furans, because PCDDs/PCDFs are undesired but unavoidable by-products of chemical synthesis, and concentrations of about 1 mg/kg PCDDs/PCDFs in PCP have been reported (Blessing, 1997). Furthermore, dibenzofurans are contaminants of PCBs, whose application also caused increased concentrations of PCDFs. Blessing (1997) has measured >20 mg/kg pentachlorodibenzofuran (PeCDF) in Clophen (see Section 1.4.2.3). In case of combustion, the formation of PCDDs/PCDFs might occur via precursors or via de novo reactions in the presence of organically bonded carbon and chlorine. It was demonstrated by Scholz-Böttcher et al. (1992) that the well-known Beilstein test, which is frequently used as a screening method for chlorine, gives ideal conditions for the formation of PCDDs/PCDFs. For the evaluation of PCDD/PCDF concentrations found in materials or urban air, toxicity equivalencies (TEQs) have been calculated in relation to 2,3,7,8-TCDD (TEQ = 1). The TEQ value is determined by multiplying the concentrations of the PCDD/PCDF isomers by the related toxicity equivalence factors followed by consecutive addition of the obtained products. Toxicity equivalence factors were established by different committees and administrative bodies (see Table 1.4-2). An international method for determination of TEQ values, which is now widely accepted, was established by NATO/CCMS (ITE values). When calculating equivalencies according to the former German Federal Health Office (BGA – Bundesgesundheitsamt), the German Federal Environmental Agency (UBA – Umweltbundesamt) and the United States Environmental Protection Agency (EPA), the summation values of the single subgroups have also to be taken into consideration. The summation values for the TCDD-HpCDDs and TCDF-HpCDFs do not include the 2,3,7,8-chlorine-substituted isomers of the groups. In Germany a PCDD/PCDF concentration <0.5 pg TEQ/m^3 (BGA/UBA) is recommended for preventive purposes (Sagunski et al., 1989; Pluschke, 1996).

Typical background concentrations of PCDDs/PCDFs (outdoor air) are in the range from <0.05 pg ITE/m^3 to 0.2 pg ITE/m^3. Päpke et al. (1989) have determined PCDDs/PCDFs indoor air concentrations in a number of kindergartens where wood preservatives had been previously used. The total concentrations of PCDDs/PCDFs ranged from 2.58pg/m^3 to 427pg/m^3 [0.050 – 1.553pg TEQ/m^3 (BGA/UBA), 0.014 – 0.185pg TEQ/m^3 (EPA)]. Blessing (1997) has investigated 27 public buildings and has measured PCDD/PCDF values from <0.1 pg ITE/m^3 to 1.6 pg ITE/m^3. In four cases the renovation guide value was exceeded (see Table 1.4-3).

Table 1.4-2. Toxicity equivalencies (TEQ) for selected PCDDs/PCDFs in relation to 2,3,7,8-TCDD according to NATO/CCMS (ITE), BGA/UBA and EPA.

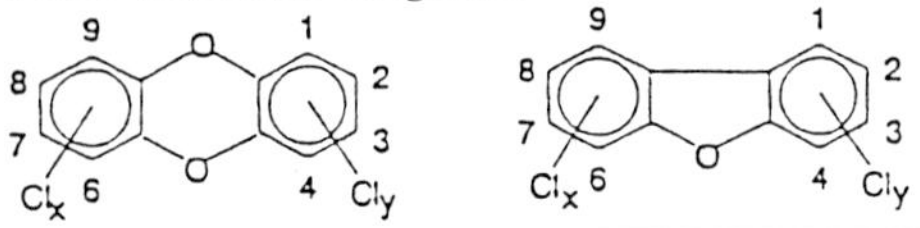

PCDD and PCDF congeners	NATO/ CCMS	BGA/UBA	EPA
2,3,7,8–tetrachlorodibenzodioxin (TCDD)	1	1	1
1,2,3,7,9-pentachlorodibenzodioxin (PeCDD)	0.5	0.1	0.5
1,2,3,4,7,8-hexachlorodibenzodioxin (HxCDD)	0.1	0.1	0.04
1,2,3,6,7,8-hexachlorodibenzodioxin (HxCDD)	0.1	0.1	0.04
1,2,3,7,8,9-hexachlorodibenzodioxin (HxCDD)	0.1	0.1	0.04
1,2,3,4,6,7,8-heptachlorodibenzodioxin (HpCDD)	0.01	0.01	0.01
Octachlorodibenzodioxin (OCDD)	0.001	0.001	-
2,3,7,8-tetrachlorodibenzofuran (TCDF)	0.1	0.1	0.1
2,3,4,7,8-pentachlorodibenzofuran (PeCDF)	0.5	0.1	0,1
1,2,3,7,8–pentachlorodibenzofuran (PeCDF)	0,05	0.1	0.1
1,2,3,4,7,8-hexachlorodibenzofuran (HxCDF)	0.1	0.1	0.01
1,2,3,6,7,8-hexachlorodibenzofuran (HxCDF)	0.1	0.1	0.1
1,2,3,7,8,9-hexachlorodibenzofuran (HxCDF)	0.1	0.1	0.1
1,3,4,6,7,8-hexachlorodibenzofuran (HxCDF)	0.1	0.1	0.1
1,2,3,4,6,7,8-heptachlorodibenzofuran (HpCDF)	0.01	0.01	0.01
1,2,3,4,7,8,9-heptachlorodibenzofuran (HpCDF)	0.01	0.01	0.01
Octachlorodibenzofuran (OCDF)	0.001	0.001	-
Σ-TCDD[1]	-	0.01	0.1
Σ-PeCDD[1]	-	0.01	0.005
Σ-HxCDD[1]	-	0.01	0.0004
Σ-HpCDD[1]	-	0.001	0.00001
Σ-TCDF[1]	-	0.01	0.001
Σ-PeCDF[1]	-	0.01	0.001
Σ-HxCDF[1]	-	0.01	0.0001
Σ-HpCDF[1]	-	0.001	0.00001

[1] Applied summation values do not include the respective 2,3,7,8-chloro-substituted isomers.

Table 1.4-3. Typical concentrations of PCDDs/PCDFs, PAHs and PCBs in different types of homes.

Compound	Unit	Range	Reference
PCDDs/PCDFs	pg ITE/m^3	< 0.1 – 1.6	Blessing, 1997
OCDD	pg/m^3	40 – 50	Blessing, 1997
PCDDs/PCDFs [1]	pg TEQ/m^3	0.050 – 1.553	Päpke et al., 1989
PCDDs/PCDFs [2]	pg TEQ/m^3	0.014 – 0.185	Päpke et al., 1989
PCDDs/PCDFs	pg/m^3	2.58 – 427	Päpke et al., 1989
PCDDs/PCDFs [1]	pg TEQ/m^3	0.5 – 2.0	Sagunski et al., 1989
Phenanthrene	ng/m^3	9.2 – 210	Wilson et al., 1991
Benzo(a)pyrene	ng/m^3	<0.01 – 4.13	Wilson et al., 1991
Chrysene	ng/m^3	0.18 – 8.61	Wilson et al., 1991
Fluoranthene	ng/m^3	2.40 – 37.37	Wilson et al., 1991
Pyrene	ng/m^3	1.4 – 18.1	Wilson et al., 1991
PCBs	ng/m^3	<60 – >3000	Pluschke, 1996
PCBs	ng/m^3	39 – 580	MacLeoad, 1981

[1] TEQ according to BGA/UBA.
[2] TEQ according to EPA.

1.4.2.2 PAHs

The distribution and fate of PAHs in the environment has been reviewed in detail (Jacob et al., 1984; 1986; 1991). Human exposure is of particular concern, since many PAHs show mutagenic and/or carcinogenic activity in screening tests and animal experiments. About 500 PAHs have been detected in the air, EPA has included 16 PAHs on the list of priority pollutants (see Table 1.4-4). In the indoor environment, the major sources of PAHs are infiltration from outdoor air and combustion products from cigarette smoking, heaters, stoves and fireplaces. Furthermore, PAHs can be released from materials containing tar oil and bitumen, which were used as glue for floor parquet. The latter is now recognized as a severe problem of indoor contamination. In homes with floor parquet, where tar oil and bitumen had been used, PAH concentrations up to 1000 mg/kg can be measured in house dust. In one case, analysis of house dust (vacuum cleaner) yielded values of 2618 mg/kg and 179 mg/kg for total PAH and benzo(a)pyrene (BaP), respectively. BaP is recommended as the target compound for evaluation of indoor contamination with PAHs. Guideline values were proposed by the German Federal Environmental Agency (UBA) for analysis of material, house dust and indoor air (UBA, 1998a;b). PAH concentrations in indoor air were investigated in 33 homes by Wilson et al. (1991). The values varied widely from a few pg/m^3 to several hundred ng/m^3. The overall levels were much higher in homes occupied by smokers. Concentration ranges are compiled in Table 1.4-3 for some target compounds. Halogenated PAHs are also of relevance for indoor pollution. Especially mono- and dichloronaphthalenes have been used as fungicides since the early 20th century. Later, chlorinated naphthalenes were replaced by pentachlorophenol (Pluschke, 1996).

Table 1.4-4. 16 PAHs classified as priority pollutants by EPA.

Compound	Formula	b.p. [°C]
Naphthalene (NAP)	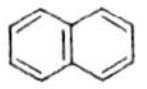	218
Acenaphthylene (ANY)	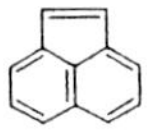	270
Acenaphthene (ANE)	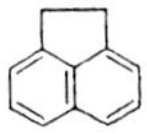	297
Fluorene (FLU)	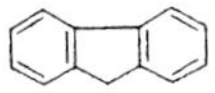	294
Phenanthrene (PHEN)	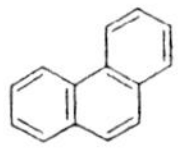	338
Anthracene (ANT)	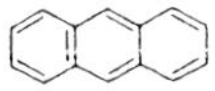	340
Fluoranthene (FLA)	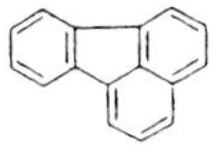	383
Pyrene (PYR)	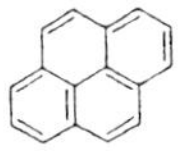	393
Benzo[*k*]fluoranthene (BkFL)	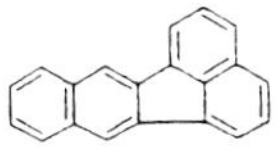	481
Benzo[*g,h,i*]perylene (BghiP)	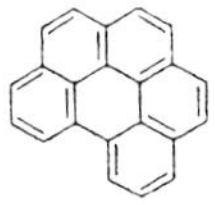	542
Benzo[*a*]anthracene (BaA)	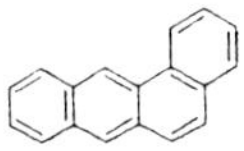	425
Chrysene (CHR)	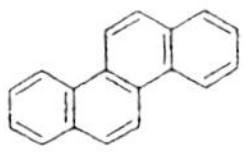	431
Benzo[*b*]fluoranthene (BbFL)	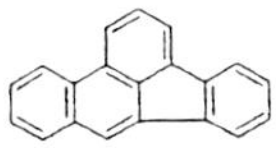	481
Benzo[*a*]pyrene (BaP)	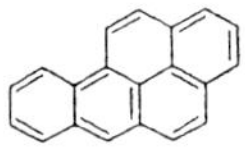	496
Dibenzo[*a,h*]anthracene (DbahA)	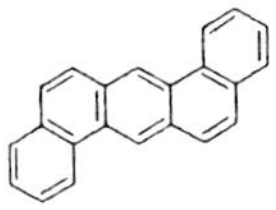	535
Indeno[1,2,3−*c,d*]pyrene (IND)	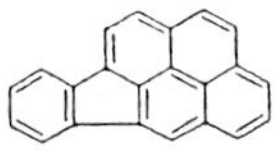	534

1.4.2.3 PCBs

Polychlorinated biphenyls (PCBs) were commercially used for many years because of their non-flammability, electrical insulating properties and chemical and thermal stability. In 1980, the overall world production since 1929 was estimated to be close to 2×10^9 kg (Ballschmiter and Zell, 1980). PCBs are ubiquitous and can be detected in rivers, oceans, arctic ice, animals and even human mother's milk. There are 209 possible PCB congeners; 20 can attain a planar configuration due to the absence of substituents in ortho-position. The most toxicologically active congeners are those having chlorine substitution at both para (4,4') and two meta positions (3,3',5,5') (Creaser et al., 1992). Hanberg et al. (1990) have proposed toxicity equivalencies for selected PCBs relative to 2,3,7,8-TCDD. Ballschmiter and Zell (1980) have suggested a systematic numbering of PCB compounds. Nowadays, their system is widely accepted and known as the *Ballschmiter nomenclature*. For practical reasons, the 6 PCB congeners with Ballschmiter nos. 28, 52, 101, 138, 153 and 180 are monitored when carrying out material and air analysis (see Table 1.4-5). The total value is then multiplied by 5 to estimate the concentration of total PCBs. Many of the products containing PCBs such as textile dyes, paints, fireproofing agents, transformers and electrical capacitors were made for indoor use (MacLeod, 1981). For primary sources, PCB contents of 1 g/kg or more are common. Distinctly lower concentrations of <1 g/kg are found when secondary sources are considered. A main emission source of PCBs indoors are polymeric sealing materials, which were especially applied in buildings constructed of concrete slabs. Such sealants contained technical PCB mixtures (e.g. Aroclor and Clophen) in amounts of 5–45 % (Pluschke, 1996). Schwerdt (1994) has determined 2480 mg/kg of the coplanar compound 3,3',4,4'-tetrachlorobiphenyl (PCB 77) in Clophen. In very few cases were PCBs used in wood protection agents. Typical PCB concentrations in indoor air are in the order of ng/m^3. MacLeod (1981) has measured levels between 39 ng/m^3 and 580 ng/m^3 when studying airborne PCBs in 14 homes. In one

Table 1.4-5. Target PCBs with Ballschmiter numbers.

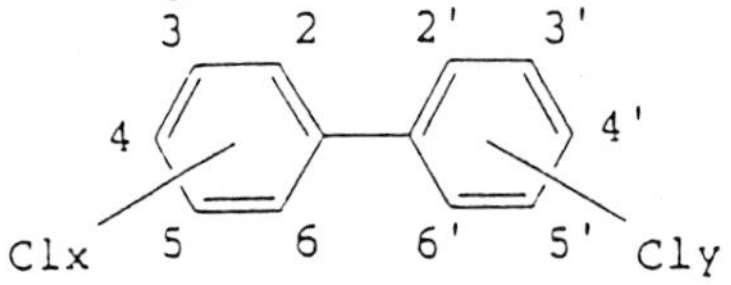

PCB congeners	Ballschmiter No.
2,4,4'-Trichlorobiphenyl	28
2,2',5,5'-Tetrachlorobiphenyl	52
2,2',4,5,5'-Pentachlorobiphenyl	101
2,2',3,4,4',5-Hexachlorobiphenyl	138
2,2',4,4',5,5'-Hexachlorobiphenyl	153
2,2'3,4,4'5,5'-Heptachlorobiphenyl	180

special case, a room contained a fluorescent tube with a defective igniter. The PCB concentration on the day of burnout was 11600 ng/m^3 and decayed to <200 ng/m^3 within 12 months. In Germany, the guideline for evaluation and sanitation of PCB-contaminated buildings gives the following recommendations. Indoor air concentrations <300 ng PCBs/m^3 (precautionary value) are to be considered as tolerable on a long-term scale. For indoor air concentrations between 300 ng/m^3 and 3000 ng/m^3 air it is recommended to locate the emission source and to remove it if the expense can be tolerated. If this is not possible, the PCB concentration should at least be reduced by intensive regular ventilation and thorough cleaning and dust removal. The target value is <300 ng/m^3. Indoor air concentrations of PCBs >3000ng/m^3 (renovation guide value) should be avoided in any case, as further pathways of human exposure are not controllable. Such high concentrations should first be verified by carrying out control analysis. If the value is confirmed, prompt measures are to be taken to reduce PCB concentrations and to avoid adverse health effects. Again, the target value is <300 ng/m^3.

1.4.3 General Aspects of Measurement Strategies

The German guideline VDI 4300, Part 1 (1995), describes general requirements relating to the measurement of indoor air pollution and the boundary conditions which are to be observed for the special substances or substance groups. Part 2 of VDI 4300 (1997) describes special measurement strategies for PCDDs/PCDFs, PAHs and PCBs. It is advisable to study both guidelines before carrying out investigations in industrial and private living spaces. Especially if SVOCs and POM are the target compounds, the determination of indoor air concentrations is not necessarily the method of choice. In a room treated with PCP-containing wood preservatives, PCDD/PCDF measurements should only be carried out if the quotient of treated wood area (A) and room volume (V) is $A/V \geq 0.2$ m^{-1} and the PCP content is higher than 50 mg/kg of wood (VDI 4300, Part 2). On sampling, the air volume per hour should not exceed 10 % of the room volume. General case-related investigation strategies should be taken into account, and it is of importance to conduct a systematic and stepwise operation. For the first step, a detailed questionnaire is a valuable tool and should specify symptoms and complaints as well as technical information (building materials, furniture, living conditions, etc.) (VDI 4300, Part 1, 1995; Maroni et al., 1995). In many cases it will be more useful to carry out analysis of material, house dust or wiped samples. Generally the sampling procedure is simple and rapid, and the results are suitable for identifying relevant organic components and emission sources. Analysis of material and wiped samples is recommended in case of fires or any other combustion processes to determine the contamination with PCDDs/PCDFs and PAHs. It was demonstrated that deposit of PAHs on surfaces is one of the reasons for the phenomenon of black staining or "Black Magic Dust" that suddenly occurs in living spaces. Fogging experiments and analysis of wiped samples might be suitable methods to study this effect (Wensing et al., 1998; Wensing and Salthammer, 1999).

1.4.4 Sampling and Analysis

PCDDs/PCDFs, PAHs and PCBs require almost identical equipment for sampling. A typical collection device includes the sampling unit (metal filter holder equipped with a particle filter backed up by solid sorbent trap) and a pumping system. A schematic diagram is shown in Fig. 1.4-1. Different types of pumps and adsorption media are briefly described in the following (see Table 1.4-6). Procedures for clean-up and analytical details are then discussed in dependence of the relevant compound group.

1.4.4.1 Pumping Systems

LIB filter apparatus (VDI 3875, Part 1, 1996): This device includes the sampling unit, a rotary pump with an output of about 16m^3/h, a gas meter with thermometer, timer, sample air cooler and silencer.

Small filter apparatus (VDI 3875, Part 1, 1996): This device includes the sampling unit, a rotary pump with an output of 2.6m^3/h to 2.8m^3/h, a separator for the abraded carbon of the rotary pump, a plate anemometer and a timer.

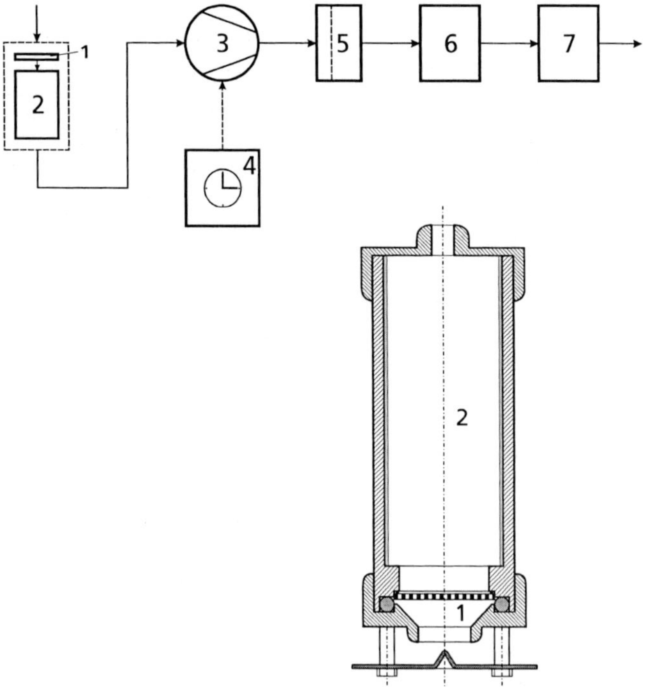

Figure 1.4-1. Scheme of a sampling device for PCDDs/PCDFs, PAHs and PCBs: *1* glass fiber filter, *2* solid sorbent (e.g. PUF or XAD-2), *3* pump, *4* timer, *5* dust filter, *6* cooling unit, *7* flow meter.

Table 1.4-6. Methods for sampling and analysis of PCDDs/PCDFs, PAHs and PCBs in indoor air.

Compound	Sampling equipment	Sampling media	Analysis	Ref.
PCDDs/PCDFs	LIB filter apparatus Small filter apparatus	glass fiber filter/PUF	GC/MS	A
PAHs	Small filter apparatus High volume pump[1]	glass fiber filter glass fiber filter/XAD-2	GC/FID GC/MS	B C,D
PCBs	LIB filter apparatus Small filter apparatus	glass fiber filter/PUF	GC/MS GC/ECD	E

[1] Air flow 15 m^3/h.
A = VDI 3498, Part 1 (1998).
B = VDI 3875, Part 1 (1996).
C = Wilson et al. (1991).
D = Chuang et al. (1990).
E = VDI 2464, Part 1 (1992).

1.4.4.2. Sampling Media

Glass fiber filter (VDI 3498, Part 1, 1998): External diameter 120 mm (LIB apparatus) or 50 mm (small filter apparatus) with a collection efficiency of 99.8 % for particles with an aerodynamic diameter >0.3 μm. Thc filter must be baked-out at 250–400 °C before use.

Polyurethane foam (PUF) (VDI 3498, Part 1, 1998): This is made of toluene diisocyanate (TDI) and polyoxypropylenetriol with a density of 25 kg/m^3 or TDI-polyether foam with a density of 33 kg/m^3 cut into cylinders of 50 mm length and 110 mm diameter. For clean-up, the foam is extracted in a Soxhlet apparatus with toluene (24 h) and acetone (24 h). The PUF is then placed in a vacuum oven and dried at room temperature under nitrogen atmosphere.

Styrene/divinylbenzene polymer (XAD-2): A batch of XAD-2 (500 μm, 20–60 mesh) is placed in a Soxhlet apparatus and extracted with dichloromethane for 16 h. Dichloromethane is replaced with fresh reagent and the procedure is repeated. Then the XAD-2 is placed in a vacuum oven and dried at room temperature for 2–4 h under nitrogen atmosphere.

1.4.4.3 Analysis of PCDDs/PCDFs

In the guideline VDI 3498, Part 1, (1998) the measurement of PCDDs/PCDFs in indoor air by using either the LIB filtering apparatus or the small filter apparatus is described. The sampling unit is equipped with one glass fiber filter and two PUF plugs. The total sampling volume is about 400 m^3. This requires a measurement time of about 27 h for the LIB apparatus and of about 150 h for the small filter apparatus. Before sampling and clean-up, the filter is spiked with known amounts of ^{13}C-marked tracers of 2,3,7,8-chloro-substituted PCDD/PCDF isomers. The tracers are used as internal standards for determination of recoveries and for quantification. Filter and PUF plugs are extracted in a

Soxhlet apparatus with toluene for 20 h. Sample clean-up is done by column chromatography (silica gel and aluminum oxide) in 2 consecutive steps. Analysis is perfomed by high-resolution gas chromatography/mass spectrometry (GC/MS) operated in the selective-ion monitoring mode (SIM). The most intense ion mass peak (base peak) is used for quantification. The method achieves a detection limit of 5 fg/m^3 for 2,3,7,8-TCDD. TEQ-values are calculated by multiplying the concentration of the PCDD/PCDF congener with the corresponding factor (see Table 1.4-2) and addition of products. Blessing (1997) has performed separate analysis of filters and PUF, and 85 % of the analyzed PCDDs/PCDFs could be recovered on the filter. The hexachloro-octachloro-substituted isomers were retained by the filter, while the tetra-penta CDDs/CDFs were detected on PUF. This result also demonstrates that PCDDs/PCDFs are mainly adsorbed on suspended dust and only a small fraction are found in the gas phase.

1.4.4.4 Analysis of PAHs

The guideline VDI 3875, Part 1 (1996), describes the measurement of PAHs. Sampling is carried out using the small filter apparatus. The suspended particulate matter on which the PAHs are adsorbed is deposited on a glass fiber filter. PAHs with a boiling point >475 °C are deposited quantitatively. The quantitative determination of PAHs with boiling points <475 °C is not described in this guideline. The total sampling volume is about 5–10 m^3. The required measurement time for the small filter apparatus is 2–4 h. After sampling and before filter preparation, a known mass of indeno[1,2,3-cd]fluoranthene is added to the sample as an internal standard. The filter is then extracted into toluene by use of a Soxhlet apparatus or an ultrasonic bath. Sample clean-up is done by column chromatography (silica gel). For the gas-chromatographic analysis and detection via flame ionisation (GC/FID), cool-on-column injection and splitless injection are suitable. Quantification is performed by the internal standard method; the detection limit should be <0.1 ng/m^3. PAHs with boiling points <475 °C and vapor pressures >10^{-8} kPa may be distributed between gas phase and particle-associated phase. These PAHs will vaporize from the filter during sampling, and back-up solid sorbents such as PUF plugs or XAD-2 are required. Keller and Bidleman (1984) describe a glass fiber filter-PUF system with analysis via GC/FID and HPLC/fluorescence, which is suitable for collection of 3- and 4-ring PAH vapors. Chuang et al. (1987; 1990) have compared the efficiency of XAD-2 and PUF with sampling volumes of about 300 m^3 and GC/MS analysis. The results show that many PAHs break through the standard PUF plugs, while 2- and 3-ring PAHs are efficiently collected and retained on XAD-2. Hach et al. (1994) also reported insufficient recoveries for PAHs when using PUF as solid sorbent.

1.4.4.5 Analysis of PCBs

For the measurement of PCBs in indoor air, the VDI guideline 2464, Part 1 (1992) is currently in preparation. Sampling can be performed on a glass fiber filter and on PUF plugs by use of the LIB filtering apparatus or a small filter apparatus. Before sampling and clean-up, the filter is spiked with known amounts of ^{13}C-marked tracers of different PCB isomers. The tracers are used as internal standards for determination of recoveries and for quantification. Filter and PUF plugs are extracted in a Soxhlet apparatus with toluene for 20 h. Sample clean-up is done by column chromatography (silica gel and aluminum oxide). Analysis can be perfomed by GC/MS or GC/ECD. The method should give detection limits between 0.1 ng/m^3 and 1 ng/m^3 for the relevant components. Ullrich (1993) has reported results of an intercomparison exercise to determine PCBs in indoor air by use of different techniques. Analytical methods for the determination of non-ortho-substituted PCBs by use of different analytical techniques were reviewed by Creaser et al. (1992). Larsen et al. (1992) have discussed congener-specific analysis of 140 PCBs by use of GC. A measurement procedure for coplanar PCBs by GC/MS and GC/ECD was described by Schwerdt (1994).

1.4.5. Conclusion

The accurate determination of PCDDs/PCDFs, PAHs and PCBs in indoor air is a demanding task, which requires sophisticated methods for sampling and analysis. Due to the complexity, only a brief discussion of occurrence, sampling strategies and recent developments in monitoring techniques could be given in this chapter. For a detailed understanding and as a precursor to any active steps in this field, a study of the original references is essential.

References

Anderson E.L. and Albert R.E. (1999): Risk assessment and indoor air quality. Lewis, Boca Raton.

Ballschmiter K. and Bacher R. (1996): Dioxine. VCH, Weinheim.

Ballschmiter K. and Zell M. (1980): Analysis of polychlorinated biphenyls (PCBs) by glass-capillary gas chromatography. Fresenius J. Anal. Chem., *302*, 20–31.

Blessing R. (1997): Polychlorierte Dioxine und Furane in öffentlichen Gebäuden. Gefahrstoffe–Reinhaltung der Luft, *57*, 305–309.

Chuang J.C., Hannan S.W. and Wilson N.K. (1987): Field comparison of polyurethane foam and XAD-2 resin for air sampling of polynuclear aromatic hydrocarbons. Environ. Sci. Technol., *21*, 798–804.

Chuang J.C., Kuhlmann M.R. and Wilson N.K. (1990): Evaluation of methods for simultaneous collection and determination of nicotine and polynuclear aromatic hydrocarbons in indoor air. Environ. Sci. Technol., *24*, 661–665.

Creaser C.S., Krokos F. and Startin J.R. (1992): Analytical methods for the determination of non-ortho-substituted chlorobiphenyls: A review. Chemosphere, *25*, 1981–2008.

Hach R., Donnevert G. and Alter E. (1994): Analyse von polyzyklischen aromatischen Kohlenwasserstoffen in Außenluft. Staub–Reinhaltung der Luft, *54*, 337–341.

Hanberg A., Waern F., Asplund L. Haglund E. and Safe S. (1990): Swedish dioxin survey: determination of 2,3,7,8–TCDD toxic equivalent factors for some polychlorinated biphenyls and naphthalenes using biological tests. Chemosphere, *20*, 1161–1164.

Jacob J., Karcher W., Belliardo J.J., Dumler R. and Boenke A. (1991): Polycylic aromatic compounds of environmental and occupational importance–their occurrence, toxicity and the development of high purity certified reference materials. Part III. Fresenius Z. Anal. Chem., *340*, 755–767.

Jacob J., Karcher W., Belliardo J.J. and Wagstaffe P.J. (1986): Polycylic aromatic compounds of environmental and occupational importance–their occurrence, toxicity and the development of high purity certified reference materials. Part II. Fresenius Z. Anal. Chem., *323*, 1–10.

Jacob J., Karcher W. and Wagstaffe P.J. (1984): Polycylic aromatic compounds of environmental and occupational importance–their occurrence, toxicity and the development of high purity certified reference materials. Part I. Fresenius Z. Anal. Chem., *317*, 101–114.

Keller C.D. and Bidleman T.F. (1984): Collection of airborne polycyclic aromatic hydrocarbons and other organics with a glass fiber filter–polyurethane foam system. Atmos. Environ., *18*, 837–845.

Larsen B., Bowardt S. and Tilio R (1992): Congener specific analysis of 140 chlorobiphenyls in technical mixtures on five narrow bore GC columns. Int. J. Anal. Chem., *47*, 47–68.

MacLeod K.E. (1981): Polychlorinated biphenyls in indoor air. Environm. Sci. Technol., *15*, 926–928.

Maroni M., Seifert B. and Lindvall T. (1995): Indoor Air Quality–a comprehensive reference book. Elsevier, Amsterdam.

Päpke O., Ball M., Lis Z.A. and Scheunert K. (1989): PCDDs and PCDFs in indoor air of kindergartens in northern W. Germany. Chemosphere, *18*, 617–626.

Pluschke P.(1996): Luftschadstoffe in Innenräumen. Springer, Berlin.

Rotard W. (1990): Risikobewertung von Dioxinen in Innenräumen. Bundesgesundheitsblatt, *3*, 104–107.

Sagunski H., Forschner S. and Kappos A.D. (1989): Indoor air pollution by dioxins in day-nurseries. Risk assessment and management. Chemosphere, *18*, 1139–1142.

Scholz-Böttcher B.M., Bahadir M. und Hopf H. (1992): Die Beilstein-Probe: Eine unbeabsichtigte Dioxin-Quelle in Routine- und Forschungslaboratorien? Angew. Chem., *104*, 477–479.

Schulz D. (1993): PCDDs/PCDFs–German policy and measures to protect man and the environment. Chemosphere, *27*, 501–507.

Schuster J. and Dürkop J. (1993): Dioxine und Furane–ihr Einfluß auf Umwelt und Gesundheit. Bundesgesundheitsblatt, Sonderheft, *3*, 1–14.

Schwerdt P. (1994): Bestimmung coplanarer PCBs in der Immission. GIT Fachz. Lab., 9/94, 907–912.

UBA–Umweltbundesamt (1998a): Belastung mit polyzyklischen aromatischen Kohlenwasserstoffen (PAK) in Wohnungen mit Parkettböden, Berlin.

UBA–Umweltbundesamt (1998b): Empfehlungen zu polyzyklischen aromatischen Kohlenwasserstoffen (PAK) in Wohnungen mit Parkettböden, Berlin.

Ullrich D. (1993): Results of an intercomparison excercise to determine PCBs in indoor air. In: Saarela K., Kalliokoski P. and Seppänen O. (eds): Proceedings of the 6th International Conference on Indoor Air and Climate, Vol. 2, 363–368, Helsinki.

VDI 2464, Part 1 (1992): Indoor air pollution measurement. Measurement of polychlorinated biphenyls (PCBs). In preparation.

VDI 3498, Part 1 (1998): Air pollution measurement. Indoor air measurement. Measurement of polychlorinated dibenzo-*p*-dioxins and dibenzofurans. Beuth, Berlin.

VDI 3875, Part 1 (1996): Outdoor air pollution measurement. Indoor air pollution measurement. Measurement of polcyclic aromatic hydrocarbons (PAHs). Gas chromatographic determination. Beuth, Berlin.

VDI 4300, Part 1 (1995): Indoor air pollution measurement. General aspects of measurement strategy. Beuth, Berlin.

VDI 4300, Part 2 (1997): Indoor air pollution measurement. Measurement strategy for polycyclic aromatic hydrocarbons (PAHs), polychlorinated dibenzofurans (PCDFs), polychlorinated dibenzodioxins (PCDDs) and polychlorinated biphenyls (PCBs) in indoor air. Beuth, Berlin.

Wensing M., Moriske H.J. and Salthammer T. (1998): Das Phänomen der „Schwarzen Wohnungen“. Gefahrstoffe–Reinhaltung der Luft, *58*, 463–468.

Wensing M. and Salthammer T. (1999): The Phenomenon of "Black Magic Dust" in housings. Proceedings of the 8th International Coference on Indoor Air and Climate, Edinburgh, in press.

Wilson K., Chuang, J.C. and Kuhlmann M.R. (1991): Sampling polycyclic aromatic hydrocarbons and related semivolatile organic compounds in indoor air. Indoor Air, *1*, 513–521.

1.5 Application of Diffusive Samplers

Derrick Crump

1.5.1 Introduction

A diffusive/passive sampler is a device which is capable of taking samples of gas or vapour pollutants from the atmosphere at a rate controlled by a physical process such as diffusion through a static air layer or permeation through a membrane, but which does not involve the active movement of air through the sampler (Brown, 1993; Moore, 1987). Compared with methods requiring pumps for active air movement, diffusive samplers are more convenient and have lower associated costs.

The need to evaluate and understand the performance of diffusive samplers resulted in the development of protocols describing tests to assess effects of analyte concentration, environmental conditions and exposure time on sampler performance as well as accuracy and reproducibility (CEN, 1995). A number of commercially available samplers have been evaluated for workplace monitoring, and, for example, the UK Health and Safety Executive (HSE, 1995) recommends a method for the determination of VOCs based on sampling with a tube type sampler (Perkin-Elmer, Beaconsfield, UK) containing a sorbent, and analysis by thermal desorption gas chromatography (TD/GC). The HSE also recommends a method for the determination of hexane that describes the use of the OVM 3500 badge sampler (3M Company, USA) and the ORSA 5 sampler (Dräger, Lübeck, Germany) which are solvent desorbed and the eluent is analysed by GC (HSE, 1992).

Compared with methods for the workplace, methods for diffusive sampling of non-occupational indoor air and ambient air have been less well developed. The range of concentrations and environmental conditions used to evaluate samplers for workplace monitoring is not directly applicable to non-occupational environments. However, diffusive monitors are finding increasing use in non-occupational environments. This chapter discusses the principles governing diffusive sampling and the factors that can influence sampler performance, and reviews studies that have applied the technique for the measurement of VOCs in indoor air.

1.5.2 Principles of Diffusive Sampling

In the ideal case, the resistance to mass transfer of a diffusive sampler is confined to the stagnant air gap between the face of the sampler and the sorbent material (Van de Hoed and Van Asselen, 1991). Then the mass flow through the sampler can be described by Ficks first law of diffusion

$$J = -DA\ \mathrm{d}c/\mathrm{d}x \qquad (1)$$

J = diffusion flux (g/s), D = air diffusion coefficient (cm^2/s), A = cross sectional area of the sampler (cm^2) and dc/dx is the concentration gradient of the compound across the stagnant air gap.

Assuming the compound is effectively trapped by the sampler, the concentration at the interface of the air gap and collecting surface (*Ce*) will be zero. Then the mass collected by the sampler (*M*) will be given by

$$M = K\, t_{(e)}\, C_o \qquad (2)$$

where $t_{(e)}$ is the exposure time, C_o is the concentration of the compound in ambient air and *K* is the uptake rate of the sampler defined by

$$K = DA/L \qquad (3)$$

where *L* is the length of the stagnant air gap (cm).

The expression *DA/L* has units of cm^3/s and represents the diffusive uptake rate of the sampler under ideal conditions.

Equation 3 enables the ideal uptake of the diffusive sampler to be calculated with knowledge of *A, L* and the diffusion coefficient for the analyte of interest. Mass transfer considerations limit the maximum sampling rate that can be achieved for a given air velocity in the environment being monitored. If the environment at the face of the sampler cannot be replenished with more of the analyte of interest at a rate greater than that demanded by the sampler, then the uptake rate will be lower than that predicted for the ideal case (Eq. 3). This can be described as starvation of the sampler and is of particular concern when undertaking stationary monitoring in indoor environments with low air speeds. Conversely, high air velocities can increase the sampling rate by causing turbulence inside the geometrical air gap of the sampler and thus reduce the effective diffusive path length of the sampler.

Equation (2) assumes a zero concentration at the interface of the air gap and collecting surface. Samplers that rely on chemisorption or other reactive trapping will maintain a zero concentration provided that the capacity of the trapping agent is not exceeded. However, if the trapping is based on reversible sorption, the perfect sink situation at the collecting surface may not be achieved and the effective sampling rate will change during the sampling period. In addition, sample loss, or reverse diffusion, will occur if the ambient concentration decreases below that which is in equilibrium with the sorbent surface.

For indoor environments the effects of changes in temperature and pressure on diffusive uptake rate will be insignificant compared with other sources of error. High humidity can affect the sorption capacity of hydrophilic sorbents such as charcoal. This will reduce the time before saturation of the sorbent occurs.

If the concentration at the face of the sampler fluctuates there is the possibility that the concentration inside the sampler air gap will not reflect the actual ambient concentration. The sampler would then not provide a true integrated response to the exposure concentrations. The response time of passive samplers is typically 1–10 s, and, provided that the total sampling time is large relative to the response time, errors will be small (Brown R.H., 1993).

It is necessary to recover the analyte from the sorbent by solvent extraction or thermal desorption. A weaker sorbent allows more efficient recovery during analyses but will not act as the perfect sink at the collecting surface. No single sampling or analytical technique is optimum for the wide range of compounds that occur in indoor air.

This consideration of the principles of diffusive sampling identifies a range of factors which may influence the performance of a diffusive sampler for monitoring of VOC concentrations in indoor air. These factors will potentially be a source of error in such measurements and add to the overall uncertainty of the result given by the measurement procedure. In addition, the amount of uncertainty will be influenced by other factors including amount and consistency of background contamination of sorbents, repeatability of analytical determination, formation of artifacts, stability of analyte on the sorbent, recovery of analyte during analyses and presence of interferents.

As well as establishing methods with acceptable accuracy and precision, the sampling method should be applied according to the strategy that ensures a representative measurement of the environment of interest. Guidance on designing a sampling strategy to meet different sampling objectives in the indoor environment has been developed by a European expert working group (ECA, 1994).

1.5.3 Studies of the Performance of Diffusive Samplers for the Measurement of VOCs in Indoor Air

The exposure periods required to obtain sufficient sensitivity to measure VOCs in low $\mu g/m^3$ concentrations using diffusive samplers are typically several days or weeks. It would be a major task to apply the laboratory and field tests required by protocols to evaluate diffusive monitors used in the workplace for these long exposure periods. At present the uncertainty associated with measurement by diffusive samplers requires further investigation, but a number of studies have shown them to be applicable to the measurement of VOCs in indoor air.

As for workplace monitoring, there have been two main types of diffusive sampler used for monitoring of indoor air: (i) badge type samplers containing a strong adsorbent such as charcoal that requires solvent desorption for GC analysis and (ii) tube type samplers with weaker adsorbents such as the porous polymer Tenax that can be thermally desorbed. The samplers most widely used have been developed for monitoring of workplace atmospheres and applied to indoor air through modification of the exposure period and the analytical method.

The badge type samplers have higher diffusive uptake rates and, because strong sorbents are used, are not prone to reverse diffusion effects. They require solvent desorption (which is not easily automated) and the use of toxic solvents, and recovery of some compounds is poor. Contaminants in the solvent can reduce sensitivity as does the dilution effect; typically 2 ml of solvent is used to desorb and 1 µl (i.e. 0.05 % of the collected mass of analyte) is used for GC analysis.

The tube type samplers have lower diffusive uptake rates and are therefore less prone to starvation effects. In consequence they require longer exposure periods to collect the same mass of analyte. The occurrence of reverse diffusion depends on the analyte-sorbent interaction, but for weaker adsorbents such as Tenax TA this does occur with more volatile compounds, e.g. compounds more volatile than toluene.

Back diffusion can be prevented by use of stronger adsorbents such as graphitised carbon, but recovery of less volatile compounds by thermal desorption is less efficient. The analysis can be automated and commercial equipment is available. The samplers can be re-used many times without loss of performance. Sensitivity of TD/GC is potentially higher because typically 10–30 % of the collected analyte is transferred to the GC column, although this has an associated disadvantage because each sampler can only be analysed once.

This section now reviews studies that have investigated the performance of diffusive monitors for measurement of VOCs in indoor air.

1.5.3.1 Badge samplers with solvent desorption

Seifert and Abraham (1983) explored the use of a badge type sampler to determine VOCs in indoor air using a 24 h exposure period. The sampler was the Gasbadge™ (National Mine Service Co, USA) and consisted of a 5.5 × 8 × 2 cm plastic casing containing a glass wool pad coated with charcoal. After exposure the VOCs were recovered using carbon disulfide and analysed by GC using a 50-m OV-101 glass capillary column. The sampler was exposed to a test gas containing 7 VOCs for periods of up to 24 h, and investigations of recovery and reproducibility were undertaken. Samini (1987) reported significant underestimation of air concentrations at air velocities less than 22 cm/s and overestimation at velocites greater than 44 cm/s.

The most widely used badge type sampler for indoor air studies has been the OVM 3500. This is a circular badge with a 1-cm diffusion length containing a charcoal wafer. Desorption of VOCs is carried out within the monitor itself by the addition of carbon disulfide. Exposure periods applied have ranged from 24 h to 3 weeks. The diffusive uptake rates reported by the manufacturer for 8-h exposure periods are about 30 ml/min, but actual values are compound specific. For the monitoring of hexane in the workplace, the diffusive uptake rate is not significantly affected by ambient air movement, provided that there is a minimum air velocity of about 0.1 m/s (HSE, 1992).

Seifert et al. (1989) undertook parallel field exposures of the Gasbadge and OVM 3500 samplers in a study of 12 homes in Berlin. For some compounds, similar standard deviations were given by the pooled data as found for duplicate OVM 3500 samplers, but for others, particularly polar compounds and terpenes, the standard deviations were higher. This study included an exposure test of OVM 3500 and Gasbadge samplers using six VOCs (benzene, toluene, *p*-xylene, *n*-hexane, *n*-heptane and 1,1,1-trichloroethane) in a chamber at concentrations of 3–60 $\mu g/m^3$ for 2 weeks. There was good agreement between the samplers and the predicted concentration, the maximum discrepancy for the OVM 3500 being 22 % for 1,1,1-trichloroethane.

Cohen et al. (1990) also investigated the performance of the OVM 3500 badge containing charcoal using a chamber experiment. They used 3-week exposure periods and investigated the effect of two relative humidities on the uptake of five VOCs (benzene, heptane, perchloroethylene, chloroform and *p*-dichlorobenzene) present at two concentrations (10 $\mu g/m^3$ and 100 $\mu g/m^3$). Results indicated that the samplers work best under conditions of high concentration with low relative humidity (27 %) and low concentration with high relative humidity (73 %). Excluding chloroform, deviations from the predicted values varied between −41 % and +22 %.

Bortoli et al. (1986) undertook an investigation of the performance of OVM 3500 samplers using a test chamber with a 4-day exposure period. Nine compounds were studied (pentanal, 1-butanol, 1,1,2-trichloroethane, 1-octene, butyl acetate, 3-heptanone, o-xylene, *α*-pinene and *n*-decane). Pentanal was substantially underestimated, but comparison between passive and active samplers was good for apolar compounds (active sampling gave 10–50 % higher values for polar compounds). Effective sampling rates of the passive device were 23–24 ml/min. Reproducibility was about 13 % on average and no information was given in the paper about air velocities in the chamber during the exposure period.

Ullrich and Nagel (1996) report further chamber tests using the same six-compound VOC mix at low $\mu g/m^3$ levels as used by Seifert et al. (1989). OVM 3500 samplers were exposed for periods of 2 days to 3 weeks in a 1-m^3 chamber with an air velocity of about 0.02 m/s and results of diffusive sampling were compared with an active sampling method using NIOSH type charcoal tubes (SKC Inc., USA). They report a mean bias ranging from +2 % to −19 % and mean coefficients of variation from ±5 % to ±17 %.

1.5.3.2 Tube type samplers with thermal desorption

The tube type diffusive sampler that has been most extensively used for ambient air monitoring is the Perkin-Elmer sampler which was specifically designed for thermal desorption (Brown et al., 1981). It consists of a 90 mm long × 6.3 mm OD (5 mm ID) steel tube within which an adsorbent is retained by a stainless steel mesh. The rear end of the tube is sealed with a brass Swagelock® fitting and PTFE ferrule. At the sampling end there is a 1.53-cm air gap between the mesh containing the adsorbent and the end of the tube. A diffusive end cap which contains a stainless steel mesh screen can also be used and this can prevent air movement within the diffusive air gap when exposed to high air velocities (Brown, 1993). Evaluation of the sampler for workplace monitoring found that the diffusive uptake rate of the sampler was not influenced by air velocities as low as 0.007 m/s.

Brown et al. (1992) measured toluene, xylenes, decane and the TVOC concentrations in the air of 100 homes using the Perkin-Elmer tube packed with Tenax TA and a 4 week exposure period. Investigations of the influence of exposure time on net diffusive uptake rate have shown that for a more volatile VOC, such as benzene, the uptake rate is lower for 6 weeks than for 3 weeks exposure periods. There was no effect of exposure period on uptake for the less volatile compounds such as undecane. Brown and Crump (1993) further examined the influence of exposure period by exposing the tube samplers for periods of up to 7 months. No further decline in uptake rate with exposure period occurred for benzene after 5 months and no significant decline in the rate for undecane occurred until exposure periods exceeded 5 months.

Diffusive uptake rates for six VOCs were determined experimentally by Brown et al. (1993) who exposed Perkin-Elmer tubes packed with Tenax TA in standard atmospheres for periods of 7–42 days. The results showed a significant decline in the effective sampling rate with exposure time for the more volatile compounds (e.g. benzene and toluene) and no significant change for compounds less volatile than xylenes. Repeatability of measurements was good, coefficients of variation for five measurements being 1.5 % for benzene, the largest being 2.5 % for decane. Environmental conditions in the exposure chamber were 23°C and 45 % RH.

Further unpublished laboratory studies have been undertaken at the Building Research Establishment to determine effective diffusive uptake rates using Perkin-Elmer tubes packed with Tenax TA. An atmosphere containing a mixture of 9 VOCs was produced using a $1-m^3$ stainless steel chamber. Diffusion- controlled emission sources were placed on the perforated chamber floor, which was set to produce a climate of 23°C and 45 % RH, air velocity 0.3 m/s and an air exchange rate of 1 h^{-1}. Seven sampling tubes (without diffusive end caps) were mounted on a steel rack near the centre of the chamber. The tubes were exposed for 14 days and the concentration of VOCs in the air leaving the chamber was monitored periodically by active sampling. Table 1.5-1 shows the results of active sampling, predicted chamber concentrations from weight loss of the sources and the calculated diffusive uptake rate (based on the active sampling concentration data). The results of active sampling and predicted concentrations are in good agreement except for decane and undecane. Repeatability of the samplers for most of the VOCs was good (RSD <4 %), the poorest being for ethylbenzene (RSD 7 %).

Table 1.5-1 also summarises the results for the 28–day exposure periods reported previously. It should be noted that different VOC mixtures and concentrations were used in the two experiments; however, the results show some differences in the diffusive sampling rates for 14 and 28 days, the greatest change being for benzene, which is the most volatile of the compounds studied. Figure 1.5-1 illustrates the change in effective diffusive uptake rate with exposure period for toluene by combining the data reported previously with the data in Table 1.5-1. This shows that it is important to use the diffusive

Table 1.5-1. Results of an experiment to measure 14 days diffusive uptake rates for 9 VOCs using the Perkin-Elmer tube with Tenax TA. (RSD = relative standard deviation).

Compound	Concentration (μg/m³)			Diffusive uptake rate (ml/min)		
	Predicted by weight loss	Measured by active method Mean	RSD	14 day uptake rate Mean	RSD	28 day uptake rate
Benzene	750	740	16 %	0.27	3 %	0.2
Toluene	420	390	15 %	0.38	1.6 %	0.32
Ethylbenzene	58	35	17 %	0.47	7 %	–
m- + *p*-Xylene	210	190	16 %	0.50	1.6 %	0.44
o-Xylene	130	100	15 %	0.52	1.3 %	0.45
Decane	140	70	12 %	0.51	3.9 %	0.51
1,2,4-Trimethylbenzene	110	90	20 %	0.62	6.4 %	0.54
Undecane	30	15	17 %	0.57	2.6%	0.53
Dodecane	50	50	4 %	0.41	2.7 %	–

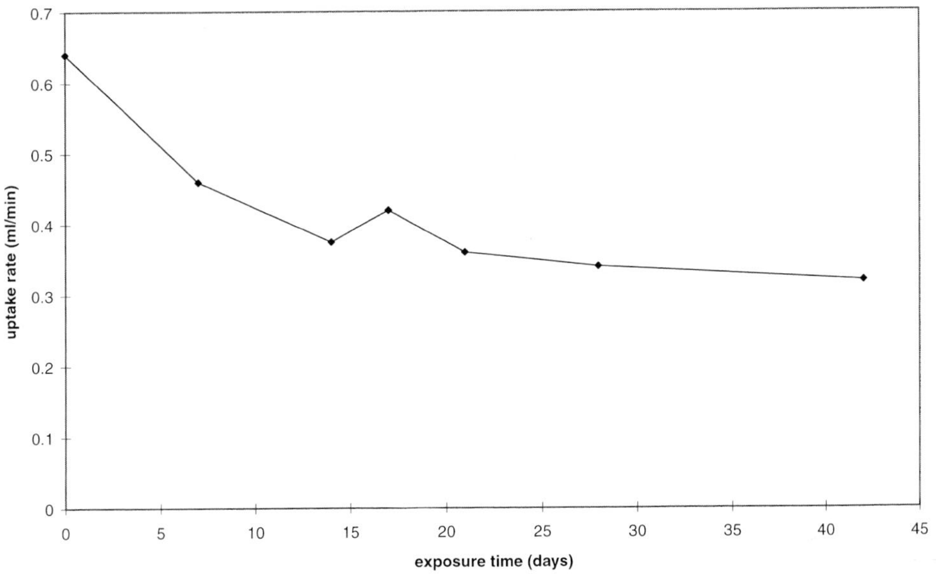

Figure 1.5-1. Experimentally determined diffusive uptake rates for toluene using different exposure periods and the Perkin-Elmer type sampler packed with Tenax TA.

uptake rate value appropriate to the exposure period when measuring analytes that exhibit non-ideal behaviour with the sorbent used.

Crump and Madany (1993) found good agreement between results of periodic active sampling in a home using Tenax tubes involving two 1-h samples per day and results of diffusive sampling using a 4 week exposure period. Cao and Hewitt (1993) measured VOC concentrations in a library using Perkin-Elmer tubes packed with either Tenax TA or Tenax GR. Calculated ideal diffusive uptake rates were applied to derive concentration values. They compared amounts of background contamination on different adsorbents and found that amounts on Tenax TA and Tenax GR were very low compared with those on Chromosorb 106 and Carbotrap.

Bradshaw and Ballantine (1995) compared the mass uptake of seven VOCs by Perkin-Elmer tubes containing Tenax TA in an indoor and an outdoor environment over an exposure period of 24 h, 7 days and 14 days. The mass uptake rate decreased with exposure time for the more volatile components such as acetone and hexane but there was little change for decane and undecane. A computer model was used to predict mass uptake rates for each analyte and each exposure period. When these predicted effective uptake rates were applied to the experimental data for benzene and other less volatile compounds there was no significant bias in the results for the different exposure periods.

1.5.4 Studies of VOCs in Indoor Air Using Diffusive Samplers

1.5.4.1 Solvent desorbed samplers

The Gasbadge and OVM 3500 samplers were applied in a study of 230 homes in western Germany (Krause et al., 1987). The exposure period for both types of badge was 2 weeks, and 57 VOCs were determined. GC analyses were conducted with two fused silica capillary columns of different polarity, and detection was by simultaneous FID/ECD. During the course of the study the commercial manufacture of the Gasbadge monitor was discontinued. Table 1.5-2 shows results of a selection of the compounds monitored that includes compounds with both significant outdoor sources such as benzene from motor vehicle exhausts and indoor sources such as solvents containing undecane.

This technique was further applied by Seifert et al. (1989) to investigate seasonal variation in VOC concentrations in 12 homes in Berlin. The TVOC concentration in the cold season was about two to three times higher than that in the warm season. Renovation activities, especially painting, caused VOC concentrations to go up by factors of between about 3 and 10.

The OVM 3500 monitor was also applied by Cohen et al. (1989) to determine VOC concentrations in a study of 35 homes in West Virginia, USA. Samplers were placed in the main bedroom at a height of approximately 1.5 m and outdoors in a specially fabricated aluminium shelter that kept the sampler dry. They were exposed for one 3 week period before analysis by solvent desorption using 1.5 ml carbon disulfide and analysis with GC/FID using a 60-m fused silica column (DB-1, J&W Scientific). Results for indoor concentrations of the 17 compounds reported are shown in Table 1.5-3.

Wolkoff et al. (1991) applied the OVM-3500 sampler in a study of 2 new homes in Denmark which were identical except that one was occupied by a three-person family

Table 1.5-2. Mean VOC Concentrations measured in homes in Germany (μg/m³).

Compound	230 homes (Krause et al., 1987)	59 homes (Ullrich et al., 1996)		43 homes, Hannover (Levsen et al., 1996)	
		Bit.	Zerbst	Suburb	City
Benzene	9.3	6.7	5.4	2.4	3.3
Toluene	76	77.8	61.3	29.1	27.8
Ethylbenzene	10	10.2	4.1	3.8	2.6
m- + *p*-Xylene	23	26.1	8.8	10.1	6.8
o-Xylene	7	8.6	3.7	2.8	2.4
1,2,4-Trimethylbenzene	11	–	–	–	–
n-Decane	14	–	–	–	–
n-Undecane	10	–	–	–	–
Limonene	25	60.2	44	–	–

Bit. = Bitterfeld

Table 1.5-3. Indoor concentrations of VOCs in 35 homes in West Virginia, USA (Cohen et al., 1989).

Compound	Concentration (μg/m³)	
	Mean	SD
Ethylene dichloride	20.9	27.0
Methyl chloroform	19.4	47.9
Benzene	6.7	10.7
Carbon tetrachloride	10.1	15.8
Trichloroethylene	9.6	14.1
Ethylene dibromide	6.1	5.9
Chlorobenzene	16.5	21.9
Tetrachloroethylene	1.3	0.9
Ethyl benzene	5.4	6.9
m- + *p*-Xylene	17.7	23.8
Styrene	1.2	0.8
o-Xylene	6.5	9.1
4-Ethyltoluene	3.0	5.0
1,2,4-Trimethlybenzene	11.5	16.2
p-Dichlorobenzene	45.5	182.8
Decane	206	39.8
1,2,4-Trichlorobenzene	12.8	30.0

and one remained unoccupied. An exposure period of 7 days was used and monitoring occurred periodically over a 12-month period. Concentrations of 21 VOCs were reported, with a strong decrease occurring over time in those originating from building materials. Comparison of VOCs found using the diffusive sampler with those given by active sampling using Tenax TA adsorbent and TD/GC showed that polar VOCs such as 1,2-propanediol are not eluted by the carbon disulfide solvent used for the diffusive sampler.

A nationwide study of indoor air concentrations of 26 VOCs was conducted in Canada in 1991 (Otson et al., 1994). An OVM-3500 sampler was exposed for 24 h on one occasion in 757 homes. Approximately equal numbers of homes were sampled in each of the four seasons. VOCs were extracted with 1.5 ml of carbon disulfide and 26 compounds were determined by GC with a mass-selective detector operated in the selected-ion monitoring mode. The GC column used was a 30-m capillary with a 0.25-μm film thickness (DB-wax, J & W). Detection limits were estimated to be in the range 1.6–5.9 μg/m^3. Mean concentrations of the 26 compounds are shown in Table 1.5-4.

Ullrich et al. (1996) used the OVM 3500 sampler in a study of indoor, outdoor and personal exposure concentrations involving 156 people in Eastern Germany. The diffu-

Table 1.5-4. Mean concentrations of 26 VOCs measured in 757 Canadian homes (Otson et al. 1994).

Compound	μg/m³	Compound	μg/m³
n-Hexane	1	*m*-Dichlorobenzene	<1
Dichloromethane	16	D-Limonene	20
Benzene	5	Pentachloroethane	<1
Trichloroethane	<1	*p*-Cymene	1
n-Decane	31	Hexachlorobenzene	<1
Tetrachloroethylene	3	*o*-Xylene	6
Toluene	41	*p*-Dichlorobenzene	19
1,2-Dichloroethane	<1	*m*-Xylene	14
Ethylbenzene	8	1,1,2,2-Tetrachloroethane	<1
1,3,5-Trimethylbenzene	3	*p*-Xylene	6
Chloroform	2	Trichloroethylene	<1
1,2,4-Trimethylbenzene	12	Styrene	<1
α-Pinene	23	Naphthalene	4

sive sampler was worn by subjects for either 2 days or 2 weeks. Samplers were also exposed simultaneously to measure VOCs in indoor and outdoor air. The results for 11 VOCs in the indoor air for a 2-week exposure period are shown in Table 1.5-2. The authors compare the results with the previous measurements undertaken in homes in Western Germany and conclude that there are no remarkable differences between the two groups studied. The indoor air results were broadly similar to the personal exposure measurements which for most VOCs were considerably higher than for the outdoor air.

Levsen et al. (1996) report the use of the ORSA (Dräger, Germany) type diffusion sampler to monitor VOCs in 42 homes in Hannover, Germany, using a 14-day exposure period. The sampler is tube-shaped and contains activated charcoal which is solvent desorbed with carbon disulfide. The diffusion area is 0.88 cm^2 and the diffusion distance 5 cm. The diffusive sampling rate applied for occupational monitoring for sampling periods of 0.5–8 h is 5–10 ml/min depending on the substance (Dräger Ltd., 1997). Table 1.5-2 shows the concentrations of VOCs measured in the homes. Analysis of the solvent eluate was by GC/MS.

Moscato et al. (1997) studied concentrations of benzene, toluene and xylene in a hospital in Italy located far away from industrial and high-traffic pollution. They undertook fixed-site and area monitoring using the 130-Radiello, radial-diffusive type cartridge filled with 530 mg of activated charcoal and an exposure period of 8 h. For analysis the charcoal was solvent desorbed with carbon disulfide and analysed by GC/FID using a Vocol 30 m/0.53 mm ID/3 μm film thickness fused silica capillary column (Supelco, USA). Indoor benzene concentrations were low (<1 μg/m^3), mean toluene concentrations ranged from 1.44 μg/m^3 (summer) to 6.04 μg/m^3 (autumn) and xylenes ranged from 10.78 μg/m^3 (summer) to 21.48 μg/m^3 (winter).

1.5.4.2 Thermally desorbed samplers

Kristensson and Lunden (1988) report the use of Perkin-Elmer type tubes in the diffusive mode to characterise air quality for identifying sick buildings. Diffusive tubes containing Tenax TA were exposed for 5–7 days and analysed by TD/GC/mass spectrometry. The air profiles produced were used as a screening test to assess whether further investigations of the air quality were warranted.

Berry et al. (1996) used Perkin-Elmer type tube samplers packed with Tenax TA to monitor VOCs in the air of 174 homes in the County of Avon in the UK during 1991/92. VOCs were monitored for up to 12 months in each home using 4-week exposure periods. The households were recruited to the study in monthly intervals in groups of approximately ten, with the first set commencing sampling in November 1990 and the last set joining in March 1992. Samplers were placed in the living room and the main bedroom in each home, and in 12 homes additional samplers were placed in the kitchen, bathroom and a second bedroom. Outside air was measured at 12 locations, the samplers being contained in a purpose-built steel open-ended box placed in back gardens. At the end of each exposure period samplers were returned to the laboratory by post and further sets of samplers were delivered to the homes by post.

Analysis of the samplers was by TD/GC. The bulk of samplers were analysed on a system composed of a Perkin-Elmer ATD50 and an 8310 GC fitted with an FID. Fifty duplicate samplers were analysed using a second TD/GC system (Perkin-Elmer ATD400/8700 GC) incorporating a Finnigan ion trap detector (ITD). All tubes were desorbed at 250°C for 10 min under a flow of helium with an outlet split of 30 : 1. The two GCs were equipped with the same column (SGE, BP10, 25m) and used the same temperature programme (40°C for 1 min, 2°C/min up to 75°C then 5°C/min up to 220°C).

TVOC concentrations were calculated using the sum of all chromatographic peaks and the FID response factor for toluene. The average annual mean TVOC concentration for the main bedrooms in the 174 homes was 415 μg/m^3, whilst the annual mean concen-

Table 1.5-5. Mean annual concentrations of VOCs measured in the main bedrooms of 174 homes in Avon, England and 14 outdoor sites.

Compound	Concentration (μg/m^3) Indoors Mean	SD	Percentile Values 10 %	50 %	75 %	95 %	Outdoors Mean	SD
TVOC	415	323	126	308	491	1064	40	30
Benzene	8	4	4	7	9	14	5	1
Toluene	40	86	14	25	39	73	12	11
Undecane	14	17	3	9	16	43	1	0
Xylenes	27	23	10	20	30	69	6	3
1,2,4-Trimethylbenzene	10	11	3	7	12	35	1	1
Decane	26	32	5	15	28	91	<1	<1

tration for the 14 outdoor sites was 40 µg/m³. The detection limit for toluene was 0.2 µg on the adsorbent tube and this is equivalent to 2 µg/m³ in air for a 4-week exposure period. The total number of VOC measurements recorded indoors in Avon was 3225. Table 1.5-5 summarises the results of measurements of TVOCs, benzene, toluene and undecane as well as results for xylenes, 1,2,4-trimethylbenzene and decane which have not been reported previously.

Brown et al. (1995) examined the range of compounds present in the air samples from Avon and found that the mean number of peaks was 97 with a range of 13 to 261 compared with a mean of 24 in the outdoor air. 48 different compounds occurred as the dominant peak on at least one occasion.

Other studies of UK homes using the same sampling and analytical procedures as reported by Berry et al. (1996) are summarised in Table 1.5-6. In addition the technique has been used for personal exposure monitoring to examine the relationship between VOC concentrations determined by fixed-site samplers in different micro-environments and those attached to the clothing of people using those micro-environments (Mann et al., 1997).

Brown and Crump (1998) report modification of the analytical method to provide a lower limit of detection for the measurement of VOCs in indoor and outdoor air. Using a Perkin-Elmer ATD 400 thermal desorber with the outlet split set at 16 : 1 and connected to an Autosystem GC, the detection limit was 2 ng of toluene on the tube which is equivalent to 0.2 µg/ m³ in air for a 4-week exposure period.

The Perkin-Elmer tube packed with Tenax was also used in a study of exposure to air pollutants of 100 office workers in Milan, Italy (Cavallo et al., 1997). Details of exposure time and analytical methods are not given in the paper. Table 1.5-7 shows the concen-

Table 1.5-6. Annual/four-week mean TVOC concentrations measured in UK buildings (µg/m³).

Study	Period	Mean	SD	Min.	Max.
173 homes in Avon; normal occupation – main bedrooms (Berry et al., 1996)	Annual	415	323	40	2051
43 buildings (36 homes); occupants concerned about IAQ (Brown et al., 1996)	4-week	1040	258	4	16537
6 homes in Hertfordshire; normal occupation (Berry et al. 1996)	Annual	232	83	125	322
40 homes in Southampton; 20 with partial mechanical ventilation and heat recovery (Brown et al., 1996)	Annual	330	301	94	1804
2 newly built timber framed houses (T1Annual and T2); unoccupied, first year (Crump et al., 1997)	Annual	1938 1603	2352 1805	635 648	8270 6472
2 newly built houses with double masonry walls (M1 and M2); unoccupied first year (Crump et al. 1997)	Annual	1429 1954	1384 1458	531 863	5179 5660

Table 1.5-7. VOC Concentrations in homes of 100 office workers in Milan, Italy (Cavallo et al., 1997).

Compound	Concentration (µg/m^3) Mean	Median	Min.	Max.
TVOC	411.3	341.5	192	2596
Benzene	28.8	23.5	0.1	135
Toluene	42.3	31.0	0.1	450
Xylenes	20.6	16.5	0.1	85

trations reported in homes. TVOC, toluene and xylene concentrations are very similar to those measured in homes in Avon, England although benzene concentrations are considerably higher.

1.5.5 Conclusion

Diffusive samplers developed for workplace monitoring can be applied to the study of VOCs in non-industrial indoor air through a modification of sampling times and analytical methods. Two main types of samplers have been used for major studies of VOCs in indoor air and these are the OVM 3500 badge which must be solvent-desorbed prior to analysis by gas chromatography and the Perkin-Elmer tube designed for thermal desorption. The performance of these techniques has not been assessed to the same extent as was undertaken for their application to workplace monitoring. This would be a major task because of the long exposure period applied and the wide range of chemicals of interest.

Available data based on limited exposure tests in chambers and comparisons with pumped sampling methods in the laboratory and field show the techniques can be used to determine mean concentrations of VOCs over periods of a day to several weeks for fixed site and personal monitoring. Problems of poor recovery at the desorption stage and possible losses by back diffusion means the investigator needs to consider carefully the choice of sampler and, in the case of thermally desorbable tubes, the optimum sorbent for the investigation.

Other issues include (a) the possibility of starvation during sampling which will depend upon the type of sampler employed and the air movement in the environment to be monitored, (b) the detection limit required, which may be limited by the amount and consistency of contaminants in unexposed samplers and (c) whether automation of procedures and avoidance of solvents are important factors. Research is required to quantify further the uncertainty associated with diffusive sampling including the stability of more reactive species on the sorbents.

Diffusive samplers have proved to be the preferred approach for large-scale surveys of indoor air quality and are being increasingly used for non-occupational personal exposure studies. The low costs and convenience compared with pumped methods ensure a growing use of the technique for health studies, building investigations and the study of trends in indoor air quality that result from changes in outdoor air quality, construction practice, building operation and occupant behaviour.

Acknowledgement

I would like to thank Veronica Brown and Chuck Yu for their work on measuring VOCs in UK homes and studies of diffusive sampler performance.

References

Berry R.W., Brown V.M., Coward S.K.D., Crump D.R., Gavin M., Grimes C.P., Higham D.F., Hull A.V., Hunter C.A., Jeffery I.G., Lea R.G., Llewellyn J.W. and Raw G.J. (1996): Indoor air quality in homes: The Building Research Establishment, Indoor Environment Study. BRE Report BR 299 and BR 300, CRC Ltd, Watford, UK.

Bortoli M.D.E., Molhave L., Thorsen M.A. and Ullrich D. (1986): European interlaboratory comparison of passive samplers for organic vapour monitoring in indoor air. Commission of the European Communities Report EUR 10487 EN. Luxembourg, Commission of the European Communities.

Bradshaw N.M. and Ballatine J.A. (1995): Confirming the limitations of diffusive sampling using Tenax TA during long term monitoring of the environment. Env. Tech., *16*, 433–444.

Brown R.H., Chalton J. and Saunders H.J. (1981): The development of an improved diffusive sampler. Amer Ind. Hyg. Assoc. J., *42*, 12, 865–869.

Brown R.H. (1993): The use of diffusive samplers for monitoring of ambient air. Pure Appl. Chem., *65*, 1859–1874.

Brown V.M. and Crump D.R. (1993): Appropriate sampling strategies to characterise VOCs in indoor air using diffusive samplers. Proceedings of the International Conference on VOCs in the Environment, London, 27–28 October 1993, 241–249.

Brown V.M. and Crump D.R. (1998): Diffusive sampling of volatile organic compounds in ambient air. Environmental Monitoring and Assessment, 52, 43–55.

Brown V.M., Crump D.R. and Gardiner D. (1992): Measurement of volatile organic compounds in indoor air by a passive technique. Environ. Technol., *13*, 367–375

Brown V.M., Crump D.R. and Yu C. (1993): Long term diffusive sampling of volatile organic compounds in indoor air. Environ. Technol., *14*, 771–777.

Brown V.M., Crump D.R. and Mann H. (1995): Concentrations of volatile organic compounds including formaldehyde in five UK homes over a three year period. Proceedings of the 2nd International Conference on VOCs in the Environment, London, 7–9 November 1995, 289–300.

Brown V.M., Cockram A.H., Crump D.R. and Mann H.S. (1996): The use of diffusive samplers to investigate occupant complaints about poor indoor air quality. Proceedings of Indoor Air 96, July 21–26, Nagoya, Japan, Vol. 2, 115–120.

Cao X.L. and Hewitt N. (1993): Passive sampling of organic vapours in indoor air. Proceedings of Indoor Air 93, Helsinki, 4–8 July 1993, Vol. 2, 227–232.

Cavallo D., Alcini D., Carrer P., Basso A., Bollini D., Lovato L., Vercelli F., Visigalli F. and Marconi M. (1997): Exposure to air pollutants in homes of subjects living in Milan. Proceedings of Healthy Buildings/IAQ 97, Washington DC, USA, September 27–October 2, 1997, Vol. 3, 141–145.

CEN (1995): Pre-Standard on Workplace Atmospheres–diffusive samplers for the determination of gases or vapours–requirements and test methods. prEN 838, 1995. European Standards Organisation (CEN).

Cohen M.A., Ryan P.B., Yanagisawa Y., Spengler J.D., Ozkaynak H. and Epstein P.S. (1989): Indoor/outdoor measurements of volatile organic compounds in the Kanawha Valley of West Virginia. JAPCA, *39*, 1086–1093.

Cohen M.A., Ryan P.B., Yanagisawa Y. and Hammond S. (1990): The validation of a passive sampler for indoor and outdoor concentrations of volatile organic compounds. J. Air and Waste Management, *40*, 993–997.

Crump D.R. and Madany I. (1993): Daily variations of volatile organic compound concentrations in residential indoor air. Proceedings of Indoor Air '93, Helsinki, 4–8 July 1993, Vol. 2, 15–20.

Crump D., Squire R. and Yu C. (1997): Sources and concentrations of formaldehyde and other volatile organic compounds in the indoor air of four newly built unoccupied test houses. Indoor Built Environ., *6*, 45–55.

Dräger Ltd (1997): Dräger Tube Handbook. Drägerwerk AG, Lübeck, Germany.

ECA (1994): Sampling strategies for volatile organic compounds (VOCs). European collaborative action–Indoor Air Quality and its Impact on Man, Report No 14, European Commission, EUR 16051 EN, Luxembourg.
HSE (1992): *n*-hexane in air-laboratory method using charcoal diffusive samplers, solvent desorption and gas chromatography. M D H S 74, Health and Safety Executive, Bootle, UK.
HSE (1995): Volatile organic compunds in air–laboratory method using diffusive solid sorbent tubes, thermal desorption and gas chromatography. MDHS 80, Health and Safety Executive, Bootle, UK.
Krause C., Mailahn, Nael R., Schulz C., Seifert B. and Ullrich D. (1987) Occurrence of volatile organic compounds in the air of 500 homes in the Federal Republic of Germany. Proceedings of Indoor Air 87, Berlin, 17–21 August 1987, Vol. 1, 102–106.
Kristensson J. and Lunden A. (1988): Air profile–a method to characterise indoor air. Proceedings Healthy Buildings 88 Stockholm, Sweden, June 1988, Vol. 3, 361–369.
Levsen K., Schimming E., Angerer J., Wickmann H.E. and Heinrich J. (1996): Exposure to benzene and other aromatic hydrocarbons: indoor and outdoor sources. Proceedings of Indoor Air 96, Nagoya, Japan, July 1996, Vol. 1, 1061–1066.
Mann H.S., Crump D.R. and Brown V.M. (1997): The use of diffusive samplers to measure personal exposure and area concentrations of VOCs including formaldehyde. Proceedings of Healthy Buildings/IAQ97, Washington, 28 September–3 October 1997, Vol. 3, 135–140.
Moore G. (1987): Diffusive sampling–a review of theoretical aspects and state of the art. In: Berlin A., Brown R.H. and Saunders H.J. (eds) Diffusive Sampling. Royal Society of Chemistry, London, 1987.
Moscato U., Volpe M. and de Belvis A. (1997): Aromatic hydrocarbons indoor pollution: results of a survey in an Italian hospital. Proceedings of Healthy Buildings/IAQ 97 , Washington, USA, September 1997, Vol. 1, 347–352.
Otson R., Fellin P. and Tran Q. (1994): VOCs in representative Canadian residences. Atmos. Environ. *28*, 22, 3563–3569.
Samini B.S. (1987): The effect of face air velocity on the rate of sampling of air contaminants by a diffusive sampler. In: Berlin A., Brown R.H. and Saunders K.J. (eds) Diffusive Sampling. Royal Society of Chemistry, London.
Seifert B. and Abraham H. (1983): Use of passive samplers for the determination of gaseous organic substances in indoor air at low concentration levels. Intern. J. Environ. Ann. Chem. *13*, 237–253.
Seifert B., Mailahn W., Schulz C. and Ullrich D. (1989): Seasonal variation of concentrations of volatile organic compounds in selected German homes. Environment International, *15*, 397–408.
Ullrich D. and Nagel R. (1996): Comparison of diffusive sampling with different sampling periods and short term active sampling at low VOC concentrations. Proceedings of Indoor Air 96, Nagoya, Japan, July 1996, Vol. 2, 703–708.
Ullrich D., Brenske K.R., Heinrich J., Hoffman K., Ung L. and Seifert B. (1996): Volatile organic compounds comparison of personal exposure and indoor air quality measurements. Proceedings of Indoor Air 96, Nagoya, Japan, July 1996, Vol. 4, 301–306.
Van de Hoed N. and Asselen O.L. (1991): A computer model for calculating effective uptake rates of tube-type diffusive samplers. Ann. Occup. Hyg., *35*, 3, 273–285.
Wolkoff P., Clausen P., Nielsen P. and Molhave L. (1991): The Danish twin apartment study. Part 1: formaldehyde and long term VOC measurements. Indoor Air, *4*, 478–490.

1.6 Real-Time Monitoring of Organic Compounds

Lars E. Ekberg

1.6.1 Introduction

Real-time monitoring of various pollutants can often be advantageous in indoor air quality investigations, especially if the monitoring is carried out simultaneously at several locations and if the monitoring period can be extended to several days or weeks. Continuous monitoring of the concentrations of organic substances indoors and outdoors can be carried out in order to reveal temporal concentration variations and/or to determine differences in concentration between various locations within a building (Bunding Lee et al., 1993 and Ekberg, 1994).

Not only indoor pollution sources can be expected to influence the concentration of organic compounds indoors, but also sources in the outdoor air may be of importance, e.g. vehicle exhaust and industrial emissions (Ekberg, 1993 and EC-European Commission, 1994). Various sources of pollution can give rise to considerable concentration variations with time. The use of real-time monitoring devices over extended periods can give detailed information about such temporal variations. Since real-time monitoring devices are generally not selective there is often also a need to use a method for identification and quantification of individual VOCs, e.g. by sampling on an adsorbent for subsequent analysis by gas chromatography and mass spectrometry. In this context, continuously monitored concentration data can be used to improve the strategy for VOC sampling with respect to sampling locations, sampling periods etc. There are various reasons for real-time monitoring of organic compounds, and the objective can be, for example:

- to reveal the presence of intermittent sources of organic compounds indoors,
- to study the influence of ventilation rate and ventilation system operation on the concentration in indoor air,
- to study the influence of the outdoor air cleanliness on the concentration indoors, and
- to study the decay of the emission rate when emissions from materials are investigated.

As mentioned, devices for real-time monitoring of concentrations of organic compounds in air are generally not selective, which means that the output signal is the sum of the instrument's response to various organic compounds present in the air analyzed. The result obtained with such an instrument is a total concentration of organic compounds, generally expressed as an equivalent of one single compound used for calibration of the instrument, e.g. toluene. Furthermore, a concentration measured with one type of instrument cannot be expected to be directly comparable to the result obtained with an instrument operating with a different detection principle, due to the fact that different detection principles show different responses for a variety of compounds (or groups of compounds).

The relevance of using the total concentration of volatile organic compounds (TVOC) for prediction of effects regarding health and comfort in non-industrial buildings has been questioned. Several scientific publications have concluded that, at present, it is not possible to use the TVOC concentration as a risk index, regardless of which of the many available definitions is used (Andersson et al., 1997 and Møhave et al., 1997). However, it has been stated that further research may, in the future, lead to the establishment of a useful risk index for VOCs. In order to realize this, it will certainly be necessary to consider the mixture of a number of specially selected compounds and to weight the concentrations with respect to the biological effects of the individual compounds. In view of these considerations, it is most unlikely that the data from real-time monitoring devices can be utilized in the context of a risk indicator. However, as indicated previously, real-time monitoring can be of substantial utility in indoor air quality investigations for other reasons, for example for screening purposes.

The purpose of the present article is (1) to point out some of the main features of various direct-reading instruments available, (2) to indicate a suitable methodology for real-time monitoring, and (3) to give examples of results obtained by real-time monitoring of organic substances.

1.6.2 Various Detection Principles and their Features

The most commonly used equipment for continuous monitoring of organic compounds at low concentrations in air (e.g. in non-industrial indoor environments) does not allow for identification of individual organic compounds, but, because of the principle of operation, the output signal is a result of the contribution of a variety of organic compounds present in the air analyzed. Furthermore, the response of the instrument is generally different for different compounds, i.e. the output signal varies not only with the concentration, but also with varying composition of the air with respect to the mix of different organic compounds. This feature, which mostly is considered as a major disadvantage of direct reading instruments, is discussed below in further detail for each detection principle listed in Table 1.6-1.

1.6.2.1 Photoacoustic Spectroscopy

Many VOCs absorb infrared (IR) light at a wavelength of 3.4 μm, and in instruments based on PAS the air sample is exposed to IR light pulses. If IR radiation is absorbed (i.e. hydrocarbons are present in the air sample), the light pulses will generate changes of pressure in the measurement cell, and the frequency of the light pulses is adjusted to

Table 1.6-1. Some principles of detection of organic compounds in air.

Mechanism	Detection principle	Compounds
IR absorption	Photoacoustic spectroscopy (PAS)	VVOC-VOC
Ionization of molecules	Flame ionization detection (FID) Photoionization detection (PID)	VVOC-VOC
Ionization of particles	Aerosol photoionizaton	Particulate PAH

make the pressure changes detectable with microphones. A detailed description of the principle of PAS has been given by Rosencwaig (1980). Instruments operating according to this principle generally are stable in time, showing only a small drift of the calibration curve. However, one disadvantage of the technique is that water vapor also absorbs IR radiation of the wavelength used for VOC detection. In order to compensate for this interference it is necessary to subtract the water vapor concentration measured at another IR wavelength. It is of major importance that the water vapor interference compensation is accurate, which in turn is a matter of a thorough calibration procedure.

Total concentrations of organic compounds measured using the IR wavelength 3.4 μm (and compensated for water vapor interference) are called $TVOC_{PAS}$ concentrations in the present paper. The $TVOC_{PAS}$ concentration may, as mentioned, represent a large number of compounds, including very volatile organic compounds. Figure 1.6-1 shows examples of the PAS detector response (at 3.4 μm) relative to toluene for various groups of VOCs, as calculated from detection limits provided by a PAS instrument manufacturer (Brüel & Kjær Innova, 1997). Empirical data presented by Hodgson (1995) are also included for comparison. It is clear that instruments for $TVOC_{PAS}$ measurements are especially sensitive for alkanes, while chlorinated compounds generally give rise only to low signals.

It should be noted that a measurement of the $TVOC_{PAS}$ concentration is sensitive for methane. The sensitivity for methane is about 2.5 times less than the sensitivity for toluene.

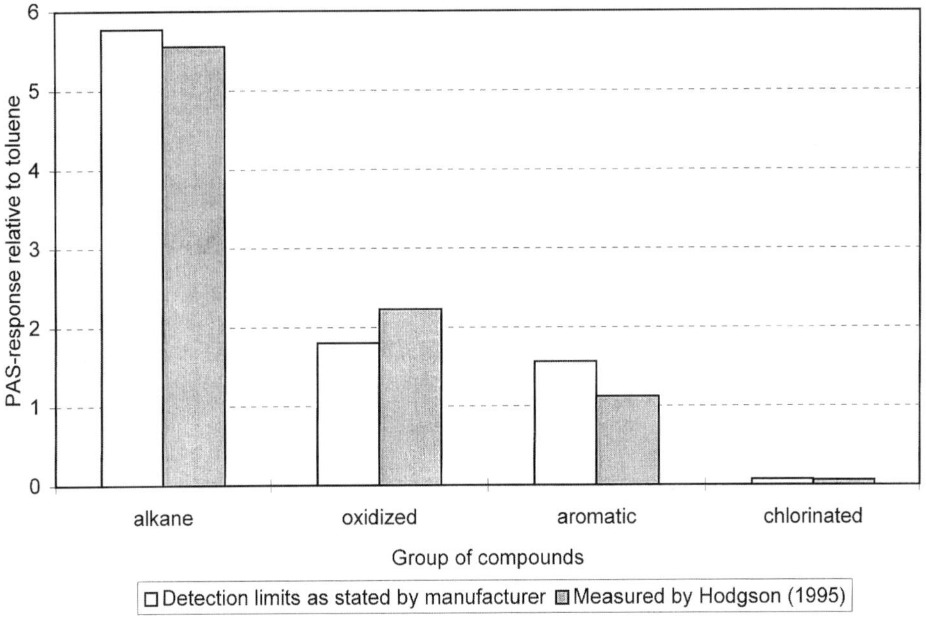

Figure 1.6-1. The PAS detector response relative to toluene for various groups of VOCs. The figure contains data calculated from detection limits provided by a PAS instrument manufacturer as well as empirical data presented by Hodgson (1995).

There is also a possibility to select the IR wavelength in order to provide enhanced sensitivity for a specific compound, for example formaldehyde. In this case the possibility of interference from other compounds must be considered.

In principle, measurement of IR absorption, e.g. by PAS, is a way of counting molecules, and therefore the measured signal is a measure of the concentration expressed in units of volume (e.g. parts per million by volume, ppm_v).

1.6.2.2 Flame Ionization Detection

In an FID the organic compounds in the sampled air are burned in a hydrogen flame, which leads to the production of ions. The ions are collected on an electrode, resulting in an electric current, which is amplified and measured. In addition to the concentration, the output signal depends on the number of carbon atoms in the molecules and also to some extent on the character of the compound. This means that the same concentration of two different compounds may cause different levels of output signal (EC-European Commission, 1997). The FID response has been shown to be about equal for alkanes and aromatic hydrocarbons, while the sensitivity for chlorinated and particularly oxidized compounds is significantly lower (Hodgson 1995). FID detectors are generally stable with time.

1.6.2.3 Photoionization Detection

In a PID also, the output signal is created by an electric current caused by ions, but the ionization of VOCs is in this case achieved by ultra-violet radiation. Some compounds, for example chlorinated VOCs, are ionized only to a limited extent, resulting in a low sensitivity for these compounds (EC-European Commission, 1997). The sensitivity for various VOCs may deviate by more than one order of magnitude. Typically, the presence of molecules with double bonds give rise to a stronger response, and the sensitivity is generally larger for large molecules than for small. PID instruments generally show a significant drift of the calibration curve, and therefore frequent recalibrations are required.

1.6.2.4 Aerosol Photoionization

The content of particulate PAHs can be measured with an instrument originally developed at ETH in Zürich (Burtscher et al., 1982). In this instrument, small soot particles that have accumulated a deposit of PAHs from combustion processes are selectively photoionized by UV light. After the electrons have been removed, positive ions are collected on an insulated paper filter, and the current from the charged paper filter is measured by means of a sensitive electrometer. The instrument can be calibrated against a PAH mixture, for example, one representative of vehicle exhaust, using the result from a gas chromatograph as reference. The instrument operates within the concentration range 1–2000 ng/m^3.

1.6.2.5 Other Types of Instruments

Other examples of instruments that can be used for real-time monitoring of VOCs are gas analyzers based on non-dispersive infra red (NDIR) detection and Fourier transform infrared (FTIR) detection. The use of FTIR technique may under certain circumstances enable identification and quantification of individual VOCs at low concentrations (Hicks et al., 1992 and Bunding Lee et al., 1993).

Finally, different types of VOC detectors based on semi-conductor techniques are available. Sensors of this type are commonly called "air quality sensors" and give an output signal scaled as 0–100 % air quality. Only limited experience of the use of such devices has been presented in the literature. However, a series of tests carried out by Fahl[illegible] al. (1992) showed that a common feature of such sensors is that they generally a[illegible] highly sensitive both to temperature and humidity.

1.6.2.6 Moni[illegible] of Inorganic Compounds

It may someti[illegible]e of interest to compare monitored VOC concentrations with the concentrations o[illegible]riety of inorganic compounds. For example, carbon monoxide (CO) could be use[illegible]n indicator for combustion products, e.g. environmental tobacco smoke or vehicle [illegible]ions.

Furthermore, ch[illegible]l reactions between pollutants in indoor air has been paid increased attention ov[illegible] last few years. When present in indoor air, for example, ozone (O_3), nitrogen oxide[illegible]$_x$) and various VOCs (e.g. unsaturated hydrocarbons) can be expected to be invol[illegible] chemical reactions (Weschler and Shields, 1997).

When studying ch[illegible] reactions in indoor air it can be of great value to obtain continuous data of a n[illegible]r of influencing parameters (Weschler and Shields, 1994). In addition to the conce[illegible]n of the reactants indoors and outdoors, the real-time monitoring should preferably i[illegible]ude also the air exchange rate, temperature and relative humidity.

Table 1.6-2 shows a lis[illegible]organic compounds that may be of interest to monitor in parallel with measurements [illegible] organic substances in indoor and outdoor air.

Table 1.6-2. Examples of inorganic compounds and detection principles suitable for direct reading instruments.

	Compound	Detection principle
Ozone	O_3	UV photometry
Nitrogen oxides	NO_x, NO and NO_2	Chemiluminescence
Carbon monoxide	CO	IR detection (e.g. PAS)
Carbon dioxide	CO_2	IR detection (e.g. PAS)

1.6.3 Air Sampling and Data Recording

Instruments for real-time monitoring are often operated with a continuous sampling air flow, but instruments with intermittent air sampling also exist. Furthermore, the output signal may be either continuous or in the form of discrete concentration values. It is common to use the gas analyzer together with a computerized monitoring system by which the measurement results are recorded in digital form, e.g. on a diskette. In practice, when using such a monitoring system, the concentration data are given as discrete values with a certain time interval, regardless of the nature of the primary signal from the gas analyzer. The maximum allowable time interval between recorded values depends on the purpose of the monitoring. If rapid concentration changes are to be studied it is generally necessary to keep the time interval small compared to the time constant for the ventilation system in the building or room where the monitoring is carried out.

In most applications of real-time monitoring it is necessary to compare concentrations measured at different locations in parallel. This is achieved if the monitoring system is complemented with a manifold system for automatic switching between different sampling tubes, e.g. by means of computer-controlled solenoid valves. The minimum time required for one measuring cycle is then determined by both the response time of the gas analyzer and the time required for purging of the sampling tube. Real-time monitoring devices typically enable sampling intervals down to a few minutes, even with a manifold system with sampling tubes connected.

1.6.4 Examples of Results from Real-Time Monitoring

In the following, examples are given of results obtained by real-time monitoring of organic compounds in both outdoor and indoor air. The purpose is to demonstrate situations where real-time monitoring can be especially valuable. The examples are collected from both field measurements and laboratory investigations.

1.6.4.1 Organic Compounds in Outdoor Air

Numerous measurements in outdoor air have shown a reasonably close agreement between the temporal variations of the concentrations of $TVOC_{PAS}$ and a variety of traffic-related pollutants, e.g. carbon monoxide (CO), nitrogen oxides (NO_x) and particulate PAHs. Previous studies have reported comparisons between $TVOC_{PAS}$ and NO with correlation coefficients between 0.69 and 0.79 (Krüger et al., 1995 and Ekberg, 1995). The same studies also reported correlation coefficients in the interval 0.81 – 0.94 for the comparison between NO and CO, while comparisons between NO and particulate PAHs gave correlation coefficients between 0.89 and 0.96. As an example of the co-variation between different compounds, Fig. 1.6-2 shows the result of simultaneous measurements of particulate PAH, $TVOC_{PAS}$ and NO at the outdoor air intake to an office building located in a traffic-influenced area.

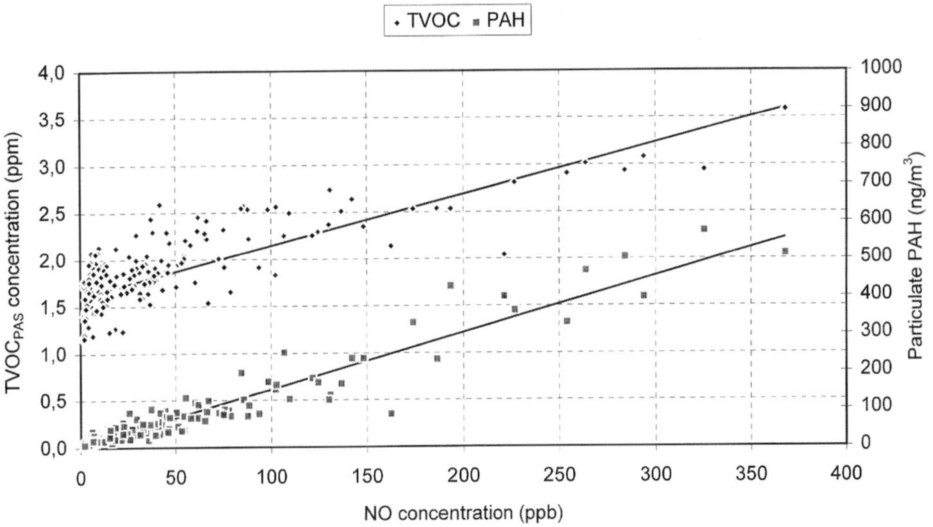

Figure 1.6-2. The concentrations of $TVOC_{PAS}$ and particulate PAHs vs the NO concentration measured at the outdoor air intake to a building located in the city of Gothenburg, Sweden. The $TVOC_{PAS}$ concentrations are reported as methane equivalents.

The measurements in the above referenced work were all carried out at the outdoor air intakes at buildings located in urban environments. Peak concentrations of $TVOC_{PAS}$ were observed in the range 4–6 ppm, mainly during periods of high traffic intensity (morning and afternoon rush hours). The background concentrations of $TVOC_{PAS}$ were typically in the range 1.5-2 ppm (methane equivalents), which corresponds quite well to the expected background concentration of methane in the atmosphere. The sensitivity to methane thus results in an unwanted interference. According to Krüger and colleagues (1995), this interference can be avoided, even if the gas analyzer used shows a pronounced sensitivity to methane. The methane concentration can be approximated by determining the VOC concentration after removal of non-methane compounds by means of a gas adsorption filter. The concentration of non-methane compounds may then be estimated as the concentration difference between an unfiltered air sample and the methane concentration.

1.6.4.2 Intermittent Indoor VOC Sources

Figure 1.6-3 shows the differences between TVOC concentrations measured in the exhaust air and in the supply air in an office building. The building is mechanically ventilated with an air flow rate corresponding to 1.1 air changes per hour (0.8 l/s per m^2 floor). The figure contains both data obtained using a PAS instrument and the results from GC/FID analysis of VOCs actively sampled on Tenax adsorbents. Both methods show a clear increase in the concentration difference, thus indicating that the internal emissions of VOCs increase over the 6-h period between 5 am and

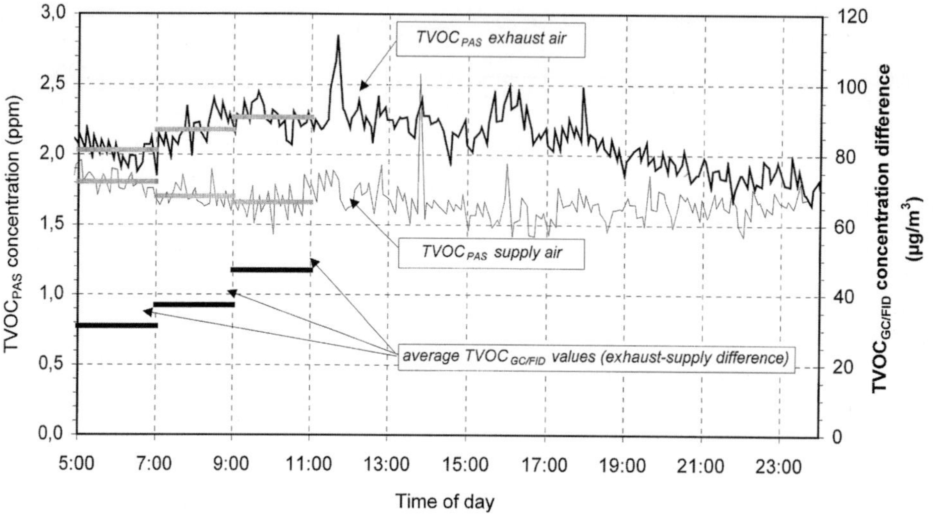

Figure 1.6-3. TVOC concentrations measured using two different methods in exhaust air and supply air in a mechanically ventilated office building. The $TVOC_{PAS}$ values are reported as methane equivalents and the $TVOC_{GC/FID}$ results as toluene equivalents.

11 am. However, as expected, according to different responses for different individual VOCs, the two methods give differing results with respect to absolute values. The magnitude of the concentration increase is greater for the PAS measurement than for the GC/FID measurement. The concentrations obtained using PAS are expressed as methane equivalents, while the GC/FID results are reported as toluene equivalents.

The GC/FID analysis also revealed that the observed increase in the TVOC concentration was mainly due to an increase in the internal emission of toluene, xylene and decane. The concentration of 2-(2-ethoxyethoxy)ethanol was also found in the exhaust air, but the concentration of this compound did not vary significantly between the three measurements. It should be noted that the air change rate was fairly constant during the period studied, and that the temperature and the relative humidity of the indoor air were found to be in the intervals 22.0–22.8 °C and 35–40 % RH, respectively.

The reader is reminded that the purpose of the present article is to indicate situations where real-time monitoring can be especially valuable, and not to present a complete evaluation of the VOC situation in certain buildings. The above presentation should be considered as an example of real-time monitoring used in order to distinguish periods with high VOC concentrations from periods with lower concentrations. Real-time monitoring carried out over a few days may reveal systematic concentration variations with time, and if the gas analyzer is used together with a multiport sampling device it should also be possible to identify locations with high VOC concentrations. By this approach real-time monitoring can be used as a tool to improve the strategy for sampling of VOCs on adsorbent tubes for subsequent identification and quantification of individual substances.

1.6.4.3 The Influence of Ventilation System Operation

The indoor concentrations of pollutants generated indoors depend, among other factors, on the outdoor airflow rate. Therefore, the temporal variation of the indoor concentration, e.g. after a change of the outdoor airflow rate, is influenced by the time constant for the ventilation system (e.g. expressed as the time for one air change). Furthermore, a variety of VOCs may show source and sink effects due to adsorption and desorption on indoor surfaces. These processes may also influence the variations with time of the indoor concentration.

In mechanically ventilated buildings it is common that the fans are in operation only during working hours. When the fans are stopped in the afternoon the ventilation rate is reduced to an infiltration rate determined by the airtightness of the building envelope, the temperature difference between indoor and outdoor air, and the wind pressure on the façades. Typical values of the infiltration rate may correspond to air change rates in the range $0.1-0.3\ h^{-1}$. During periods with reduced ventilation rate the indoor concentrations of pollutants generated indoors will increase, e.g. VOCs emitted from building materials and furniture.

Figure 1.6-4 gives an example of the difference between the $TVOC_{PAS}$ concentrations measured indoors and outdoors in an office building during the first hours after a night with the fans out of operation. The figure shows that the concentration difference decreases to approximately ¼ of the value observed immediately before the fans were started. After about 1.5 h the concentration difference has reached a minimum of about

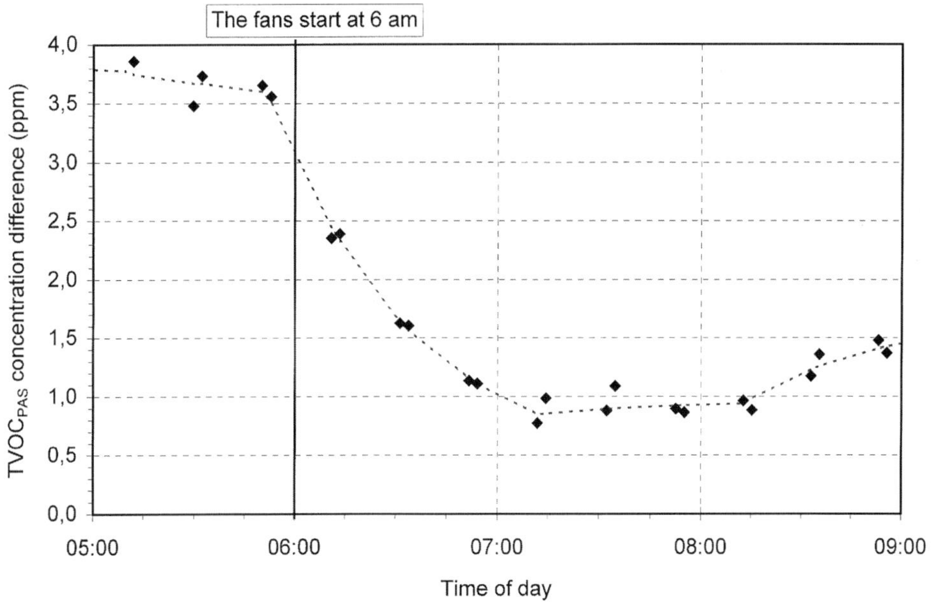

Figure 1.6-4. The difference between $TVOC_{PAS}$ concentration indoors and outdoors after starting the fans in an office building which has no mechanical ventilation during night time. The concentrations are reported as methane equivalents.

1.0 ppm. When the fans are in operation the ventilation rate corresponds to 1.4 air changes per hour.

1.6.4.4 Laboratory Investigations of Emissions from Building Products

Building products may be investigated in test chambers with respect to the emission of organic compounds. In this context, it is often necessary to consider the possibility that the emission rate is time dependent. Consequently, there may be a need to take several air samples at appropriate times after the start of the emission test in order to make an accurate determination of the emission decay. In such investigations real-time monitoring may provide guidance for the strategy for VOC sampling aiming at identification and quantification of specific VOCs. The signal from the real-time monitoring device may reveal changing characteristics of the emission, e.g. caused by drying of a paint newly applied to a substrate (Afshari, 1995).

Another example of real-time VOC monitoring in laboratory experiments has been presented by Vejrup and Wolkoff (1995), who demonstrated a method of measuring, with high time resolution, the emission profile of cleaning agents. The authors concluded that the use of a direct-reading instrument can be a powerful tool to complement emission testing by air sampling for subsequent analysis by gas chromatography. The use of real-time monitoring has also been demonstrated for the determination of the toluene emission from rotogravure printed brochures (Jensen et al., 1995).

1.6.5 Concluding Remarks

As demonstrated in the present article, real-time monitoring of organic compounds can be used as a valuable tool in indoor air quality investigations carried out both in the field and in laboratories. Appropriately used, such methods can aid the identification of indoor air pollution sources and facilitate studies of various factors influencing the concentration of organic compounds in indoor air.

However, it is of crucial importance that such methods should be used with an awareness of their limitations, and cautious interpretation of the measured data is required. It is often necessary to limit the conclusions to observations of concentration variations with time and concentration differences between different locations. Not too much attention should be paid to the absolute values of TVOC concentrations monitored. Generally, comparisons between such data obtained in different situations should be avoided.

In order to avoid results that are difficult to interpret it may be necessary to monitor the concentrations at several locations in parallel, e.g. by the use of an automatically controlled manifold system. In most situations it is of interest to study variations in the outdoor air and indoor air concentration simultaneously. However, the need for such arrangements may depend on the purpose of the investigation.

With regard to the reporting of data, it is good practice always to specify what type of instrument has been used and under what circumstances the monitoring has been carried out.

Acknowledgements

The author wishes to thank Professor Ove Strindehag at Chalmers University of Technology for his valuable comments and suggestions.

References

Afshari A. (1995): Clean room for characterizing indoor sources. Environ. Eng., *8*, 10–16.

Andersson K., Bakke J.V., Bjørseth O., Bornehag C.G., Clausen G., Hongslo J.K., Kjellman M., Kjærgaard S., Levy F., Mølhave L., Skerfving S. and Sundell J. (1997): TVOC and health in non-industrial indoor environments–Report from a Nordic scientific consensus meeting at Långholmen in Stockholm, 1996. Indoor Air, *7*, 78–91.

Brüel & Kjær Innova (1997): Gas detection limits. Brüel & Kjær Innova Air Tech Instruments, Nærum, Denmark.

Bunding Lee K.A., Clobes A.L., Hood A.L., Schroeder J.A. and Hawkins L. (1993): Methods of continuous monitoring of volatile organic compounds in indoor environments. Proceedings of Indoor Air '93, Helsinki, Finland, Vol 2., 245–250.

Burtscher H., Scherrer L., Siegman, H.C., Schmidt-Ott A. and Federer B. (1982): Probing aerosols by photoelectric charging. J. Appl. Phys., *53*, 3787–3791.

EC-European Commission (1994): Sampling strategies for volatile organic compounds (VOCs) in indoor air. Indoor Air Quality and its Impact on Man, Report No. 14, Luxembourg.

EC-European Commission (1997): Total volatile organic compounds (TVOC) in indoor air quality investigations. Indoor Air Quality and its Impact on Man. Report No. 19, Luxembourg.

Ekberg, L.E. (1993): Sources of volatile organic compounds in the indoor environment. Proceedings of the International Conference–Volatile Organic Compounds, London, UK, 573–579.

Ekberg, L.E. (1994): Volatile organic compounds in office buildings. Atmos. Environ., *28*, 3571–3575.

Ekberg, L.E. (1995): Concentrations of NO_2 and other traffic related contaminants in office buildings located in urban environments. Build. Environ., *30*, 293–298.

Fahlén P., Andersson H. and Ruud S. (1992): Demand controlled ventilating systems: Sensor tests. The Swedish National Testing and Research Institute, SP Report No. 1992 : 13, Borås, Sweden.

Hicks J.B., Winegar E.D. and Sim C.S. (1992): The application of FTIR remote sensing for indoor air quality monitoring. Proceedings of IAQ '92: Environments for People, San Francisco, American Society of Heating, Refrigerating and Air-Conditioning Engineers, Inc. Atlanta, USA, 251–256.

Hodgson A.T. (1995): A review and a limited comparison of methods for measuring total volatile organic compounds in indoor air. Indoor Air, *5*, 247–257.

Jensen B., Olsen E. and Wolkoff P. (1995): Toluene in rotogravure printed brocures–High speed emission testing with the FLEC, and comparison with chamber exposure data. Proceedings of Healthy Buildings '95, Milan, Italy, Vol. 2, 965–970.

Krüger U., Kraenzmer M. and Strindehag O. (1995): Field studies of the indoor air quality by photoacoustic spectroscopy. Environ. Int., *21*, 791–801.

Mølhave L., Clausen G., Berglund B., De Ceaurriz J., Kettrup A., Lindvall T., Maroni M., Pickering A.C., Risse U., Rothweiler H., Seifert B. and Younes M. (1997): Total volatile organic compounds (TVOC) in indoor air quality investigations. Indoor Air, *7*, 225–240.

Rosencwaig A. (1980): Photoacoustics and photoacoustic spectroscopy. In: Elving P.J., Winefordner J.D. and Kolthoff I.M. (eds) Chemical Analysis, Vol. 57. Wiley, Chichester, UK.

Vejrup K.V. and Wolkoff P. (1995): High-speed emission testing of cleaning agents with the FLEC–Part I: Method development. Proceedings of Healthy Buildings '95, Milan, Italy, Vol. 2, 983–988.

Weschler C.J. and Shields H. (1994): Indoor chemistry involving O_3, NO and NO_2 as evidenced by 14 months of measurements at a site in southern California. Environ. Sci. Technol., *28*, 2120–2132.

Weschler C.J. and Shields H. (1997): Potential reactions among indoor pollutants. Atmos. Environ., *31*, 3487–3495.

1.7 Assessment Methods for Bioaerosols

Peter S. Thorne and Dick Heederik

1.7.1 Sampling Strategy for Bioaerosols

A variety of methods for sampling and analysis of bioaerosols have been developed over the past 50 years. Much of the early research was carried out because of military concerns regarding the use of biological warfare. While airborne, bioaerosols have few readily measurable characteristics that distinguish them from other dry aerosols from the standpoint of detection. Thus, bioaerosol measurements are rarely performed in real time. Instead, they rely upon the collection of a sample into or onto solid, liquid, or agar media with subsequent microscopic, microbiologic, biochemical, immunochemical or molecular biological analysis. It is helpful to divide the discussion of microbial bioaerosol assessment into "culture-based methods" that rely upon culturing of microorganisms and "non-culture methods" or total bioaerosol methods that attempt to enumerate organisms without regard to viability. Analysis of other bioaerosol constituents relies upon bioassays, immunoassays, or chemical assays. Analysis of endotoxins, mycotoxins, or aeroallergens in bioaerosol samples generally requires air filtration and elution of collected material into a liquid for subsequent analysis.

1.7.1.1 Sample Collection

It is often difficult to construct an accurate bioaerosol exposure assessment. Decisions regarding sampling duration and location as well as occupant activities before and during sampling may depend upon the goals of the study and the purpose of sampling. In the occupational health arena there is a strong preference for air sampling during normal operations, since that is often the basis for promulgating health standards. Indoor air studies in homes, office buildings, and day care centers have most commonly employed air sampling under normal use conditions. However, some studies have preceded collection of short-term air samples by aggressive re-aerosolization of settled dust (Pope et al., 1993; Rylander et al., 1992). For studies of aeroallergens such as house dust mites and animal dander, vacuum collection of dust from floors, carpets, upholstery, or pillows may be more predictive of occupant health status since these aeroallergens are typically associated with larger particles that settle rapidly. Thus, one may find indoor aeroallergen levels expressed per cubic meter of air, per square meter of surface, or per milligram of collected dust.

1.7.2 Assessment Methods for Microbial Bioaerosols

1.7.2.1 Culture-Based Methods

A variety of devices for microbial bioaerosol sampling have been developed and are listed in Table 1.7-1. These devices and their use have been described previously (Buttner and Stetzenbach, 1991; Buttner and Stetzenbach, 1993; Chang et al., 1994; Jensen et al., 1992; Juozaitis et al., 1994; Nevalainen et al., 1992; Thorne et al., 1992). There are three standard approaches to active sampling of culturable bioaerosols: impactor methods, liquid impinger methods, and air filtration methods. Settle plates have been used for passive sampling but do not allow quantitative determination of airborne concentration. With all three of the active sampling approaches a calibrated pump draws bioaerosol-laden air into the sampling device that strips the bioaerosols from the air stream. After sample collection organism colonies are grown on culture media held at a defined temperature in an incubator. Colonies are counted manually or with the aid of image analysis as they emerge usually over a 3 to 7 day period.

With impactor sampling, bioaerosols moving in the air stream pass through a round jet or a slit and impact onto agar in a Petri dish or on some other type of nutrient surface where they are enumerated. Correction of the counts is generally required with this method of enumeration to adjust for the probability of more than one viable organism passing through a single jet and being counted as a single colony. The chance of this occurring increases as the concentration of viable bioaerosols increases. This adjustment of colony counts is usually referred to as positive hole correction and has been described elsewhere (Andersen, 1958; Leopold, 1988; Macher, 1989). Multistage devices typically have 100 to 400 jets per stage and allow size discrimination by sequentially increasing the velocity through the jet and decreasing the jet-to-plate spacing. Larger organisms deposit on the higher stages and spores as small as 0.6 μm appear on the final stage. Some slit-to-agar impaction samplers slowly rotate the collection media (held in a Petri dish) during sampling to allow longer sampling times.

Liquid impingers, such as the all-glass impinger (AGI), collect microorganisms by impinging the airstream onto the agitated surface of the liquid collection fluid. Most microorganisms and spores will be retained in the collection fluid such that, when this fluid is plated onto a suitable agar in Petri dishes, colonies will emerge. Serial dilutions of the collection fluid into sterile isotonic solutions are generally required prior to plating to allow the numbers of colonies on the plates to be in a quantifiable range (30 to 300 is desirable). Many different solutions have been used and several investigators have tested the efficacy of one fluid over another (Marthi and Lighthart, 1990; Thorne et al., 1994). The most commonly used fluids are peptone water (1 % peptone in distilled water with 0.01 % Tween 80 and 0.005 % antifoam A), betaine water (5 mM betaine with Tween 80 and antifoam A), phosphate-buffered saline with Tween 80, or tryptic soy broth with antifoam A. The multistage liquid impinger (MSLI) works in an analogous fashion to the AGI but allows fractionation of the bioaerosol into three size fractions: less than 4 μm, 4 to 10 μm, and greater than 10 μm (Lange et al., 1997; May and Harper, 1957).

Table 1.7-1. Microbial Bioaerosol Sampling Devices.

Viable Sampling Methods	
Stationary Jet-to-Agar Impactors	
Andersen Microbial Sampler (AMS)[1]	Models: N-6 1-stage, 2-stage, 6-stage
Burkard Portable Air Sampler for Agar Plates[2]	
MicroBio Air Sampler[3]	Models: MB1, MB2
Rotating Media Slit-to-Agar Impactors	
Slit-to-Agar Biological Air Sampler[4]	Models: STA-101, STA-203, STA-303
Mattson/Garvin Air Sampler[5]	Models: 220, P-320
Casella Airborne Bacteria Sampler[6]	Model: MK-II
Centrifugal Agar Impaction Samplers	
Centrifugal Air Sampler[7]	Models: Standard RCS, RCS PLUS
Liquid Impingers	
All-Glass Impinger (AGI)[8]	Models: AGI-30, AGI-4
Multi-Stage Liquid Impinger (MSLI)[2]	
Air Filtration Methods	
Nuclepore Filtration and Elution Method[9]	
Inverted Filter Method[9]	
Non-Viable Sampling Methods	
Stationary Slit-to-Microscope Slide Spore Samplers	
Burkard Personal Volumetric Air Sampler[2]	
Moving Slit-to-Microscope Slide/Tape Spore Samplers	
Recording Air Sampler[2]	
Recording Volumetric Spore Trap[2]	Models: 7-day, 24-hr
Allergenco Air Sampler[10]	Model: MK-3
Rotating Impaction Surface Sampler	
Rotarod Spore Sampler[11]	
Virtual Impactor Spore Sampler	
High Throughput Jet Spore and Particle Sampler[2]	
Analysis Methods for Liquid Impinger Solutions or Air Filter Samples	
Immunoassay Methods for Microbial Surface Antigens	
Non-Specific Staining Method with Enumeration by Microscopy or Flow Cytometry	
Fluorescence *In Situ* Hybridization Specific Labeling with Enumeration by Microscopy or Flow Cytometry	
Polymerase Chain Reaction Analysis	

Suppliers for the bioaerosol samplers:

1. Graseby Andersen-Nutech, 500-T Technology Ct., Smyrna, GA 30082, USA. 770-319-9999 *www.graseby.com*
2. Burkard Manufacturing Co., Woodcock Hill, Rickmansworth, Hertfordshire WD3 1PU, UK. +44 1923 773134 *www.burkard.co.uk*
3. Spiral Biotech, Inc., 7830 Old Georgetown Rd., Bethesda, MD 20814, USA. 301-657-1620 *www.spiralbiotech.com*
4. New Brunswick Scientific Co., Inc., 44 Talmadge Rd., Edison, NJ 08817. 732-287-1200 *www.nbsc.com*
5. Barramundi Corp., P.O. Box 4259, Homosassa Springs, FL 34447, USA. 352-628-0200 (no website)
6. Casella London Ltd., Lee Valley Technopark, Ashley Road, Tottenham Hale, London N17 9LN England +44-1818-804560 *www.casella.co.uk/consultancy/bedata.htmlhealth*
7. Biotest Diagnostics Corp., 66 Ford Rd., Denville, NJ 07834, USA. 201-625-1300 *www.biotest.com*
8. Ace Glass Co., 1430 N. West Blvd., Vineland, NJ 08360, USA. 609-692-3333 *www.aceglass.com*
9. For assorted filter holders: BGI, Inc., 58 Guinan St., Waltham, MA 02154, USA. 617-891-9380 *www.bgiusa.com*
10. Allergenco/Blewstone Press, P.O. Box 8571, Wainwright, TX 78208, USA. 210-822-4116 *www.txdirect.net/corp/allergen*
11. Sampling Technologies, Inc., 10801 Wayzeta Blvd., Suite 340, Minnetonka MN 55305, USA. 612-544-1588 *www.rotarod.com*

Several sampling methods in common use rely upon air filtration to remove bioaerosols from a sampled air volume. Generally the aerosols are removed by a combination of impaction on the filter surface as well as interception, electrostatic attraction, and diffusion deposition. After sampling, filters are agitated in a solution to re-entrain the microbes and this solution is then serially diluted and plated onto appropriate culture media.

With all of the viable bioaerosol assessment methods described above, the culture medium is selected to test for broad-spectrum bacteria or fungi or to select for specific groups, genera, or species. The most commonly used media and incubation temperatures are: trypticase soy agar (TSA) or R2A with cycloheximide for broad spectrum mesophilic bacteria at 22–35 °C, blood agar for mesophilic bacteria of human origin at 37 °C, TSA for thermophilic bacteria at 50–55 °C, eosin methylene blue agar (EMB) or MacConkey's medium (MAC) for Gm- bacteria at 22–35 °C, and malt extract agar (MEA), rose bengal streptomycin (RBS) agar, or TG-18 for fungi at 22–30 °C. Recipes for these media appear in standard texts (Difco, 1984; Gerhardt et al., 1994), and dry mixes are commercially available. If one is sampling to isolate a particular genus, then media, antimicrobials, and growth conditions must be tailored to support the growth of that genus and to exclude competing organisms.

1.7.2.2 Non-Culture Methods

1.7.2.2.1 Direct Count Methods

Non-specific, non-culture methods are based on the observation that particular fluorochromes can attach to or intercalate into resident nucleic acids (in DNA or RNA), thereby labeling the microorganisms in a sample. Indoor air samples are collected by air filtration followed by fixation and staining of the organisms. Counting is performed by microscopy under epifluorescence illumination to assess total counts of the microbes in the air. Several recently developed techniques have improved the sensitivity of this bioaerosol direct count method. Scanning confocal microscopy provides higher resolution and enhanced sensitivity over traditional methods. New fluorochromes have been developed that improve the enumeration process. Monomeric and dimeric cyanines can be successfully used to stain the bacteria or bacterial cell walls with reduced background fluorescence from foreign materials, bacterial degradation debris, and other autofluorescing dust particles collected during the bioaerosol sampling. Acridine orange (AO) has been routinely used as the staining agent for quantifying bacteria by counting individual organisms under fluorescence microscopy using the direct count method (Eduard and Aalen, 1988; Eduard et al., 1990; Lange et al., 1997; Palmgren et al., 1986; Thorne et al., 1992). This dye is widely used because it is inexpensive and has a characteristic narrow excitation band and relatively low background illumination. However, at the bacterial size-exclusion detection limits, selective identification and counting of cells from the fluorescing dust particles is subjective and tedious. In addition, AO binds to DNA in a non-specific manner and autofluorescence of organic debris can lead to over-estimation of bacterial counts. Acriflavine dye staining methods have been suggested to give more uniform and reproducible counting of bacteria (Bergstrom et al., 1986). For other nucleotide chemistry applications, several dyes have been developed for the analysis of nucleic acids by intercalation-fluorescence enhancement techniques. These polyfunctional intercalating dyes have been used effectively at much lower concentration levels than monomeric dyes for intercalation with DNA (Ogura et al., 1994). It has recently been demonstrated that labeling with 4'-6-diamidino-2-phenylindole (DAPI) and counting organisms by flow cytometry can improve the sensitivity of this method by several orders of magnitude (Lange et al., 1997). This makes direct counting a practical method for use in the evaluation of bioaerosols in indoor air when information on organism taxa is not needed.

1.7.2.2.2 Molecular Biology-Based Techniques

Polymerase chain reaction (PCR) techniques allow one to amplify small quantities of target DNA typically by 10^4 to 10^6 times and determine in a semi-quantitative manner the presence of specific microorganisms. PCR analysis can be automated and provide turnaround times on the order of hours rather than the 3 to 5 days required for culture techniques (Atlas, 1991). Quantitation of PCR is accomplished by most probable number methods or by calibration with the function: $C_{\text{final}} = C_{\text{initial}} \times 2^n$, where C is the concentration (number of copies of DNA per unit volume) and n is the number of thermal cycles through which the sample has been processed. PCR has been shown to be useful for analysis of water samples for microbial contamination (Bej et al., 1990) and clinical specimens (Eisenach et al., 1991; Kolk et al., 1992). Viral particles of both RNA and DNA viruses are assayed primarily by PCR and related molecular methods. The application of PCR to sampling the airborne environment has been demonstrated (Alvarez et al., 1994; Khan and Cerniglia, 1994) but is less well established.

Advances in molecular biology have made *in situ* hybridization of labeled oligonucleotide probes with ribosomal ribonucleic acids (rRNA) from microorganisms a potentially useful method for characterization of organisms in bioaerosols at the family, genus, or species level, regardless of their viability. Oligonucleotide probes can now be made quickly and economically, there are publicly accessible data bases of rRNA sequences, and a variety of nonradioactive fluorescent labels can be used, which avoids the inherent handling problems associated with autoradiography. Potential advantages of fluorescent *in situ* hybridization (FISH) techniques include identification of nonculturable organisms, highly sensitive and species-specific enumeration, and automated analysis. Oligonucleotide probes are short, single-stranded, synthetic deoxynucleotide base sequences. The identification of single microorganisms by FISH uses probes that are complementary to unique sequences of 16s rRNA contained within every ribosome. rRNA consists of phylogenetically conserved and variable nucleotide regions allowing determination of taxon at any desired level. Each individual active bacterial cell contains approximately 10^4 to 10^5 ribosomes with the same number of identical copies of rRNA. The number of rRNA copies within a bacterium is proportional to the rate of growth. Thus, the rRNA represents a natural amplified target site for probe hybridization. Detection of hybridization has been accomplished with epifluorescence microscopy (DeLong et al., 1989), scanning confocal microscopy for higher sensitivity (Bauman et al., 1990), and flow cytometry for automated enumeration (Amann et al., 1990). These techniques were developed using pure cultures at high concentrations at the optimal growth stage. Recently, quantification of bioaerosols using FISH and flow cytometry has been demonstrated for laboratory-generated bioaerosols and air samples collected in the field using both the AGI and the MSLI (Lange et al., 1997). Specific quantitation was accomplished using three oligonucleotide probes with different fluorophores: a general probe complementary to all eubacteria, a specific probe complementary to all species within the genus *Pseudomonas*, and a nonsense probe that controlled for non-specific binding. This produced differential labelling of pseudomonads from other eubacteria and facilitated specific enumeration by flow cytometry. The application of FISH to environmental bioaerosol samples is presently limited because of the lower numbers of rRNA targets in these cells compared to cultured organisms captured in the logarithmic growth phase. Work is in progress to develop methods to amplify the DNA *in situ* using PCR and then to introduce labeled probes specific for the DNA. New techniques for specific quantitation of bioaerosols regardless of viability using molecular biology methods are likely to produce major advances in exposure assessment to bioaerosols.

1.7.2.3 Problems with Accurate Assessment of Microbial Bioaerosols

There are a number of significant problems with measuring bioaerosol exposures, and these are summarized in Table 1.7-2.

There are large temporal, seasonal and spatial variations in bioaerosols. These can relate to the specific activities that disperse the bioaerosols as well as those conditions that lead to sporulation of organisms. Some commonly used sampling methods allow only short-term sampling because of their relatively low upper limits of detection. With these methods exposures may be over- or under-estimated by an order of magnitude.

In 1877, Pasteur discovered that some microorganisms could produce substances that would kill or inhibit other species. This was termed antibiosis. More recently it has been found that fungi produce mixtures of mycotoxins that synergistically produce interference competition to impede other prokaryotic and eukaryotic organisms. Multiple organisms may cohabitate a building thriving in separate amplification sites but then become intermingled as aerosols. When those bioaerosols are collected in a sampler, competitive organisms may inhibit each other. While positive hole correction can adjust the total counts from a jet-to-agar or slit-to-agar sampler, it does not allow one to account for the presence of the inhibited species in that environment. For viable bioaerosol sampling, this means that the full range of airborne organisms may not be realized, and the numbers of viable bioaerosols will be under-predicted. In a similar fashion, the choice of media on which to culture the organisms can also affect the emergence of colonies. Some organisms such as obligate plant pathogens, some anaerobes, and many mushroom spores, will not grow under standard culture conditions. For those organisms that can be cultured, the media may not yield colonies of the organisms in the same proportion by genus or species as existed in the environment that was sampled. The collection medium used in liquid impingers may differentially favor or disfavor the cul-

Table 1.7-2. Problems with Assessment of Bioaerosol Concentrations Indoors.

General:	
	Large short-term temporal variability due to activity
	Large seasonal variability
	Large variability from area to area
	Few methods available to assess total bacterial concentration and taxa
	Methods are labor intensive and tedious
Viable bioaerosol sampling methods:	
	For some methods, relatively short sampling times due to sampler limitations
	Loss of small organisms from some samplers through re-entrainment or wall losses
	Physicochemical trauma to organisms resulting in loss of viability
	Viability loss through bombardment or solubilization of toxic vapors
	Antibiosis and colony masking leads to lower colony yield
	Selection of media and culture conditions affects colony growth
	Viability may be irrelevant to the toxicity
Non-culture bioaerosol sampling methods:	
	Quantitative identification of organisms is difficult
	Debris may be falsely enumerated as microorganisms
	Large variability if method uses microscopic enumeration
	Viability may be required for the disease process
Molecular biology assessment methods:	
	Dependent upon the access of the oligonucleotide probe to the organismal nucleic acids
	FISH methods dependent upon sufficient rRNA in the bioaerosol
	Can not determine viability of the organisms
Immunoassay aeroallergen assessment methods:	
	Dependent upon the specificity of the antibody probe for the antigen
	No standard antigens for many aeroallergens
	Non-specific binding of the antibody with sample debris

turability of bioaerosols in the sampling environment. Many methodologic issues in this debate remain to be resolved.

It is traumatic for microorganisms to undergo the process of sampling in any active sampling device. Trauma can reduce culturability for viable sampling and impede identification for non-viable methods. Most sampling devices require impaction on a solid or liquid surface at a high enough velocity to cause the aerosol to exit airflow streamlines and deposit. Volumetric flowrates vary from 12 to 200 l/min in commercial viable microbial air samplers with 28.3 l/min common. The impaction velocity is dependent upon the gas velocity, U, at the exit from the impactor nozzle or jet. This velocity for two common devices, the AGI and the AMS in the N6 configuration, is 26500 cm/sec and 2400 cm/sec, respectively (Nevalainen et al., 1993). When they strike the surface of the Petri dish or enter the collection fluid, their microenvironment changes in terms of temperature, hydration, osmolarity, nutrients, and oxygen tension. There is differential resilience to these shocks across bacterial and fungal species. In addition, some fungal spores are hydrophobic while others can trap air on their surface making them buoyant. Trapping of these spores in a liquid impinger may be difficult because of re-entrainment into the air stream and loss from the sampler. Thus, whether one is using a slit-to-agar sampler, a jet-to-agar sampler, or a single or multistage liquid impinger, only a portion of the viable organisms present in the sampling environment will be detected.

VOC or SVOC in the air of some buildings may concentrate in the collection media and differentially affect the viability of organisms. For samplers that impact organisms onto agar with fixed agar plates, an organism collected early in the sampling period may continue to be bombarded for the rest of the sampling period. For liquid impingers, compounds highly soluble in water may concentrate in the impinger solution. With air filtration, sampling in dry environments leads to desiccation of the microbes, reducing their culturability.

Non-culturable organisms often represent more than 95 % of the total viable plus non-viable bioaerosols. For a number of bioaerosol-induced diseases, dead organisms are as potent causative agents as viable organisms. This is true for hypersensitivity diseases associated with microbial or plant allergens and is true for diseases caused by endotoxins or other microbial toxins. While many fungal spores can be classified by morphology, it is difficult to speciate non-culturable bacteria and yeast. The direct-count method using intercalating dyes has relatively poor reproducibility and a high lower limit of detection. The use of FISH technology, PCR methods, and immunoassays has opened new possibilities for detection and speciation regardless of whether or not the organisms are culturable.

1.7.3 Assessment Methods for Aeroallergens

Antibody-based immunoassays, particularly enzyme-linked immunosorbent assays (ELISA) are widely used for the measurement of aeroallergens and allergens in settled dust in buildings. These assays use antibodies with specificity for the target aeroallergen to bind the allergen in the air or settled dust sample. Detection is provided via reaction of a substrate with an enzyme-linked second antibody that immunologically recognizes and binds to the anti-aeroallergen primary antibody. The antibody-bound enzyme reacts with the substrate to produce a spectral shift or color change that can be measured in a microplate reader. Specific antibodies for these assays may be collected from a seropositive patient or a group of patients (a patient pool) or may be raised in rabbits immunized repeatedly with crude or purified antigen accompanied by an adjuvant. Monoclonal antibodies can be produced in vitro by cloned cells derived from a single antibody-producing mouse lymphocyte. Monoclonal antibodies offer enhanced specificity over polyclonal antibodies. Samples are generally assayed at multiple dilutions, and a titer is determined based on the highest dilution where the optical density exceeds a defined threshold. Standard allergens are used as a basis for assigning units to the assay. ELISAs can also be run as inhibition assays where standard allergen is added to the sample at defined concentrations and the loss of antibody binding is assessed. In radioimmunoassays (RIA), radiolabeling is used for detection rather than enzymatic reaction with a substrate. The principals of these methods are described in standard clinical immunology texts and their use in aeroallergen detection has been reviewed (Pope et al., 1993). To date, the antigens *Der p* I, *Der f* I, and *Der p/f* II have been most widely investigated and the methods have been well described (Platts-Mills and Chapman, 1987; Platts-Mills and de Weck, 1989; Price et al., 1990). Methods for assessment of exposure to allergens from animals (Eggleston et al., 1989; Schou et al., 1991; Swanson et al., 1985; Virtanen et al., 1986; Wood et al., 1988), cockroaches (Pollart et al., 1991), storage mites (Hollander et al., 1997; Iverson et al., 1990; van Hage-Hamsten, 1992), and latex rubber (Miguel et al., 1996) have also been published.

1.7.3.1 Problems with Accurate Assessment of Aeroallergens

Little is known about the validity and comparability of immunoassays that have been used to measure allergen levels in the air. Two studies clearly indicate that differences in the extraction procedure and the use of extraction buffers contribute to differences between assay protocols up to at least a factor 10 (Zock et al., 1996; Hollander et al., 1998). The most important determinants that contribute to differences between assays seem to be the antibody source (monoclonals vs polyclonals), type of immunoassay (inhibition vs sandwich) and the standard used (purified allergens vs crude extract). Although few comparative studies exist, the available evidence suggests that while the correlation of allergen levels obtained with different assays is very good, large systematic differences seem to occur (Hollander et al., 1998; Renström et al., 1998). The available literature is limited to a few allergens and generalizations can not be made. Validation and comparison of different assays by intralaboratory comparisons are important issues that have not received the attention needed.

1.7.4 Assessment Methods for Endotoxins

Analytical chemistry methods for quantification of LPS have been developed employing gas chromatography-mass spectrometry (GC-MS) (Sonesson et al., 1988; Sonesson et al., 1990). These methods require special LPS extraction procedures and have not been widely used. Immunoassays have also been developed but have not been widely adopted. Although the biological activity rests with the Lipid A portion of the molecule, the antibodies produced against LPS recognize the polysaccharide part (Rietschel et al., 1996). More importantly, since the biological activity of endotoxins and LPS from various organisms varies depending upon their chemical structure and bioavailability, chemical and immunochemical methods may not represent the potency as well as a bioassay. Since LPS is a pyrogenic macromolecule (induces fever), the earliest assessments of endotoxin potency evaluated the lowest dilution of a material that could induce fever in rabbits. The discovery by Bang in 1956 that Gm- bacteria could induce gelation of lysate prepared from blood cells of the horseshoe crab, *Limulus polyphemus* (Bang, 1956), led to the development of the *Limulus* amebocyte lysate (LAL) gel clot assay for endotoxin (Levin and Bang, 1968). The LAL contains a mixture of proenzymes that are activated by endotoxin in a cascading set of reactions (Iwanaga, 1993).

Modern versions of this assay are based on the activation of a clotting enzyme (via factor C) present in mixed lysates of hemolymph from a large number of *Limulus polyphemus*. In addition to LPS, β(1-3)-glucan, which is present in the cell walls of some plants and fungi, is also capable of triggering the LAL coagulation system (via factor G) (Iwanaga, 1993; Rylander et al., 1994) unless that clotting pathway is deliberately blocked. Cell wall peptidoglycans of Gm+ bacteria can also activate the LAL system via factor G, though orders of magnitude higher concentrations are required (Brunson et al., 1976; Mikami et al., 1982; Morita et al., 1981; Tanaka et al., 1991). Interference by LAL-reactive glucans and peptidoglycans may differ for assay kits from various suppliers. With the LAL test, predominantly cell wall-dissociated endotoxins are detectable (Sonesson et al., 1990). In contrast, animal inhalation experiments have shown that cell-bound endotoxins have similar or even increased toxicity compared to free endotoxin (Burrell and Shu-Hua, 1990; Duncan et al., 1986). Different Gm- bacterial species or even strains have shown differing degrees of inducing the LAL coagulation reaction. However, endotoxin activity in the LAL assay is highly correlated to the pyrogenicity in the rabbit (Baseler et al., 1983; Ray et al., 1991; Weary et al., 1980) and to pulmonary effects in the guinea pig, mouse and human (Deetz et al., 1997; Gordon et al., 1992; Michel et al., 1997; Rylander, 1994; Schwartz et al., 1994; Schwartz et al., 1995; Thorne et al., 1996). Thus, the LAL assay is a highly sensitive bioassay with physiologic relevance.

Several LAL-based assays have been developed including turbidimetric and chromogenic methods (Iwanaga, 1979; Teller et al., 1979). The turbidimetric method measures an increase in light scattering in proportion to endotoxin concentration, whereas the chromogenic methods measure an increase in chromophore release from synthetic substrates with increasing endotoxin concentrations. The turbidimetric techniques are most frequently used in the kinetic mode, where the rate of increase in turbidity is measured over time (Remmilard et al., 1987). The chromogenic methods traditionally used acid-quenched endpoint measures, where the absorbance at a given wavelength following a specific incubation period is compared to standard endotoxin concentrations treated in the same manner (Dunér, 1993). There are now well-developed commercial LAL-based endotoxin assays using kinetic chromogenic types of measurements in microtiter plates (Cohen and McConnell, 1984; Milton et al., 1992). These have been evaluated as to their robustness (Douwes et al., 1995; Hollander et al., 1993; Milton et al., 1997; Thorne et al., 1997). A new biochemical method different from the LAL assay is currently under development (Associates of Cape Cod). This is a direct fluorescence method that uses a labeled tracer reactive with endotoxin. The tracer is a recombinant DNA-prepared LPS binding protein. No published studies are yet available that have evaluated this method against the LAL assay.

The LAL assay is at present the most widely accepted assay for endotoxin measurements, having been adopted as the standard assay for endotoxin detection by the United States Food and Drug Administration (FDA) in 1980. The more recent kinetic chromogenic versions of the LAL assay are very sensitive and have a broad measurement range (0.01 – 100 Endotoxin Units (EU)/ml ≈ 1 pg/ml – 10 ng/ml). The detection limit for airborne endotoxin measurements is approximately 0.05 EU/m^3 (5 pg/m^3). Since different test batches may give different results an internal standard must be used. The U.S. FDA has established a Reference Standard Endotoxin (currently RSE: *E. coli*-6 {EC-6}) as part of their standardization procedures. All endotoxin analyses must therefore be referenced to RSE:EC-6 to be considered valid. Large differences in both the hydrophilic and to a lesser extent the lipid A moiety between endotoxins of different species or strains make a comparison on weight basis almost meaningless (Cooper, 1985). The RSE:EC-6 is based on purified LPS from *E. coli* and is expressed in Endotoxin Units (EU) which is a measure for LAL activity. Since the RSE is expensive and in limited supply, most laboratories use a Control Standard Endotoxin (CSE) that is standardized to the RSE. The CSE is normally included in commercial LAL assay kits and allows results to be expressed in EU.

Environmental monitoring of endotoxins is usually performed by sampling airborne dust and a subsequent aqueous extraction. Dust is sampled on filters using pumps to draw air through the filters. Several types of filters are commonly used for endotoxin sampling: glass fiber, polycarbonate, cellulose, polyvinyl chloride, polytetrafluoroethylene, or polyvinylidene difluoride. No standard method exists for sample storage and extraction. All glassware used during extraction, storage and analyses should be rendered pyrogen free by heating at 190 °C for 4 h. Most laboratories use pyrogen-free water or buffers like Tris or phosphate triethylamine, with or without detergents such as Tween-20, Tween-80, triton-x-100 and saponin. The use of buffers and dispersing agents may be beneficial in the LAL assay in case of deviation from the optimal pH (6.5) or increased ion strength of the extract. The most common method of extraction is rocking or sonication of filters in extraction media or a combination of both. Although studies have been published on optimization of filter choice, filter extraction methods, extraction buffers and choice of glassware (Douwes et al., 1995; Gordon et al., 1992; Olenchock et al., 1989; Milton et al., 1990; Novitsky et al., 1986; Thorne et al., 1997), no generally accepted protocol exists.

1.7.4.1 Problems with Accurate Assessment of Endotoxins

A number of studies have investigated interferences with the LAL assay and have attempted to optimize the assay (Douwes et al., 1995; Hollander et al., 1993; Thorne et al., 1997). These studies have demonstrated that results may vary depending upon the sample matrix, the extraction method and the assay method. Other constituents present in the sample may interfere with the LAL assay and cause inhibition or enhancement of the test or aggregation and adsorption of endotoxins, resulting in under- or over-estimation of the concentration. Techniques such as spiking with known quantities of purified endotoxin and analysis of dilution series of the same sample have been described to deal with these interferences (Hollander et al., 1993; Milton et al., 1990; Milton et al., 1992; Whitakker, 1988). Studies in the laboratories of the authors of this chapter have demonstrated within-laboratory coefficients of variation between 15 % and 20 % in routine assay work. When extra care is taken to optimize precision this can be reduced to under 5 % in the endpoint chromogenic assay (Thorne, unpublished data). Several interlaboratory comparison studies have been performed and demonstrate much greater variability. One in-depth comparison of two laboratories experienced in the LAL assay, compared results of assays performed on sets of 12 simultaneous samples under varying conditions (Thorne et al., 1997). Regression analyses demonstrated variability attributable to the assay method (kinetic vs endpoint chromogenic LAL assay), air-sampling filter type (glass fiber vs polycarbonate) and interactions between the assay method and filter type. Intralaboratory correlation coefficients were 0.92 across extraction methods and 0.98 across filter type. Reliability coefficients within methods for analysis of 12 paired filters were all greater than 0.94. Two multilaboratory round robin studies have been peformed. The first included six laboratories (Reynolds et al., 1998). Excluding one laboratory that was an outlier, the remaining laboratories agreed to within ½ log unit, on the average, for air samples generated from swine and chicken barn dust. The correlation coefficients for these five were all in excess of 0.85. A second round robin test was performed with 8–10 laboratories analyzing endotoxin in air samples from generated cotton dust (Chun et al., 1998). When laboratories used their typical extraction and analysis methods the results were within 0.83 log unit (again excluding one outlier). When the extraction method was harmonized in a subsequent round of testing, the results still spanned 0.77 log unit. In spite of the above-mentioned difficulties, it is generally accepted that a harmonized LAL-based endotoxin assay method will allow adequate assessments of exposure for standard setting. This harmonization will facilitate better epidemiologic and experimental health effects data on endotoxin.

1.7.5 Assessment Methods for Other Bioaerosol Components

1.7.5.1 Glucans

There are currently three principal methods in use for the assay of β(1-3)-glucans. Two are based upon the bioactivity of this molecule in the factor G-mediated *Limulus* coagulation pathway (Iwanaga, 1993; Morita et al., 1981). The third method is an immunoassay (Douwes et al., 1996). The amebocyte lysate from *Limulus polyphemus* or its Asian relatives *Tachypleus tridentatus, T. giga,* or *T. rotundicauda* coagulates upon contact with LPS via factor C and β(1-3)-glucans via factor G. Thus, an LAL assay for glucans can be designed by either disabling the factor C pathway (Roslansky and Novitsky, 1991) or by extraction and purification of components of the factor G pathway from the lysate (Nagi et al., 1993; Obayashi et al., 1995). Commercial kits using this approach are now available (Maruha Corp., Tokyo, Japan; Seikagaku Corp., Tokyo, Japan). An additional approach is under study that utilizes a monoclonal antibody against the LPS-factor C complex in the LAL assay to inactivate the factor C pathway allowing the glucans to stimulate the cascade via factor G. Recently, a protein that specifically binds β(1-3)-glucans was isolated from the amebocyte lysate of *T. tridentatus* (Tamura et al., 1997). This *Tachypleus*-glucan binding protein (T-GBP) was used to construct a sandwich ELISA. In this assay, the microplates are first coated with monoclonal mouse anti-{β(1-3)-glucan} IgG-κ antibody. The sample or glucan standard is then introduced and allowed to bind. After washing, the biotinated T-GBP is added to the well and the ELISA is developed via addition of peroxidase-strepavidin and the enzyme substrate. An inhibition ELISA for β(1-3)-glucans that is totally independent of the horseshoe crab hemolymph has been developed recently (Douwes et al., 1996). In this assay, β(1-3)-glucans in the test sample inhibit the binding of affinity-purified rabbit anti-glucan antibodies to the glucans coating the microtiter plate. Quantitation is achieved by labeling the rabbit antibodies with an enzyme-linked anti-rabbit antibody. There should continue to be advancements in methods for the analysis of β(1-3)-glucans since they may serve as important markers for indoor fungal contamination.

1.7.5.2 Mycotoxins

Mycotoxins possess distinct chemical structures and reactive functional groups that include primary and secondary amines, hydroxyl or phenolic groups, lactams, carboxylic acids and amides. Standard analytical methods for mycotoxin analysis from grains, feeds and food products have been developed and thoroughly reviewed by professional associations and international agencies, and results from inter-laboratory studies have been compared (FAO, 1990; Van Egmond, 1989). A combination of analytical methodologies must be used to evaluate the wide spectrum of mycotoxins. These include thin-layer chromatography (TLC), high-performance liquid chromatography (HPLC), solid-phase extraction (SPE), gas chromatography (GC), UV spectrophotometry, nuclear magnetic resonance spectroscopy (NMR), GC-mass spectrometry (MS), and LC-MS. Accepted methods for the analysis of mycotoxins are documented in the official methods book of the Association of Official Analytical Chemists (Scott, 1990) and the International Agency for Research on Cancer (Van Egmond, 1991). Simultaneous determination methods for aflatoxins, ochratoxin A, sterigmatocystin, and zearalenone have been evaluated, and detection limits for multianalytes have been determined (Soares and Rodriguez-Amaya, 1989). Several analytical methods such as TLC, HPLC, GC-MS, radio immunoassay, and enzyme-linked immunosorbant assay (ELISA) have been reviewed for the analysis of mycotoxins such as aflatoxins, deoxynivalenol, fumonisins, zearalenone and their metabolites (Bauer and Gareis, 1989; Richard et al., 1993). These immunologic techniques may be particularly useful for screening large numbers of samples for an array of mycotoxins. There have been only a small number of methods developed specifically for sampling mycotoxins in indoor environments (Jarvis, 1990).

1.7.5.3 EPS

Nearly all soil biomass is produced by microorganismal degradation of higher organic structures into humus. Both bacteria and fungi are important, but fungi contribute the greatest proportion (Schlegel, 1993). As fungi enzymatically degrade material such as cellulose, starch and pectin they release polysaccharides. Some of the material consumed is incorporated into the cell wall of the fungi in the form of cell wall polysaccharides. These include chitins, glucans, and mannans. Two very important molds that appear indoors and induce allergy are *Aspergillus* and *Penicillium*. These genera are particularly important in cases where water damage is evident. These organisms are ascomycetes that have a characteristic conidial stage that gives rise to allergenic spores. Their cell walls consist primarily of α(1-3)-glucan and β(1-5)-galactan (Leal et al., 1992). Since these organisms represent food storage molds there has been research into methods for monitoring food for their presence. Antibodies were developed against certain extracellular polysaccharides from these organisms (β-D-galactofuranosyl residues from mycelial galactomannans) (Kamphuis et al., 1992; Notermans and Soentoro, 1986) and were then available to develop an inhibition ELISA to detect EPS-*Aspergillus/Penicillium* in house dust samples (Douwes et al., 1998). Epidemiological studies are underway in which the EPS-*Aspergillus/Penicillium* assay is being used as an exposure variable. Chemical methods for the analysis of fungal polysaccharides have been developed for the food and animal feed industry, but these have not found wide application in indoor air studies.

1.7.6 Summary

Methods for sampling culturable microorganisms are well developed but grossly under-represent the exposure levels for total airborne microorganisms. Non-culture methods are available but generally provide little information regarding microorganism species. Molecular biology-based methods are under development and offer the possibility for quantitation by species regardless of culturability. Exposure estimates for risk assessment can use markers such as cell wall components and cell surface antigens. Methods are currently available for endotoxins, mycotoxins, glucans, and extracellular polysaccharides. Aeroallergen assays are widely used to study mite, fungal, animal, and Latex aeroallergens. Standard antigens are needed to allow comparison across studies and in order to establish exposure guidelines. Development and harmonization of rigorous exposure assessment methods for indoor bioaerosols will facilitate a better understanding of the role of these agents in indoor air pollution.

References

Alvarez A.J., Buttner M.P., Toranzos G.A., Dvorsky E.A., Toro A., Heikes T.B., Mertikas-Pifer L.E. and Stetzenbach L.D. (1994): Use of solid-phase PCR for enhanced detection of airborne microorganisms. Appl. Environ. Microbiol., *60*, 374–376.

Amann R.I., Binder B.J., Olson R.J., Chisholm S.W., Devereux R. and Stahl D.A. (1990): Combination of 16S rRNA-targeted oligonucleotide probes with flow cytometry for analyzing mixed microbial populations. Appl. Environ. Microbiol., *56*, 1919–1925.

Andersen A.A. (1958): New Sampler for the collection, sizing, and enumeration of viable airborne particles. J. Bacteriol., *76*, 471–484.

Associates of Cape Cod, Endo-Fluor Assay, Woods Hole, MA, USA.

Atlas R.M. (1991): Environmental applications of the polymerase chain reaction. ASM News, *57*, 630–632.

Bang, F.B. (1956): A bacterial disease of *Limulus polyphemus*. Bull. John Hopkins Hosp., *98*, 325–350.

Baseler M.W., Fogelmark B. and Burrell R. (1983): Differential toxicity of inhaled gram-negative bacteria. Infect. Immun., *40*, 133–138.

Bauer J. and Gareis M. (1989): Analytical methods for mycotoxins. DTW–Deutsche Tierarztliche Wochenschrift., *96*, 346–350.

Bauman J.G.J., Bayer J.A. and Van Dekken H. (1990): Fluorescent in-situ hybridization to detect cellular RNA by flow cytometry and confocal microscopy. J. Microscopy, *157*, 73–81.

Bej A.K., Steffan R.J., DiCesare J., Haff L. and Atlas R.M. (1990): Detection of coliform bacteria in water by polymerase chain reaction and gene probes. Appl. Environ. Microbiol., *56*, 307–314.

Bergstrom I., Heinnanen A. and Salonen K. (1986): Comparison of acridine orange, acriflavine, and bisbenzimide stains for enumeration of bacteria in clear and humid waters. Appl. Environ. Microbiol., *51*, 664–667.

Brunson K.W. and Watson D.W. (1976): Limulus amebocyte lysate reaction with streptococcal pyrogenic exotoxin. Infect. Immun., *14*, 1256–1258.

Burrell R. and Shu-Hua Y. (1990): Toxic risks from inhalation of bacterial endotoxin. Brit. J. Ind. Med., *47*, 688–691.

Buttner M.P. and Stetzenbach L.D. (1991): Evaluation of four aerobiological sampling methods for the retrieval of aerosolized *Pseudomonas syringae*. Appl. Environ. Microbiol., *57*, 1268–1270.

Buttner M.P. and Stetzenbach L.D. (1993): Monitoring airborne fungal spores in an experimental indoor environment to evaluate sampling methods and the effects of human activity on air sampling. Appl. Environ. Microbiol., *59*, 219–226.

Chang C.-W., Hwang Y.-H., Grinshpun S.A., Macher J.M. and Willeke K. (1994): Evaluation of counting error due to colony masking in bioaerosol sampling. Appl. Environ. Microbiol., *60*, 3732–3738.

Chun D.T., Chew V., Bartlett K., Gordon T., Jacobs R.R., Larsson B.-M., Larsson L., Lewis D.M., Liesivuori J., Michel O., Milton D.K., Rylander R., Thorne P.S., White E.M. and Brown M.E. (1998): Report on the results of a round robin endotoxin assay study using cotton dust. In press.

Cohen J. and McConnell J.S. (1984): Observations on the measurement and evaluation of endotoxemia by a *Limulus* lysate microassay. J. Infect. Dis., 150, 916–924.

Cooper J.F. (1985): Ideal properties of a LAL reagent for pharmaceutical testing. In: Bacterial endotoxins; structure, biomedical significance, and detection with the *Limulus* amebocyte lysate test. Alan R. Liss, Inc., New York, 241–249.

Deetz D.C., Jagielo P.J., Thorne P.S., Bleuer S.A., Quinn T.J. and Schwartz D.A. (1997): The kinetics of grain dust-induced inflammation of the lower respiratory tract. Am. J. Respir. Crit. Care Med., *155*, 254–259.

DeLong E.F., Wickman G.S. and Pace N.R. (1989): Phylogenetic stains: Ribosomal RNA-based probes for the identification of single cells. Science, *243*, 1360–1363.

Difco Manual, 10th edn., (1984): Difco Laboratories, Ann Arbor, MI.

Douwes J., Versloot P., Hollander A., Heederik D. and Doekes G. (1995): Influence of various dust sampling and extraction methods on the measurement of airborne endotoxin. Appl. Environ. Microbiol., *61*, 1763–1769.

Douwes J., Doekes G., Montijn R., Heederik D. and Brunekreef B. (1996): Measurement of β(1→3)-glucans in occupational and home environments with an inhibition enzyme immunoassay. Appl. Environ. Microbiol., *62*, 3176–3182.

Douwes J., van der Sluis B., Doekes G., van Leusden F., Wijnands L., van Strien R., Verhoeff A. and Brunekreef B. (1999): Fungal extracellular polysaccharides in house dust as a marker for exposure to fungi: Relations with culturable fungi, reported home dampness and respiratory symptoms. J. Allergy Clin. Immunol., *103*, 494–500.

Duncan R.L., Jr., Hoffman J., Tesh V.L. and Morrison D.C. (1986): Immunologic activity of lipopolysaccharides released from macrophages after the uptake of intact *E. coli* in vitro. J. Immunol., *136*, 2924–2929.

Dunér K.I. (1993): A new kinetic single stage LAL method for the detection of endotoxin in water and plasma. J. Biochem. Biophys. Meth., *26*, 131–142.

Eduard W. and Aalen O. (1988): The effect of aggregation on the counting precision of mould spores on filters. Ann. Occup. Hyg., *32*, 471–479.

Eduard W., Lacey J., Karlsson K., Palmgren U., Ström G. and Blomquist G. (1990): Evaluation of methods for enumerating microorganisms in filter samples from highly contaminated occupational environments. Am. Ind. Hyg. Assoc. J., *51*, 427–436.

Eggleston P.A., Newill C.A., Ansari A.A., Pustelnik A., Lou S.R., Evans III R.I., Marsh D.G., Longbottom J.L. and Corn M. (1989): Task related variation in airborne concentrations of laboratory animal allergens: Studies with *Rat n* I. J. Allergy Clin. Immunol., *84*, 347–352.

Eisenach K.B., Sifford M.D., Cave M.D., Bates J.H. and Crawford J.T. (1991): Detection of *Mycobacterium tuberculosis* in sputum samples using a polymerase chain reaction. Am. Rev. Respir. Dis., *144*, 1160–1163.

FAO. FAO Food and Nutrition Paper. (1990): Manuals of food quality control. 10. Training in mycotoxins analysis, *14*, 1–113.

Gerhardt P., Murray R.G.E., Wood W.A. and Krieg N. (eds) (1994): Methods for general molecular bacteriology. American Society for Microbiology Press, Washington, DC.
Gordon T. (1994): Role of the complement system in the acute respiratory effects of inhaled endotoxin and cotton dust. Inhal. Tox., *6*, 253–66.
Gordon T., Galdanes K. and Brosseau L. (1992): Comparison of sampling media for endotoxin contaminated aerosols. Appl. Occup. Environ. Hyg., *7*, 427–436.
Hollander A., Heederik D., Versloot P. and Douwes J. (1993): Inhibition and enhancement in the analysis of airborne endotoxin levels in various occupational environments. Am. Ind. Hyg. Assoc. J., *54*, 647–653.
Hollander A., Heederik D. and Doekes G. (1997): Respiratory allergy to rats: exposure-response relationships in laboratory animal workers. Am. J. Crit. Care Med., *155*, 562–567.
Hollander A., Gordon S., Renström A., Thissen J., Doekes G., Larsson P.H., Malmberg P., Venables K. and Heederik D. (1999): Comparison of methods to assess airborne rat and mouse allergen levels I: Analysis of Samples. Allergy, *54*, 142–149.
Iversen M., Korsgaard J., Hallas T.E. and Dahl R. (1990): Mite allergy and exposure to storage mites and house dust mites in farmers. Clin. Exp. Allergy, *20*, 211–219.
Iwanaga S. (1979): Chromogenic substrates for horseshoe crab clotting enzyme, its application for the assay of bacterial endotoxins. Haemostasis, *12*, 183–188.
Iwanaga S. (1993): The *Limulus* clotting reaction. Curr. Opinion Immunol., *5*, 74–82.
Jarvis B.B. (1990): Mycotoxins and indoor air quality. In: Morey P.R., Feeley J.C. and Otten J.A. (eds) Biological contaminants in indoor environments, ASTM STP 1071, ASTM, Philadelphia, PA, 201–214.
Jensen P.A., Todd W.F., Davis G.N. and Scarpino P.V. (1992): Evaluation of eight bioaerosol samplers challenged with aerosols of free bacteria. Am. Ind. Hyg. Assoc. J., *53*, 660–667.
Juozaitis A., Willeke K., Grishpun S.A. and Donnelly J. (1994): Impaction onto a glass slide or agar versus impingement into a liquid for the collection and recovery of airborne microorganisms. Appl. Environ. Microbiol., *60*, 861–870.
Kamphuis H.J., Ruiter de G.A., Veeneman G.H., Boom van J.H., Rombouts F.M. and Notermans S.H.W. (1992): Detection of *Aspergillus* and *Penicillium* extracellular polysaccharides (EPS) by ELISA: Using antibodies raised against acid hydrolyzed EPS. Antonie van Leeuwenhoek, *61*, 323–332.
Khan A.A. and Cerniglia C.E. (1994): Detection of *Pseudomonas aeruginosa* from clinical and environmental samples by amplification of the exotoxin A gene using PCR. Appl. Environ. Microbiol., *60*, 3739–3745.
Kolk A.H.J., Schuitema A.R.J., Kuijper S., van Leeuwen J., Hermans P.W.M., van Embden J.D.A. and Hartskeerl R.A. (1992): Detection of *Mycobacterium tuberculosis* in clinical samples by using polymerase chain reaction and a nonradioactive detection system. J. Clin. Microbiol., *30*, 2567–2575.
Lange J.L., Thorne P.S. and Lynch N.L. (1997): Application of flow cytometry and fluorescent in situ hybridization for assessment of exposures to airborne bacteria. Appl. Environ. Microbiol., *63*, 1557–1563.
Leal J.A., Guerrero C., Gómez-Miranda B., Prieto A. and Bernabé M. (1992): Chemical and structural similarities in wall polysaccharides of some *Penicillium, Eupenicillium* and *Aspergillus* species. FEMS Microbiol. Lett., *90*, 165–168.
Leopold S.S. (1988): "Positive hole" statistical adjustment for a two stage, 200-hole-per-stage Andersen air sampler. Am. Ind. Hyg. Assoc. J., *49*, 88–89.
Levin, J. and Bang F.B. (1968): Clottable protein in *Limulus*: Its localization and kinetics. Thromb. Diath. Haemorrh., *19*, 186–197.
Macher J.M. (1989): Positive-hole correction of multiple-jet impactors for collecting viable microorganisms. Am. Ind. Hyg. Assoc. J., *50*, 561–568.
Marthi B. and Lighthart B. (1990): Effects of betaine on enumeration of airborne bacteria. Appl. Environ. Microbiol., *56*, 1286–1289.
May K.R. and Harper G.J. (1957): The efficiency of various liquid impinger samplers in bacterial aerosols. Br. J. Ind. Med., *14*, 287–297.
Michel O., Nagy A.M., Schroeven M., Duchateau J., Neve J., Fondu P. and Sergysels R. (1997): Dose-response relationship to inhaled endotoxin in normal subjects. Am. J. Respir. Crit. Care Med., *156*, 1157–64.
Miguel A.G., Cass G.R., Weiss J. and Glovsky M.M. (1996): Latex allergens in tire dust and airborne particles. Environ. Health Perspec., *104*, 1180–1186.

Mikami T., Nagase T., Matsumoto S., Suzuki S., Suzuki M. (1982): Gelation of *Limulus* amoebocyte lysate by simple polysaccharides. Microbiol. Immunol., *26*, 403–409.
Milton D.K., Gere R.J., Feldman H.A. and Greaves I.A. (1990): Eendotoxin measurement: aerosol sampling and application of a new *Limulus* method. Am. Ind. Hyg. Assoc. J., *51*, 331–227.
Milton D.K., Feldman H.A., Neuberg D.S., Bruckner R.J. and Greaves I.A.(1992): Environmental endotoxin measurement: the kinetic limulus assay with resistant-parallel-line estimation. Envir. Res., *57*, 212–230.
Milton D.K., Johnson D.K., Park J.-H. (1997): Environmental endotoxin measurement: interference and sources of variation in the *Limulus* assay of house dust. Am. Ind. Hyg. Assoc. J. *58*, 861–867.
Morita T.S., Tanaka S., Nakamura T., Iwanaga S. (1981): A new (1–3)-ß-Glucan-mediated coagulation pathway found in *Limulus* amebocytes. FEBS Lett., *129*, 318–21.
Nagi N., Ohno N., Adachi Y., Aketagawa J., Tamura H., Shibata Y., Tanaka S. and Yadomae T. (1993): Application of *Limulus* test (G pathway) for the detection of different conformers of (1→3)-β-D-glucans. Biol. Pharm. Bull., *16*, 822–828.
Nevalainen A., Pastuszka J., Liebhaber F. and Willeke K. (1992): Performance of bioaerosol samplers: Collection characteristics and sampler desing considerations. Atmos. Environ., *26A*, 531–540.
Nevalainen A., Willeke K., Liebhaber F., Pastuszka J., Burge H. and Henningson E. (1993): Bioaerosol sampling. In: Willeke K. and Baron P. (eds) Aerosol measurement–principles, techniques, and applications. Van Nostrand Reinhold, New York, 471–492.
Notermans S. and Soentoro P.S.S. (1986): Immunological relationship of extracellular polysaccharide antigens produced by different mould species. Antonie van Leeuwenhoek, *52*, 393–401.
Novitsky T.J., Schmidt-Gegenbach J. and Remillard J.F. (1986): Factors affecting recovery of endotoxin absorbed to container surfaces. J. Parenteral Sci. Tech., *40* 284–286.
Obayashi T., Yoshida M., Takeshi M., Goto H., Yasuoka A., Iwasaki H., Teshima H., Kohno S., Horiuchi A., Ito A., Yamaguchi H., Shimada K. and Kawai T. (1995): Plasma (1→3)-β-D-glucan measurement in diagnosis of invasive deep mycosis and fungal febrile episodes. Lancet, *345*, 17–20.
Ogura M., Koo K. and Mitsuhashi M. (1994): Use of the fluorescent dye YOYO-1 to quantify oligonucleotides immobilized on plastic plates. Biotechniques, *16*, 1032–1033.
Olenchock S.A., Lewis D.M. and Mull J.C. (1989): Effects of different extraction protocols on endotoxin analysis of airborne grain dusts. Scand. J. Work Environ., *15*, 430–435.
Palmgren U., Strom G., Blomquist G. and Malmberg P. (1986): Collection of airborne microorganisms on nuclepore filters, estimation and analysis–CAMNEA method. J. Appl. Bacteriol., *61*, 410–406.
Platts-Mills T.A.E. and Chapman M.D. (1987): Dust mites: immunology, allergic diseas, and environmental control. J. Allergy Clin. Immunol., *80*, 755–775.
Platts-Mills T.A.E. and de Weck A.L. (1989): Dust mite allergens and asthma–a worldwide problem. J. Allergy Clin. Immunol., *83*, 416–427.
Pollart S.M., Smith T.F., Morris E.C., Gelber L.E., Platts-Mills T.A.E. and Chapman M.D. (1991): Environmental exposure to cockroach allergens: analysis with monoclonal antibody-based enzyme immunoassays. J. Allergy Clin. Immunol., *87*, 505–510.
Pope A.M., Patterson R. and Burge H. (eds) (1993): Indoor allergens: assessing and controlling adverse health effects. National Academy Press, Washington, DC, 1993.
Price J.A., Pollock I., Little S.A., Longbottom J.L. and Warner J.O. (1990): Measurement of airborne mite antigen in homes of asthmatic children. Lancet, *336*, 895–897.
Ray A., Redhead K., Seikirk S. and Poole S. (1991): Variability in LPS composition, antigenicity and reactogenicity of phase variants in *Bordetella pertussis*. FEMS Microbiol. Lett., *63*, 211–217.
Remillard J.F., Case Gould M., Roslansky P.F. and Novitsky T.J. (1987): Quantitation of endotoxin in products using the LAL kinetic turbidimetric assay. In: Detection of bacterial endotoxins with the *Limulus* amebocyte lysate test. Alan R. Liss, Inc., New York, 197–210.
Renström A., Gordon S., Hollander A., Spithoven, J., Larsson P., Venables K., Heederik D. and Malmberg P. (1999): Comparison of methods to assess airborne rat and mouse allergen levels: II Factors influencing allergen detection. Allergy, *54*, 150–157.
Reynolds S.J., Thorne P.S., Donham K.J., Croteau E., Kelley K., Milton D.K., Lewis D.M., Whitmer M., Heederik D., Douwes J., Connaughton I., Koch S., Malmberg P. and Larsson B.M. (1999): Interlaboratory comparison of endotoxin assays using agricultural dusts. In press.
Richard J.L., Bennett G.A., Ross P.F. and Nelson P.E. (1993): Analysis of naturally occurring mycotoxins in feedstuffs and food. J. Animal Sci., *71*, 2563–2574.

Rietschel E.T., Brade H., Holst O., Brade L., Muller-Loennies S., Mamat U., Zahringer U., Beckmann F., Seydel U., Brandenburg K., Ulmer A.J., Mattern T., Heine H., Schletter J., Loppnow H., Schonbeck U., Flad H.D., Hauschildt S., Schade U.F., Di Padova F., Kusumoto S. and Schumann R.R. (1996): Bacterial endotoxin: Chemical constitution, biological recognition, host response, and immunological detoxification. Curr. Topics Microbiol. Immunol., *216*, 39–81.

Roslansky P.F. and Novitsky T.J. (1991): Sensitivity of *Limulus* amebocyte lysate (LAL) to LAL-reactive glucans. J. Clin. Microbiol., *29*, 2477–2483.

Rylander R. (1994): Endotoxins. In: Rylander R. and Jacobs R.R., (eds) Organic dusts: exposure, effects and prevention. Lewis Publishers, Boca Raton, Fla.

Rylander R., Persson K., Goto H., Yuasa K. and Tanaka S. (1992): Airborne β(1–3)-glucan may be related to symptoms in sick buildings. Indoor Environ., *1*, 263–267.

Rylander R., Goto H., Yuasa K., Fogelmark B. and Polla B. (1994): Bird droppings contain endotoxin and (1–3)-β-D-glucans. Internat. Arch. Allergy Immunol., *103*, 102–104.

Schlegel H.G. (1993): General microbiology, 7th edn, translated by M. Kogut, Cambridge Univ. Press, Cambridge, UK, 447.

Schou C., Svendsen U.G. and Lowenstein H. (1991): Purification and characterization of the major dog allergen, *Can f* I. Clin. Exper. Immunol., *21*, 321–328.

Schwartz D.A., Thorne P.S., Jagielo P.J., White G.E., Bleuer S.A. and Frees K.L. (1994): Endotoxin responsiveness and grain dust-induced inflammation in the lower respiratory tract. J. Appl. Physiol. *267* (Lung Cell. Mol. Physiol. *11*), L609–L617.

Schwartz D.A., Thorne P.S., Yagla S.J., Burmeister L.F., Olenchock S.A., Watt J.L. and Quinn T.J. (1995): The role of endotoxin in grain dust-induced lung disease. Am. J. Respir. Crit. Care Med., *152*, 603–608.

Scott P.M. (1990): Natural poisons. In: AOAC Official Methods of Analysis. 1990, Ch. 49, 1185.

Soares L.M. and Rodriguez-Amaya D.B. (1989): Survey of aflatoxins, ochratoxin A, zearalenone, and sterigmatocystin in some Brazilian foods by using multi-toxin thin-layer chromatographic method. Journal of AOAC Intl., *72*, 22–26.

Sonesson A., Larsson L., Fox A., Westerdahl G. and Odham G. (1988): Determination of environmental levels of peptidoglycan and lipopolysaccharide using gas chromatography-mass spectrometry utilizing bacterial amino acids and hydroxy fatty acids as biomarkers. J. Chromatogr. Biomed. Appl., *431*, 1–15.

Sonesson A., Larsson L., Schütz A., Hagmar L. and Hallberg T. (1990): Comparison of the *Limulus* amebocyte lysate test and gas chromatography-mass spectrometry for measuring lipopolysaccharides (endotoxins) in airborne dust from poultry processing industries. Appl. Environ. Microbiol., *56*, 1271–78.

Swanson M.C., Agrawal M.K. and Reed C.E. (1985): An immunochemical approach to indoor aeroallergen quantitation with a new volumetric air sampler: studies with mice, roach, cat, mouse, and guinea pig antigens. J. Allergy Clin. Immunol., *76*, 724–729.

Tamura H., Tanaka S., Ikedo T., Obayashi T. and Hashimoto Y. (1997): Plasma (1→3)-β -D-glucan assay and immunohistochemical staining of (1→3)-β-D-glucan in the fungal cell walls using a novel horseshoe crab protein (T-GBP) that specifically binds to (1→3)-β-D-glucan. J. Clin. Lab. Anal., *11*, 104–109.

Tanaka S., Aketagawa J., Takahashi S., Shibata Y., Tsumuraya Y. and Hashimoto Y. (1991): Activation of a *Limulus* coagulation factor G by (1–3)-ß-D-glucans. Carbohyd. Res., *218*, 167–174.

Teller J.D. and Key K.M. (1979): A turbidimetric LAL assay for the quantitative determination of Gram negative bacterial endotoxins. In: Biomedical applications of the horseshoe crab. Alan R. Liss, Inc., New York, 423–433.

Thorne P.S., Kiekhaefer M.S., Whitten P. and Donham K.J. (1992): Comparison of bioaerosol sampling methods in barns housing swine. Appl. Environ. Microbiol., *58*, 2543–2551.

Thorne P.S., Lange J.L., Bloebaum P.D. and Kullman G.J. (1994): Bioaerosol sampling in field studies: Can samples be express mailed? Am. Ind. Hyg. Assoc. J., *55*, 1072–1079.

Thorne P.S., DeKoster J.A. and Subramanian P. (1996): Pulmonary effects of machining fluids in guinea pigs and mice. Am. Ind. Hyg. Assoc. J., *57*, 1168–72.

Thorne P.S., Reynolds S.J., Milton D.K., Bloebaum P.D., Zhang X., Whitten P. and Burmeister L.F. (1997): Field evaluation of endotoxin air sampling assay methods. Am. Ind. Hyg. Assoc. J., *58*, 792–799.

Van Egmond H.P. (1989): Current situation on regulations for mycotoxins. Overview of tolerances and status of standard methods of sampling and analysis. Food Add. Contam., *6*, 139–188.

Van Egmond H.P. (1991): Methods for determining ochratoxin A and other nephrotoxic mycotoxins. IARC Scientif. Publ., *115*, 57–70.
van Hage-Hamsten M. (1992): Allergens of storage mites. Clin. Exp. Allergy, *22*, 429–431.
Virtanen T., Louhelainen K. and Montyjarvi R. (1986): Enzyme-linked immunosorbent assay (ELISA) inhibition method to estimate the level of airborne bovine epidermal antigen in cowsheds. Int. Arch. Allergy Appl. Immunol., *81*, 253–257.
Weary M.E., Donohue G., Pearson F.C. and Story K. (1980): Relative potencies of four reference endotoxin standards as measured by the *Limulus* amoebocyte lysate and USP rabbit pyrogen tests. Appl. Environ. Microbiol., *40*, 1148–1151.
Whittaker Bioproducts, Inc. (1988): LAL review. Walkersville, MD., 4.
Wood R.A., Eggleston P.A., Lind P., Ingemann L., Schwartz B., Graveson S., Terry D., Wheeler B. and Adkinson N.F., Jr. (1988): Antigenic analysis of household dust samples. Am. Rev. Respir. Dis., *137*, 358–363.
Zock J.-P., Hollander A., Doekes G. and Heederik D. (1996): The influence of different filter elution methods on the measurement of airborne potato antigens. Am. Ind. Hyg. Assoc. J., *57*, 567–570.

1.8 Standard Test Methods for the Determination of VOCs and SVOCs in Automobile Interiors

Helmuth Bauhof and Michael Wensing

1.8.1 Introduction

A wide range of materials exist for the interiors of cars. These materials vary in their composition as well as in their internal chemical structure.

To obtain their necessary and desirable characteristics and to facilitate their production the use of a large variety of organic chemical substances is required. After completion of the production process, these materials emit chemical compounds with a wide range of volatilities into the air inside the car: these are roughly divided into volatile organic compounds (VOCs) and semivolatile organic compounds (SVOCs). High temperatures encourage those emissions, which can cause undesirable effects in the car's interior. On the one hand these can cause car users to complain of annoying odors or even health problems, whilst on the other hand the semivolatile part of the compounds has been observed to condense as a "fogging" film on the inner side of the windscreen. Together with soot and dust particles, this film dims the screen. The driver's view is further impaired if the screen has a small inclination (Eisele, 1987; Möhler and Schönherr, 1992; Munz et al., 1994).

To reduce these disadvantageous effects and to eliminate them as far as possible, comprehensive and reliable information about the types of organic compounds in the indoor air of automobiles and also their concentration and their origin is required. Earlier measurements of this kind covered *N*-nitrosoamines (Fine et al., 1980; Smith and Baines, 1982; Dropkin, 1985) and other VOCs (Zweidinger et al., 1982; Ullrich et al., 1992). The measurements were carried out under various conditions: outdoors or indoors, as well as in motion. In some cases the temperature inside the car was increased using a heater. The samples of air were simply taken directly from the car's interior or through a covered-up window-opening by using an adsorption tube and a small pump. The results were analyzed by gas chromatography (GC) with various detectors.

The procedure for examining VOCs followed a similar procedure to that for the semi-volatile "fogging" substances (SVOCs). The vehicles were either in use or parked outdoors with their windows closed and exposed to the sun. As a rule the fogging film which developed on the inside of the screen was scraped off by means of a clean razor blade and was examined subsequently by gas chromatography and infrared spectroscopy (Carter et al., 1987; Nranian et al., 1987; Munz et al., 1994).

Compared to the above-mentioned simple methodical approaches to the measurement of VOCs and SVOCs in indoor air of automobiles, the approach of the Technischer Überwachungs-Verein Nord (TÜV Nord), Hamburg, is fundamentally different. In a test room, a stationary test stand reserved exclusively for this purpose was constructed where the air inside cars could be tested by following a uniform standardized programme every time (Bauhof and Frels, 1985). This test stand was used frequently for some time

(Wensing and Schwampe, 1989–1998; Lüssmann-Geiger and Schmidt, 1995; Schmidt and Lüssmann-Geiger, 1996) and further developed in a 5-year research and development project. The discussions in the following sections are based on the work with this test stand (TÜV Nord, 1996).

1.8.2 Conditioning of the Car's Interior

The concentration of VOCs and SVOCs in the car's interior is essentially influenced by the following factors:

- the strength of the emissions from numerous individual sources in the interior equipment and also from the hollow spaces linked to the interior of the car,
- decomposition reactions in the form of ad- and absorption on the material's surface including chemical conversions (e.g. formation of salt from base and acid combinations, oxidation of aldehydes, etc.),
- gas exchanges between the passenger's compartment and the atmosphere outside the car,
- contamination by external sources near the car, from extraneous disturbing emissions from building materials and furnishings in the test hall, and from the outside atmosphere.

For a representative analysis of the emissions, it is necessary to carry out these measurements only when the condition of the interior atmosphere is stable. This is the case if certain effects that increase or decrease the concentration are balanced and are not subjected to any further change. This is why the precise conditioning of the car's interior is of utmost importance for adjusting and preserving a defined state of equilibrium.

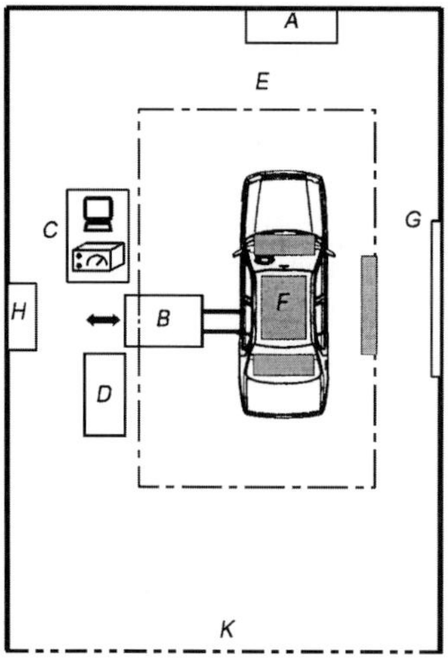

A: gas supply

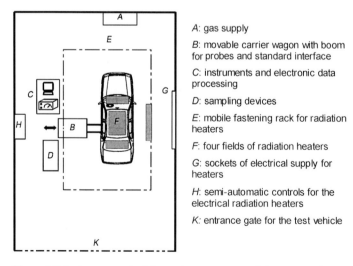

B: movable carrier wagon with boom for probes and standard interface

C: instruments and electronic data processing

D: sampling devices

E: mobile fastening rack for radiation heaters

F: four fields of radiation heaters

G: sockets of electrical supply for heaters

H: semi-automatic controls for the electrical radiation heaters

K: entrance gate for the test vehicle

Figure 1.8-1. Arrangement of the main components of the test stand.

These requirements are met by standardizing the experiment's set-up and procedure. All vehicles are tested on a test stand and always in the same location. Only a standardized interface consisting of an aluminium sheet is used for the insertion of probes into the car's interior for the sampling of air. The sheet is fitted into the car's open window and sealed. All connections leading into the interior are airtight. As a rule the set-up includes the following: several sensors to register the temperature at different places inside the interior and to measure the humidity, a mechanical system for air circulation, a glass probe to extract several samples of air simultaneously (analysis of VOCs), a device to take samples of phthalates directly, and a device to enrich the fogging precipitate (SVOCs).

The plan of the experiments usually depends on the particular problem. The essential conditions for the surroundings, however, are as follows:

- The temperature of the inside air has to be adjusted to 23 °C (ambient air temperature), 45 °C and 65 °C which are defined as standard temperatures of the inside air.
- The air has to be kept in constant circulation by mechanical means and thus homogenized.
- The heating of the interior is done by radiators from outside the car.
- If large samples of air are taken from the interior of the car, ultra-pure gas has to be added to balance the draw off.

Fig. 1.8-1 is a diagram of the test stand, showing:

- provision of ultra pure gas (A) with defined humidity,
- semi-automatic heating-up device with its control (H), its provision areas (G) and the four radiator fields (F) which are fixed to a mounting support above the front and the rear window, the roof and the right-hand side window
- vehicle (B) to carry probes with an arm and standardized interface,

Figure 1.8-2. Partial view of the test stand.

- various measuring instruments and data recording equipment (C) for continuous recording of measured signals (e.g. for 11 spots for temperature measurements, 3 spots for humidity measurements, 1 FID signal and signals of other measuring instruments),
- system for sampling of air samples,
- gate (K) for the entry and exit of the test car.

Figure 1.8-2 shows a partial photographic view.

Detailed data about the test stand and the standard test cycle are included in TÜV Nord (1996).

1.8.3 Measurement Procedure

In the interior of a modern car, the presence of a large variety of individual VOCs and SVOCs can be demonstrated. They can roughly be categorized as follows:

- VOCs responsible for the smell of brand-new cars: alkanes, aromatic hydrocarbons, carbonyl compounds, residual monomers, alcohols, esters, ethers, halogenated hydrocarbons, terpenes, nitrogen and sulfur compounds.
- SVOCs responsible for the "fogging" effect: paraffins, higher fatty acids and esters, phthalates, phosphoric acid esters, organic silicon compounds, halogenated hydrocarbons, oxygen, nitrogen and sulfur compounds.

The concentrations range from a few hundred $\mu g/m^3$ to less than 1 $\mu g/m^3$ of inside air. High standards of sensitivity and selectivity are required for the analytical methods. The analysis is complicated by water vapor which is released from water stored in the textile surfaces when the vehicle heats up. This can have a disturbing effect on sampling and analysis. Thus it is essential that procedures are employed that have been specially validated and practically tested for the analysis of the air inside cars.

1.8.3.1 Quantitative Determination

Table 1.8.1 summarizes the procedure for important substance categories.

The results of the quantitative measurements are given as concentrations of substances, expressed in mass per volume unit (e.g. $\mu g/m^3$, standardized for gas in the following conditions: temperature 20 °C, pressure 1013hPa, dry). As a rule these measured values relate only to the time span of the sampling and the condition of the car's interior at that time. If it is required to determine the potential emission of a substance *i*, the emission rate ER has to be calculated by using the measured concentration:

$$\mathrm{ER} = Q_{\mathrm{tot}} C_i \ (\mu\mathrm{g/h}) \tag{1.8.1}$$

Q_{tot} (m^3/h) is the volume flow that is used to exchange the interior atmosphere while measuring the concentration. It is determined experimentally by tracer gas methods (e.g. TÜV Nord, 1996).

Table 1.8.1. Standard procedures for determination of important substance classes.

Aromatic hydrocarbons
Sampling: charcoal tubes, type NIOSH
Sample preparation: desorption with carbon disulfide
Analysis: GC/MS, SIM-Mode; column: DB5, 60 m × 0.25 mm × 0.25 μm

Glycol ethers
Sampling: charcoal tubes, type NIOSH
Sample preparation: desorption with dichloromethane/methanol
Analysis: GC/MS, SIM mode; column: DB-WAX, 60 m × 0.25 mm × 0.25 μm

Aldehydes / ketones
Sampling: derivatization with 2,4-dinitrophenylhydrazine (DNPH) in acetonitrile
Analysis: HPLC with ODS-Hypersil/RP8e and LiChrospher 100RP-8e
UV detection at 360 nm

Phthalic acid esters
Sampling: glass fiber filter with Florisil tubes
Sample preparation: desorption with acetonitrile
Analysis: HPLC, ODS-Hypersil
UV detection at 234 nm

Amines
Sampling: silica gel tubes
Sample preparation: derivatization with 9-fluorenylmethylchloroformate (FMOC-CL)
Analysis: HPLC, ODS Hypersil
UV detection at 263 nm

Nitrosamines
Sampling: Thermosorb N tubes
Sample preparation: desorption with dichloromethane/methanol
Analysis: GC/MS, SIM mode, PCI
Column: DB-WAX, 60 m × 0.25 mm × 0.25 μm

1.8.3.2 Semi-quantitative Determination of VOCs (TVOC)

The VOCs in the inside air are enriched by means of active sampling on different sorption tubes, which are thermodesorbed or desorbed with a solvent. After adding internal standards, the analysis is done by capillary GC/MS. The hundred compounds of the chromatogram with the most intensive signals are identified by mass spectroscopy, and are then semi-quantitatively evaluated with toluene as the substance of reference. The toluene equivalents are summed (Wensing, 1996a), and this result serves as a semi-quantitative estimation of the total VOC concentration (TVOC).

1.8.3.3 Qualitative Determination of VOCs (Identification)

The taking of samples and their preparation is identical to that of the semi-quantitative analysis. However, to evaluate the complex spectra of the substances, the multi-dimensional GC technique (GCGC/MS) is employed, which consists of two series-connected GCs combined with an MS detector. As a rule, the process is adjusted in such a way that

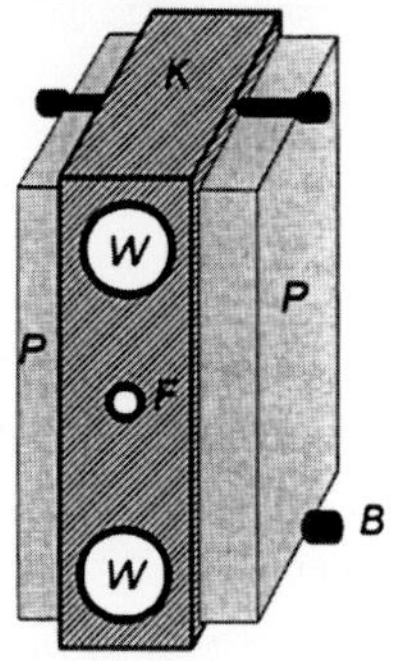

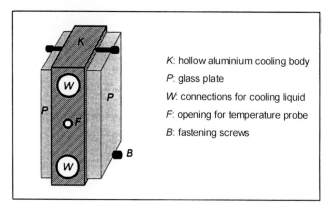

Figure 1.8-3. Apparatus for the identification of SVOCs.

polar compounds containing heteroatoms are extracted from the nonpolar hydrocarbon-matrix (of no interest) and identified. The laborious work of identification is facilitated by a special powerful data base program that works on the principle of standardized retention time indices (Wensing, 1996b).

1.8.3.4 Identification of SVOCs (Fogging Precipitate)

The samples are taken from the surface of two cooled glass plates of a special apparatus (Fig. 1.8-3), which is exposed to the air inside the car at the standardized interface. The glass plates (P) are mounted on both sides of the cooling body (K). The body is hollow inside with a coolant running through. The temperature of the cooling body is registered by a rod-shaped measuring sensor which is introduced into the drilled hole (F) and reaches right into the center of the plate. While the samples are taken, the vehicle is heated to increase emissions and thus to speed up the collection of substances. Conceptually, the procedure leans on the sample-taking technique of DIN 75201 (1992). After the sample is taken, the film which built up on the glass plates is washed off with organic solvent. The solution is concentrated and analysed by GC/MS.

1.8.3.5 Measurement of the Sum of Organic Substances (ΣVOC)

For each experiment on a vehicle at the test stand, a flame ionization detector (FID) is used for the sum determination of total organically fixed carbon in the inside air. In this way it is possible to follow the relative dependence of the total concentration of VOCs on various influences, and at the same time the state of the inside air can be continuously documented.

1.8.4 Quantitative and Qualitative Results from Brand-New Cars

In an experiment lasting several years, six brand-new cars were tested at the test stand for 40 days. For 8 h each day the inside air was heated up to 65 °C (artificial ageing). At the beginning, and after 20 and 40 days the air was characterized by means of the standardized measurement procedures described above (TÜV Nord, 1996).

As sum-of-VOCs results (FID) in Fig. 1.8-4 show, the measured values decrease rapidly. The shape of the curve follows a simple exponential function. Accordingly, the decrease in the intensities of the individual signals and the clear change in their pattern can be seen on the VOC gas chromatogram (Fig. 1.8-5).

The calculation of measured values of TVOC (toluene equivalents, TE) from the gas chromatogram of the six vehicles results in the following spans:

- new condition: 7000–24 000 μgTE/m^3,
- after 20 days of ageing: 2500–10 000 μgTE/m^3,
- after 40 days of ageing: 1000–4500 μgTE/m^3.

For a selection of VOCs, measured values of emission rates ER are given in Table 1.8.2.

Generally, the emission decreases from high measured values at the beginning to lower ones. Because of these effects, the conclusion can be drawn that the processes are controlled by evaporation. Substances at the surface or in layers near the surface of the materials of the interior furnishings are desorbed rapidly. However, there are exceptions to these reactions, e.g. the aldehydes. Here, diffusion-controlled emissions can be assumed. In a slow and steady process, diffusions from the inside of the material take

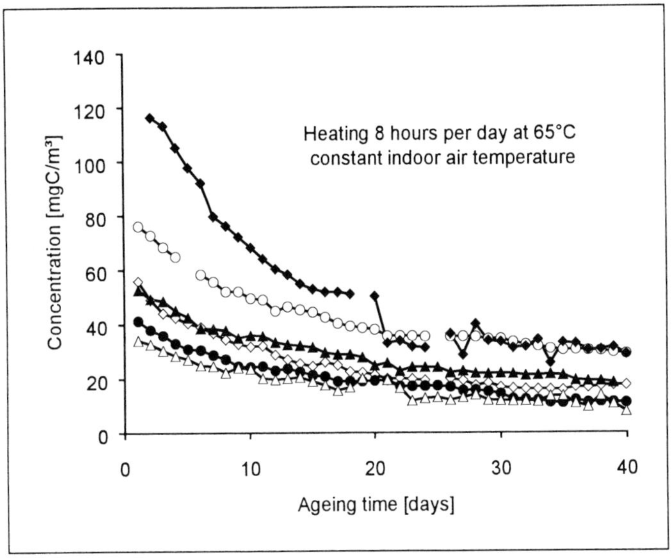

Figure 1.8-4. Decrease in the sum of VOCs for six brand-new cars during artificial ageing at 65°C.

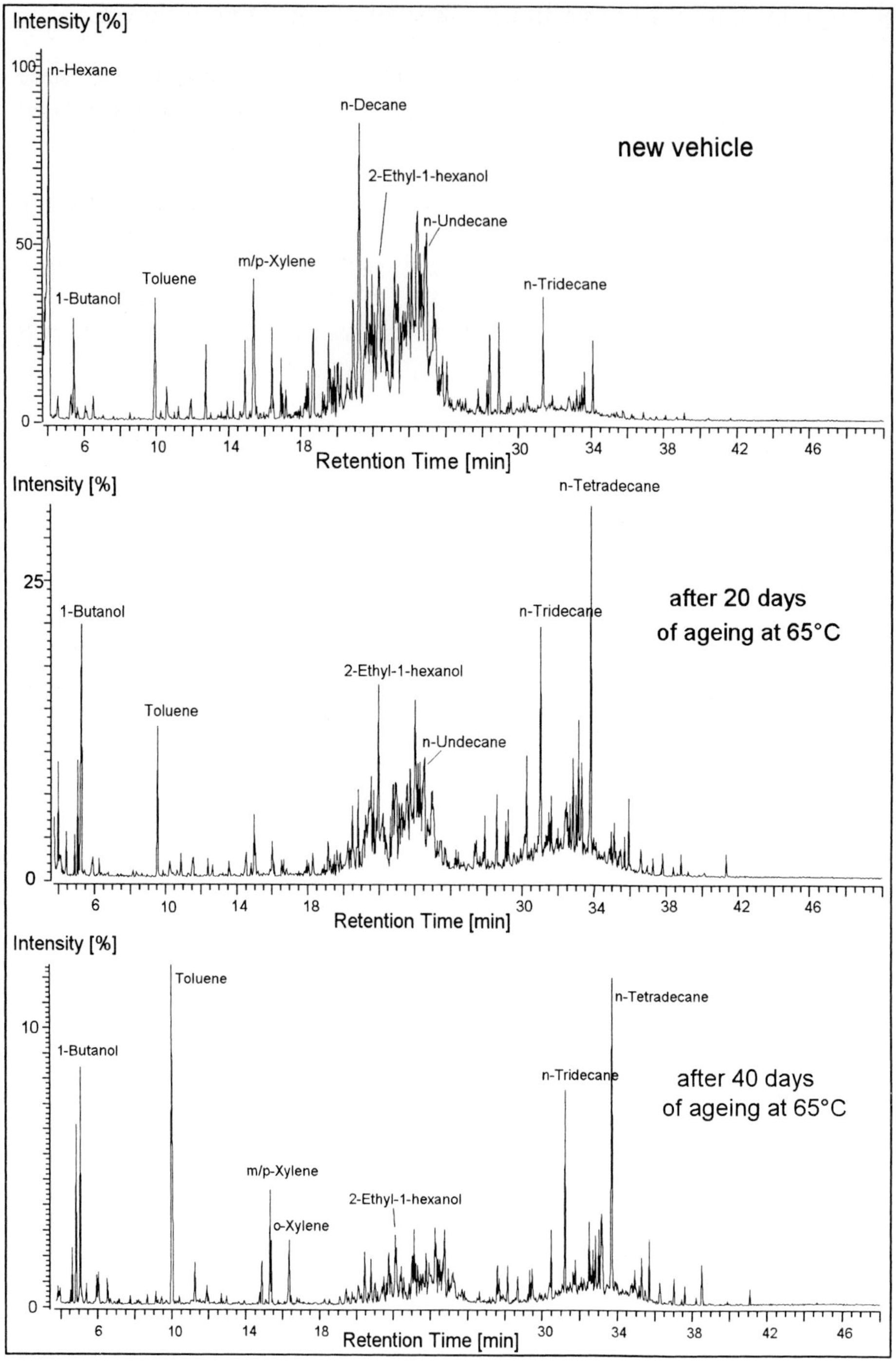

Figure 1.8-5. Changes in the VOC gas chromatogram for brand-new cars during artificial ageing.

Table 1.8.2. Emission rates of selected VOCs: arithmetic mean of six different new cars.

Substance	Emission rate ER [μg/h]		
	New vehicle	After 20 days of ageing	After 40 days of ageing
Benzene	24	22	27
Toluene	275	78	73
Ethylbenzene	434	38	18
m/p-Xylene	965	95	46
o-Xylene	272	37	20
Styrene	369	88	68
2-Methoxyethanol	3.4	0.7	0.8
2-Ethoxyethanol	2.6	0.7	0.7
2-Butoxyethanol	68	12	5.6
2-Ethoxyethyl acetate	4.4	1.4	1.3
2-Butoxyethyl acetate	25	8.6	9.5
1-Methoxypropyl acetate	104	24	6.8
Formaldehyde	40	52	43
Acetaldehyde	44	30	29
Propanal	15	9.5	13
n/iso-Butyraldehyde	19	7.2	7.2
Pentanal	29	14	16
Hexanal	42	25	13
Benzaldehyde	10	34	11
Acetone	361	143	116
Methylethyl ketone	85	13	7.5
Methylisobutyl ketone	755	78	47
Di-*n*-butyl phthalate	0.7	0.8	2.0
Dimethylamine	9.0	3.1	2.5
Diethylamine	9.6	8.1	5.7
Di-*n*-butylamine	54	14	59
N-nitrosodimethylamine	0.20	0.12	0.07

place which die off only after a longer period of time. In both cases, textiles and textile (laminated) compound materials play an essential role as storage media and as a reversible interim store for chemical substances because of their strong sorption ability (Ehrler et al., 1994).

For the accumulation of SVOCs, the apparatus described in Fig. 1.8-3 was used. The accumulation capacity during the experiments amounted to 10 μg film mass per glass plate per hour at a constant inside air temperature of 65 °C. Examples of substances found in fogging films are listed in Table 1.8.3.

Table 1.8.3. Examples of SVOCs in fogging films.

Hydrocarbons: Alkanes, *n*-Alkanes (C13–C32); Branched alkanes
Alcohols: 2-Ethylhexanol, Octadecanol.
Amines: 1,4-Diazabicyclo-2,2,2-octane; Dicyclohexylamine, Methyldicyclohexylamine; Bisdimethylaminodiethyl ether; *N,N*-Dimethylpentadecylamine.
Amides: Tridecylcycloacetamide; Pentadecylcycloacetamide.
Aromatic carbonic acid esters: Dibutyl, Dioctyl, Di-(2-ethylhexyl) phthalate; Dioctyladipate, Dioctylsebacate, Trihexyltrimellitate.
Fatty acids: Lauric acid; Stearic acid.
Fatty acid esters: Palmitic acid butyl ester; Palmitic acid 2-ethylhexyl ester; Stearic acid 2-ethyhexyl ester; Dibutyl adipate.
Phenols: 2-(1,1-dimethylethyl)-phenol; 2,6-Di-*tert*-butyl-4-methylphenol (BHT); 2,6-Di-*tert*-butyl-4-ethylphenol; 2,6-Di-*tert*-butyl-4-methoxymethylphenol.
Phosphates: Tris-(2-chloroethyl) phosphate; Tris-(1-chloro-*iso*-propyl) phosphate; Tris-(1,3-dichloro-*iso*-propyl) phosphate.
Other compounds: Benzoic acid; 2-Ethylhexaneacid; Erucic acid amide; Siloxanes.

1.8.5 Conclusion

The existing test stand method and the standardized analytical process are powerful routine methods to characterize emissions in car interiors in a reliable way. The effects on the IAQ of changes in the interior furnishing can be followed precisely. In combination with measurements on single parts of the interior in emission chambers, products can be constantly improved and the quality of the materials for interior furnishings secured (Bauhof et al. 1996).

References

Bauhof H. and Frels C. (1985): Untersuchungen von Pkw-Innenräumen auf flüchtige organische Verbindungen. Bericht Nr. 128CO00550, TÜV Nord, Hamburg.

Bauhof H., Wensing M., Zietlow J. and Möhle K. (1996): Prüfstandsmethoden zur Bestimmung organisch-chemischer Emissionen des Pkw-Innenraums. ATZ Sonderheft 25 Jahre FAT, 37–42.

Carter R.O., Jensen T.E. and McCallum J.B. (1987): Chemical characterization of automobile window film. SAE Technical Paper Series 870314, Society of Automotive Engineers, Warrendale.

DIN 75201 (1992): Bestimmung des Fogging-Verhaltens von Werkstoffen der Fahrzeuginnenausstattung. Beuth, Berlin.

Dropkin D. (1985): Sampling of automobile interiors for organic emissions. EPA/600/9, US Environmental Protection Agency, Research Triangle Park.

Ehrler P., Schreiber H. and Haller S. (1994): Emission textiler Automobil-Innenausstattung: Ursachen und Beurteilung des Kurzzeit- und Langzeit-Foggingverhaltens. Textilveredelung, *29*, 254–260.

Eisele D. (1987): Geruch und Fogging von Automobil-Innenausstattungsmaterialien. Melliand Textilberichte, *3*, 206–215.

Fine D.H., Reisch J. and Rounbehler D.P. (1980): Nitrosamines in new automobiles. IARC Sci. Publ. 31 (N-Nitroso Compd.: Anal., Form., Occurrence) 541–554.

Lüssmann-Geiger H. and Schmidt H.J. (1995): Emissionen aus Polymerwerkstoffen im Pkw-Innenraum. In: Kunststoffe im Automobilbau: Verbundsysteme, Verfahren, Anwendungen; Tagung Mannheim 29.-30.3.1995. Verein Deutscher Ingenieure, VDI-Gesellschaft Kunststofftechnik; VDI-Verlag Düsseldorf.

Möhler H. and Schönherr D. (1992): Untersuchung zum Foggingverhalten von polymeren Werkstoffen der Kfz-Innenausstattung unter zusätzlicher Berücksichtigung des UVA-Anteils der Sonnenstrahlung. Kautschuk + Gummi, Kunststoffe, *45*, 103–105.

Munz R., Faas U., Kurzmann P., Leitz R. and Meister C. (1994): Das Foggingproblem: Messmethoden, Wege und Erfolge. ATZ, *96*, 238–246.

Nranian M., McCallum J.B. and Kelly, M. (1987): An overview of automotive interior glass light scattering film. SAE Technical Paper Series 870313; Society of Automotive Engineers, Warrendale.

Schmidt H.J. and Lüssmann-Geiger H. (1996): Verunreinigungen der Fahrzeuginnenraumluft–Quellen und Gegenmassnahmen, Gefahrstoffe–Reinh. Luft, *56*, 43–46.

Smith R.L. and Baines Th.M. (1982): Nitrosamines in vehicle interiors. SAE Technical Paper Series 820785. Society of Automotive Engineers, Warrendale.

TÜV Nord (1996): Entwicklung und Erprobung von Standard-Messverfahren für die Bewertung des fahrzeugeigenen Beitrages zu organischen Luftverunreinigungen in Fahrgasträumen von Personenkraftwagen. Report Vol. I-III. Hamburg.

Ullrich D., Seifert B. and Nagel R. (1992): Concentrations of volatile organic compounds inside new cars. International Environmental Management. Papers from the 9th World Clean Air Congress, Vol. 7, IU-12A.02, Montreal.

Wensing M. (1996a): Bestimmung gasförmiger Emissionen von Bauprodukten–Anforderungen an die Prüfkammermethode und exemplarische Untersuchungsergebnisse. In: Aktuelle Aufgaben der Messtechnik in der Luftreinhaltung, VDI-Bericht 1257, p.405–438. VDI, Düsseldorf.

Wensing M. (1996b): In: E. Bagda, Emissionen aus Beschichtungsstoffen. Expert-Verlag, Kontakt & Studium, Vol. 478.

Wensing M. and Schwampe W. (1989–1998): Untersuchung des Pkw-Innenraumes auf flüchtige organisch-chemische Emissionen. Untersuchungsberichte, TÜV Nord, Hamburg.

Zweidinger R.A., Bursey J.T., Castillo N.C., Keefe R. and Smith D. (1982): Organic emissions from automobile interiors. SAE Technical Paper Series 820784. Society of Automotive Engineers, Warrendale.

1.9 Nomenclature and Occurrence of Glycols and their Derivatives in Indoor Air

Peter Stolz, Norbert Weis and Jürgen Krooss

1.9.1 Introduction

Since the 1970s a trend to use water-based coating systems and glues instead of solvent-based products has been apparent. Most water-based paints and lacquers use polyurethane or acrylate as binders. For technical reasons the systems still contain organic cosolvents as well as pigments, biocides and surfactants (Hansen et al., 1987).

As cosolvents, bifunctional alcohols (glycols) and their derivatives (alkyl ethers and esters of organic acids) are mainly used next to hydrocarbons and alcohols. The processing of water-based systems in coatings technology makes special demands on the dispersibility of the lacquer resins in water-based solutions. Water as a pure solvent shows disadvantages that have to be compensated by use of solutizers. Solutizing effects including increased cross-linkability, increased workability, film-forming and storage stability are important properties. Optimal effects concerning these properties were shown by some glycol derivatives such as 2-butoxyethanol (butylglycol) and 2-(2-butoxyethoxy)-ethanol (butyldiglycol) due to their distinct hydrophilic behavior (Woitowitz and Knecht, 1992).

1.9.2 Indoor Application Range of Glycol Derivatives

A very complex nomenclature is typical for glycol derivatives. Up to five synonyms commonly exist for one compound. In order to simplify this nomenclature for practical purposes, a systematic nomenclature according to the old IUPAC rules for nomenclature was proposed (Stolz et al., 1998). The system is based on the structural description of the parent glycol derivative according to Falbe and Regitz (1990) and can therefore also be used for new glycol derivatives. An important part of this nomenclature consists of sensible and systematic abbreviations simplifying the naming of these compounds. For the frequently applied substance butyldiglycol acetate, the term DEGMBA (diethylene glycol monobutyl ether acetate) can be derived. Abbreviations for 42 relevant compounds are given in Table 1.9-1a (glycols) and 1.9-1b (acrylates). Glycol derivatives are colorless and of different volatilities. Boiling points range from 85 °C (EGDM) to 273 °C (T3PG). Most glycols are soluble in water at room temperature. Their application in commercial products is now widespread. They are used as solvents for pharmaceutical chemicals, foodstuffs, cosmetic additives and inks, and as solvents for textile and leather dyes (Marchl, 1998). As mentioned earlier, glycol derivatives have special importance as ingredients of paints, lacquers, glues and paint removers (see Table 1.9-2).

Table 1.9-1a. Nomenclature of commonly occurring glycol derivatives and proposed simplifying abbreviations (Stolz et al., 1998).

Glycol derivative (IUPAC name)	Abbre-viation	CAS Number	Common synonyms
*E*thylene *g*lycol (1,2-ethanediol)	EG	107-21-1	glycol ethylene alcohol
*E*thylene *g*lycol *mono**m*ethyl ether (2-methoxyethanol)	EGMM	109-86-4	methyl glycol methyl cellosolve
*E*thylene *g*lycol *mono**e*thyl ether (2-ethoxyethanol)	EGME	110-80-5	ethyl glycol cellosolve
*E*thylene *g*lycol *mono*-*n*-*pr*opyl ether (2-propoxyethanol)	EGMPr	2807-30-9	propyl glycol propyl cellosolve
*E*thylene *g*lycol *mono*-*iso*-*pr*opyl ether (2-methylethoxyethanol)	EGMiPr	109-59-1	isopropyl glycol isopropyl cellosolve
*E*thylene *g*lycol *mono*-*n*-*b*utyl ether (2-butoxyethanol)	EGMB	111-76-2	butyl glycol butyl cellosolve
*E*thylene *g*lycol *mono**h*exyl ether (2-hexoxyethanol)	EGMH	112-25-4	hexyl glycol hexyl cellosolve
*E*thylene *g*lycol *mono**p*henyl ether (2-phenoxyethanol)	EGMP	122-99-6	phenyl glycol phenyl cellosolve
*E*thylene *g*lycol *d**im*ethyl ether (1,2-dimethoxyethane)	EGDM	110-71-4	glycol dimethyl ether glyme
*E*thylene *g*lycol *d**ie*thyl ether (1,2-diethoxyethane)	EGDE	73506-93-1	glycol diethyl ether diethyl cellosolve
*E*thylene *g*lycol *mono**m*ethyl ether *a*cetate (2-methoxyethyl acetate)	EGMMA	110-49-6	methyl glycol acetate methyl cellosolve acetate
*E*thylene *g*lycol *mono**e*thyl ether *a*cetate (2-ethoxyethyl acetate)	EGMEA	111-15-9	ethyl glycol acetate cellosolve acetate
*E*thylene *g*lycol *mono*-*n*-*b*utyl ether *a*cetate (2-butoxyethyl acetate)	EGMBA	112-07-2	butyl glycol acetate butyl cellosolve acetate
*Di*ethylene *g*lycol *mono**m*ethyl ether (2-(2-methoxyethoxy)-ethanol)	DEGMM	111-77-3	methyl diglycol methyl carbitol
*Di*ethylene *g*lycol *mono**e*thyl ether (2-(2-ethoxyethoxy)-ethanol)	DEGME	111-90-0	ethyl diglycol carbitol
*Di*ethylene *g*lycol *mono*-*n*-*b*utyl ether (2-(2-butoxyethoxy)-ethanol)	DEGMB	112-34-5	butyl diglycol butyl carbitol
*Di*ethylene *g*lycol *mono*-*h*exyl ether (2-(2-hexoxyethoxy)-ethanol)	DEGMH	112-59-4	hexyl diglycol hexyl carbitol
*Di*ethylene *g*lycol *d**im*ethyl ether (1-methoxy-2-(2-methoxyethoxy)-ethane)	DEGDM	111-96-6	dimethyl carbitol diglyme
*Di*ethylene *g*lycol *d**ie*thyl ether (1-ethoxy-2-(2-ethoxyethoxy)-ethane)	DEGDE	112-36-7	diethyl carbitol ethyl diglyme
*Di*ethylene *g*lycol *mono**m*ethyl ether *a*cetate (2-(2-methoxyethoxy)-ethyl acetate)	DEGMMA	629-38-9	methyl diglycol acetate methyl carbitol acetate

Table 1.9-1a. (Continue).

Glycol derivative (IUPAC name)	Abbre-viation	CAS Number	Common synonyms
*Di*ethylene *g*lycol *mono**e*thyl ether *a*cetate (2-(2-ethoxyethoxy)-ethyl acetate)	DEGMEA	112-15-2	ethyl diglycol acetate carbitol acetate
*Di*ethylene *g*lycol *mono**b*utyl ether *a*cetate (2-(2-Butoxyethoxy)-ethyl acetate)	DEGMBA	124-17-4	butyl diglycol acetate butyl carbitol acetate
1,2-*P*ropylene *g*lycol (1,2-propanediol)	1,2-PG	57-55-6	propylene glycol α-propylene glycol
1,2-*P*ropylene *g*lycol *mono**m*ethyl ether (1-methoxy-2-propanol)	1,2-PGMM	107-98-2	propylene glycol methyl ether Dowanol 33 B
2,1-*P*ropylene *g*lycol *mono**m*ethyl ether (2-methoxy-1-propanol)	2,1-PGMM	1589-47-5	
Propylene glycol mono-*n*-butyl ether (mixture of 1,2 / 2,1-isomers)	PGMB	5131-66-8[1]	butyl propylene glycol
1,2-*P*ropylene *g*lycol *mono**m*ethyl ether *a*cetate / (1-methoxy-2-propyl acetate)	1,2-PGMMA	108-65-6	methoxypropyl acetate-2
2,1-*P*ropylene *g*lycol *mono**m*ethyl ether *a*cetate / (2-methoxy-1- propyl acetate)	2,1-PGMMA	70657-70-4	methoxypropyl acetate-1
1,3-*P*ropylene *g*lycol (1,3-propanediol)	1,3-PG	504-63-2	β-propylene glycol trimethylene glycol
1,3-*P*ropylene *g*lycol *mono**e*thyl ether (3-ethoxy-1-propanol)	1,3-PGME	111-35-3	trimethylene glycol me-thyl ether
*Di**p*ropylene *g*lycol (mixture of 3 isomers)	DPG	110-98-5	dipropylene glycol
*Di**p*ropylene *g*lycol *mono**m*ethyl ether (mixture of 4 isomers)	DPGMM	34590-94-8	dipropylene glycol methyl ether
1,4-*B*utylene *g*lycol (1,4-butanediol)	1,4-BG	110-63-4	tetramethylene glycol g-butylene glycol
*Tri*ethylene *g*lycol (1,2-bis(2-hydroxyethoxy)ethane)	T3EG	112-27-6	triglycol trigol
*Tri*ethylene *g*lycol *mono-n-b*utyl ether (2-(2-(2-butoxyethoxy)ethoxy)ethanol)	T3EGMB	143-22-6	butyl triglycol butyl trigol
Tripropylene glycol (mixture of isomers)	T3PG	24800-44-0	tripropylene glycol
2,4,7,9-*T*etra*m*ethyl-*d*ec-5-*y*ne-4,7-*d*iol	TMDYD	126-86-3	
Texanol / (2,2,4-trimethylpentane-1,3-diol monoisobutyrate)	Texanol[2]	74367-33-2	
TXIB / (2,2,4-trimethylpentane-1,3-diol diisobutyrate)	TXIB[2]	6846-50-0	

[1] Refers to the pure 1,2-PGMB.

[2] Here the known trivial names are used instead of abbreviations.

Table 1.9-1b. Nomenclature of commonly occurring acrylates and simplifying abbreviations.

Acrylate (IUPAC name)	Abbre-viation	CAS Number	Common synonyms
*E*thylene *g*lycol *mono**ph*enyl ether *acr*ylate (2-phenoxyethyl acrylate)	EGMPhAcr	48145-04-6	Phenyl glycol acrylate
1,6-*H*exylene *g*lycol *diacr*ylate (1,6-hexanediol diacrylate)	1,6-HGDAcr	13048-33-4	Hexamethylene glycol diacrylate
*Triprop*ylene *g*lycol *diacr*ylate (mixture of isomers)	T3PGDAcr	42978-66-5	Tripropylene glycol acrylate

Table 1.9-2. Uses of glycol derivatives in paints, lacquers and paint removers (Rühl, 1997).

Product type	Remarks / information	Used glycol derivatives (abbrev.)
Disperse colorants	Resin dispersions with synthetic resin binders (e. g. alkyd resins, polyurethane resins or polyacrylates), water content 35–45 %, organic solvent mixtures 5–10 %, additional other hydrocarbons	1,2-PG, EG, DPGMM, DEGMM, EGMB, DEGMB
Colorless primers, water thinnable	Primer coatings of mineral basements (e.g. plastering, concrete), up to est. 3 % organic solvents, additional other solvents	Not declared
Primers with pigments, water thinnable	Dispersions of synthetic resins, glycol derivatives used as film formers or solvents	Not declared
Clear lacquers, wood glazings, water thinnable	Resin dispersions from water-thinnable alkyd resins, polyurethane resins and polyacrylates, containing 35–85 % water, 5–10 % organic solvents, additional other solvents	DEGMB, PGMM
Clear lacquers, wood glazings, solvent thinnable, free from aromatic hydrocarbons	40–50 % solvents, "high solid lacquers" up to 25 %, additional other solvents	1,2-PGMMA, 1,2-PGMM, EGMBA
Blue stain-preventing paints, water thinnable	Fungicide-containing alkyd resin or acryl resin dispersions, est. 3 % organic solvents, additional other solvents	Not declared
Silicon resin paints, water thinnable	Mainly for exterior house coatings, glycol derivatives (est. 3 %) used as film formers	1,2-PGMM, 1,2-PG
Paint removers	Commonly containing "asymmetrical" diesters, aliphatic and aromatic hydrocarbons	DPGMM, DEGMB, DEGME, DEGMH, DEGMM

Other solvents = solvents other than glycol derivatives.

The indoor air concentration of glycol derivatives depends on the type of application, the loading factor and the climatic parameters. High short-term exposures result during the processing of glycol derivatives, while long-term exposures result from emissions from building products. The high short-term burdens mainly affect the workers processing these products. This applies especially to paints and lacquers, as shown in Table 1.9-2.

In comparison to paints containing organic solvents, the application of water-based paints causes considerably lower VOC concentrations in indoor air. Processing water-based systems results in TVOC concentrations of only ca. 10 % of the TVOC values caused by processing solvent-based systems (Norbäck et al., 1995). In this study, concentrations within the range 0.04–12.7 mg/m^3 were measured.

Water-based paints and lacquers contain glycols in amounts up to 7 % (Hansemann, 1996), and the release of solvent residues from finished products may contribute to indoor air pollution. This is demonstrated in Fig. 1.9-1 for the emission of diethyleneglycol monoethylether (DEGME) from a freshly-manufactured furniture surface. The experiment was performed in a 1-m^3 test chamber; experimental details are given in the figure captions (Fuhrmann and Salthammer, 1998). One hour after loading, a high chamber value of 1858 μg/m^3 was apparent. Within 4 weeks, the chamber concentration decayed to 218 μg/m^3.

During the application of water-based coating systems in real living spaces, the concentrations of individual solvent components found were usually in the order of 10 mg/m^3. Moreover, the total concentration of all lacquer components was not significantly higher than this. An increasingly occurring problem is human exposure to high indoor concentrations of glycol derivatives during and after renovation work. Further problems

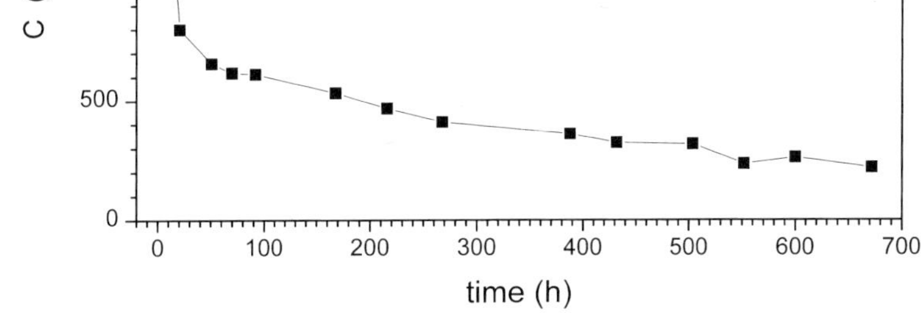

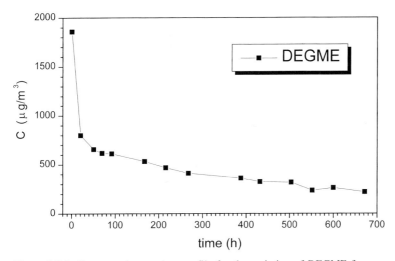

Figure 1.9-1. Concentration vs time profile for the emission of DEGME from a water-based lacquer system for furniture coating. The substrate was particle board covered with oak veneer. Test chamber conditions: 23 °C, 45 % rel. humidity, 1 h^{-1} air exchange, 1m^2/m^3 loading (Fuhrmann and Salthammer, 1998).

arise from slowly decaying emission sources such as floor glues and other building products, which may cause increased indoor concentrations over months or even years.

The application of water-based systems has also led to an increased use of biocides. Though toxicological properties of the active substances are known, health effects caused by glycol derivatives are not sufficiently investigated (Norbäck et al., 1995).

Indoor air pollution often coincides with considerable problems due to unpleasant odors, and, in unpublished investigations of such events, measured concentrations of glycol derivatives were found to range from a few to several hundreds of $\mu g/m^3$. Analogous problems can be caused by wall paints containing glycol derivatives applied over large areas (Gehrig et al., 1993). The total amounts absorbed by the human organism are normally rather low in this case compared with those resulting from coatings manufacture, but it has to be assumed that the human organism will be affected continuously (see below). In climate test chamber investigations on water-based products, concentrations up to 200 mg/m^3 were found for PG and EGMB within the first few days. Texanol concentrations of 57 $\mu g/m^3$ were found 1 year later (Clausen et al., 1990).

1.9.3 Analytical Procedure

Different techniques using sampling on appropriate adsorption media followed by solvent desorption or thermodesorption and GC-analysis are suitable for the quantitative determination of glycols in indoor air.

For solvent desorption, adsorbents such as modified silica containing cyanopropyl groups with acetone elution and GC-FID/GC-MS detection (Woitowitz and Knecht, 1992) or synthetic polymers such as XAD 2 or XAD 7 with dichloromethane elution and GC-FID/GC-MS detection (Norbäck et al., 1995) are used.

The use of non-modified activated charcoal (established as an adsorbent for non-polar VOCs) is ruled out because of low recovery rates for glycol derivatives (Woitowitz and Knecht, 1992).

TENAX TA is the recommended adsorbent for active sampling of glycols in indoor air. In combination with thermal desorption and GC/MS- or GC/FID-analysis, this method shows sufficient selectivity and sensitivity (Clausen et al., 1990; Hansen et al., 1987). However, the identification of higher glycols via low-resolution MS is somewhat sophisticated, because the mass peaks are not characteristic. In case of 1,2-PGMM and 1,2-PG the mass spectra are almost identical, with relevant *m/z*-values of 59, 45, 103 and 73. Therefore, retention indices and standard compounds are required for an exact classification (Uhde et al., 1999).

1.9.4 Toxicologic Properties of Glycol Derivatives

Human exposure to glycol derivatives may occur through inhalation, ingestion and skin contact. In the indoor environment, inhalation is the main path for intake, although glycols are also readily absorbed through skin. Toxic amounts may be resorbed dermally without showing a significant irritating effect.

Table 1.9-3. Indoor air concentration values of glycol derivatives (μg/m^3) (Plieninger, 1998).

Substance	1,2-PGMM	1,2-PG	EGMB	1,2-PGMB	DEGMB	EGMP	Texanol
Average value	4.5	7.7	7.0	6.6	1.3	1.4	1.6
90 % quantile	7.5	17.5	14.0	3.0	1.6	3.5	4.5
Maximum	118.9	69.3	102.6	298.6	37.8	15.4	14.6

Though not all glycol derivatives, (e.g. propylene glycol), are considered as harmful, health impairment caused by other glycol derivatives cannot be excluded. Many glycol derivatives irritate the mucous membranes of eyes and respiratory tract. Moreover, they cause headache and, in higher concentrations, show narcotic effects (Kühn and Birett, 1994). Glycol derivatives are also suspected to cause disturbances in tissues with high cell division rates. In animal testing, neural malfunctions and disturbances within thymus, kidney, liver and lung were observed. Nevertheless, too little information from animal testing and too few conclusions from occupational medicine are available. Thus it is not possible to describe the risk assessment of glycols in indoor air on a toxicological basis.

In general, indoor concentrations of glycol derivatives in the range 10–1000 μg/m^3 have to be assumed under living conditions during long exposure times (B.A.U.C.H., 1994). Some typical values are given in Table 1.9-3. For the evaluation of health impacts, the metabolism of this class of substances in the human body has to be considered. Depending on the substrate, harmless products may be formed, (e.g. 1,2-propylene glycol oxidizes to lactic acid, a substance naturally occurring in the body). On the other hand, some glycols are transformed to kidney-affecting products (e.g. oxalic acid) or various alkoxyacetic acids showing blood or reproductive toxicity. Due to rapid

Table 1.9-4. Proposed guideline values for some glycol derivatives (Marchl, 1998).

Abbreviation	Substance name	Berlin guideline values
EG	Ethylene glycol	25 ppb / 60 μg/m^3
EGMM	Ethylene glycol monomethyl ether	10 ppb / 30 μg/m^3
EGME	Ethylene glycol monoethyl ether	25 ppb / 90 μg/m^3
EGMB	Ethylene glycol mono-*n*-butyl ether	25 ppb / 120 μg/m^3
EGMMA	Ethylene glycol monomethyl ether acetate	10 ppb / 55 μg/m^3
EGMEA	Ethylene glycol monoethyl ether acetate	25 ppb / 170 μg/m^3
EGMBA	Ethylene glycol mono-*n*-butyl ether acetate	25 ppb / 170 μg/m^3
DEGMB	Diethylene glycol mono-*iso*-butyl ether	No value[1]
1,2-PG	1,2-Propylene glycol	No value[2]
1,2-PGMM	1,2-Propylene glycol 1-monomethyl ether	No value[2]
Sum concentration	EGMMA, EGMBA, EGMEA, EGMM, EGMB, EGME	25 ppb

[1] Too little data, no value deducible [2] Poor toxicity, value unnecessary.

resorption and metabolization of glycol derivatives and their conversion to metabolites such as alkoxyacetic acids, which in many cases are slowly excreted, accumulation of the latter may result. Continuous exposure without "recovery phases" may therefore lead to increased internal levels and to health risks.

Guideline values for indoor air were proposed by Marchl (1998) (see Table 1.9-4). According to present knowledge, exposure to levels at or below these values will not impair the health, even in case of long-term exposures and even for persons whose health is not good. Indoor air guideline levels have also been proposed by Nielsen et al. (1998) for 2-ethoxyethanol, 2-(2-ethoxyethoxy)ethanol and 2-(2-butoxyethoxy)ethanol.

1.9.5 Conclusion

The replacement of solvent-based products by water-based products for interior use has led to decreasing VOC emissions during processing. On the other hand, ingredients of low volatility such as glycols and their derivatives can be detected in indoor air over several months or even years. The permanent exposure to glycols may lead to health impairment. As a consequence, guideline concentration values have been proposed. For practical purposes, a nomenclature for glycols based on systematic abbreviation is introduced.

References

B.A.U.C.H. (1994): Vorkommen von Estern und Ethern mehrwertiger Alkohole in der Raumluft. Beratung und Analyse Verein für Umweltchemie e.V., Sachbericht, Eigenverlag, Berlin.

Clausen P.A., Wolkoff P. and Nielsen P.A. (1990): Long term emissions of volatile organic compounds from waterborne paints in environmental chambers. Proceedings of Indoor Air '93, Helsinki, Finland, Vol. 3, 557–562.

Falbe J. and Regitz M. (1990): Römpp Chemie Lexikon. Thieme-Verlag, Stuttgart.

Fuhrmann F. and Salthammer T. (1998): private communication.

Gehrig R., Hill C., Zellweger C. and Hofer, P. (1993): VOC emissions from wall paints – a test chamber study. Proceedings of Indoor Air '93, Helsinki, Finland, Vol. 2, 431–436.

Hansemann W. (1996): Wood, Surface Treatment. Ullmanns Encyclopedia of Industrial Chemistry, Vol. 28A, 385–393. VCH, Weinheim.

Hansen M.K, Larsen M. and Cohr K.-H. (1987): Waterborne paints – a review of their chemistry and toxicology and the result of determinations made during their use. Scand. J. Work Environ. Health, 13, 473–485.

Kühn and Birett (eds) (1994): Merkblätter Gefährliche Arbeitsstoffe 77. Erg.Lfg. 11/94.

Marchl D. (1998): Raumluftbelastungen durch Glykolverbindungen. In: Diel F., Feist W., Krieg H-U. and Linden W. (eds) Ökologisches Bauen und Sanieren. Verlag C.F. Müller, Heidelberg, 71–77.

Nielsen G.D., Hansen L.F., Nexr B.A. and Poulsen O.M. (1998): Indoor air guideline levels for 2-(2-ethoxyethoxy)ethanol, 2-(2-butoxyethoxy)ethanol and 1-methoxy-2-propanol. Indoor Air, Supplement no. 5/98, 37–54.

Norbäck D., Wieslander G. and Edling C. (1995): Occupational exposure to volatile organic compounds (VOCs) and other air pollutants from the indoor application of water-based paints. Ann. Occup. Hyg., 39, 783–794.

Plieninger, P. (1998): Glykolverbindungen, Phthalate und Siloxane in der Innenraumluft. Vortragsveranstaltung: Belastung des Innenraumes – Gefahren für die Gesundheit. Akademie für Öffentliches Gesundheitswesen in Düsseldorf. March 1998, Hamburg.

Rühl R. (ed) (1997): Handbuch der Bauchemikalien. Ecomed, Landsberg.

Stolz P., Weis N., Köhler M, Krieg H.U., Plieninger P., Santl H. (1998): Zur Nomenklatur der Glykolverbindungen. In: Diel F., Feist W., Krieg H-U. and Linden W. (eds): Ökologisches Bauen und Sanieren. C.F. Müller, Heidelberg, 63–71.

Uhde E., Fuhrmann F., Klare S. and Salthammer T. (1999): Identification of glycol derivatives by use of GC/MS-analysis. Proceedings of Indoor Air '99, Edinburgh, in press.

Woitowitz H.J. and Knecht U. (1992): Emissionen organischer Verbindungen bei der handwerklichen Verarbeitung von Lacken und Farben. Forschungsbericht Nr. 104 08 152 im Auftrag des Umweltbundesamtes.

2 Environmental Test Chambers and Cells

2.1 Environmental Test Chambers

Michael Wensing

2.1.1 Introduction

Building products, furnishings and household commodities often emit volatile chemical compounds. Mostly these are so-called volatile organic compounds (VOCs) or semivolatile organic compounds (SVOCs). Solvents, residual monomers, plasticizers, fireproofing agents, auxiliary agents for processing, preservatives (biocides), and reaction and decomposition products are responsible for the occurrence of these emissions. With respect to the indoor air quality (IAQ), building products (e.g. Gustafsson, 1992) and furniture (e.g. Salthammer, 1997) are of major importance as a possible source of emissions of air-polluting substances. This is also valid for products and materials of "natural" origin (e.g. Horn et al., 1998).

The essential requirement no. 3 "Hygiene, Health and the Environment" of the European Council Directive 89/106/EEC for construction products (EEC, 1989) states that a healthy indoor environment can be achieved by controlling the sources and by eliminating or limiting the release of pollutants into the air (CEC, 1993). Only those (low-emitting) building products which do not influence the indoor air quality in a negative way should be used in a building. This demand on a product necessitates standardized methods of examination with which emissions of building products in laboratory conditions can be tested and compared.

To characterize these emissions by measuring techniques, so-called environmental test chambers have been constructed in the past. Today these are used for the following tasks inter alia (Tichenor, 1989):

- to provide compound-specific data on various organic sources to guide field studies and to assist in evaluating indoor air quality in buildings,
- to provide data on emissions to develop and verify models used to predict indoor concentrations of organic compounds,
- to develop data useful to manufacturers and builders for assessing product emissions and for developing control options or to improve products,
- to rank various products and product types according to their emission profiles.

As another method of examination for the tasks listed above, the so-called emission test cell, e.g. the FLEC (field laboratory emission cell; see Chapter 2.2) can be employed to determine surface emissions.

2.1.2 Principle of the Environmental Test Chamber Method

The product/material to be examined is tested with regard to temperature, relative humidity, air exchange rate, air velocity and product loading factor (ratio of exposed surface area of the test specimen to the free emission test chamber volume). It is exam-

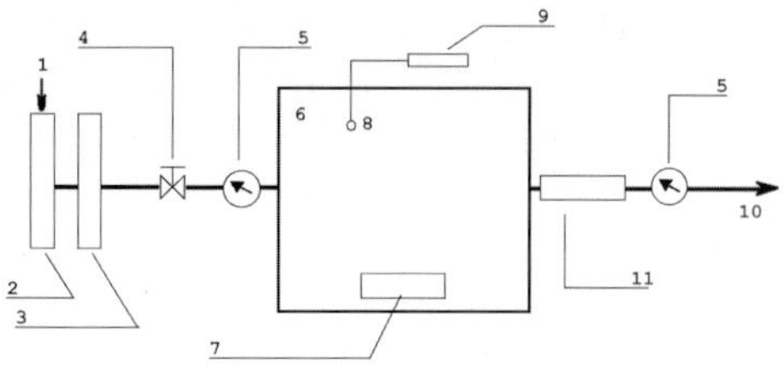

Figure 2.1-1. General arrangement of an emission test chamber.

1	Air inlet	7	Device to circulate air and to control air velocity
2	Air filter	8	Temperature and humidity sensors
3	Air conditioning	9	Monitoring system for temperature and humidity
4	Air flow regulator	10	Exhaust outlet
5	Air flow meter	11	Manifold for air sampling
6	Test chamber		

ined in standardized test conditions in an environmental test chamber that can be hermetically sealed off from the outside atmosphere. The air exchange is regulated by ultrapure gas (synthetic air) that runs through the test chamber. At the exit of the test chamber an enriching sampling with collecting phases takes place to determine the component substances of the outlet air stream. When the test chamber atmosphere is completely mixed, this is then a representative sample of it. By chemical analysis, the component substances of a collecting phase are determined so that, taking into consideration the volume of the sampling and the air exchange rate, a specific emission rate (SER) can be calculated.

Table 2.1-1. Advantages and disadvantages of small and large scale chambers (EC, 1989).

	Advantages	Disadvantages
Small-scale chambers	Relatively low cost of construction and operation	Need for replicate tests (sample inhomogenity)
	Increased flexibility (at about the same cost a larger number of chambers can be run simultaneously)	Need for additional sample preparation (sealing of edges)
	Less test material needed	Sample size limited
	More likely to be widely applied	
Large scale chambers	Possibility to simulate real life situations	Increased cost for construction and operation (energy, clean air and water)
	Reduced influence of sample inhomogeneity	Large amount of sample material needed (transport costs)
	Large-size products can be tested as a whole (e.g. furniture, construction components)	supply of clean air may become a problem
	Smaller influence of chamber surfaces	VOC test chamber background concentration may become a problem

The SER describes the product-specific emission behavior for selected chemical compounds (VOC_X) or the sum of the emissions (TVOC), e.g. as an area-specific emission rate (SER_A) measured in μg/(m^2h), or as a piece-related unit-specific emission rate (SER_U) measured in μg/(unit h).

With the aid of SER, various products can be compared ("ranking") and especially low-polluting products can be identified ("labelling") (EC, 1997).

In Fig. 2.1-1, the general arrangement of an environmental test chamber is shown. Basically, two types of environmental test chambers can be distinguished: small scale chambers with volumes ranging from a few litres to a few cubic metres, and room-size large scale chambers of the "walk-in" type. Both types of chambers have advantages as well as disadvantages (see Table 2.1-1) (EC, 1989).

The examination of particle board for formaldehyde emissions (EN 717-1, 1996) is a typical application of large scale chambers (Colombo et al., 1994). On account of the circumstances mentioned in Table 2.1-1, however, the use of small-scale chambers has spread extensively for examinations of VOC emissions of building products.

2.1.3 Standardization of Environmental Test Chamber Measurements

The results of examinations in test chambers is influenced by a large number of different factors, (see Table 2.1-2). To obtain comparable data from corresponding examinations the three stages of the procedure, preparation of the sample, examination in the test chamber and chemical analysis, have to be standardized. This is also shown by comparative measurements with test chambers (EC, 1993; EC, 1995).

Table 2.1-2. A large number of different factors influence the result of a test chamber examination.

Product history	Age of product Transport to the test laboratory and storage time Sample preparation Climatic conditioning prior to testing
Chamber measurement	Supply air quality Background concentration Sink effects/recovery Air exchange ratio Air tightness of the environmental test chamber Internal air mixing Air velocity Accuracy of temperature, RH and air exchange ratio Product loading factor
Sampling of test chamber air and chemical analysis	Choice of sorbent tube Recovery of specific VOC Detection limit Precision Accuracy

Table 2.1-3. Quality requirements of the future EN.

Parameter	Quality requirement
Temperature	23 °C ± 1 °C
Relative humidity	50 % ± 5 %
Air tightness	Leak rate: < 1 % of the supply airflow rate
Background concentration	< 10 μg/m³, TVOC < 2 μg/m³, single compound
Recovery	≥ 80 % toluene, *n*-dodecane and product-specific target VOCs
Air mixing	± 10 % of the theoretical perfectly mixed model
Air velocity	0.1–0.3 m/s
Air exchange rate	Accuracy: ± 3 %

In connection with the European Council Directive 89/106/EEC on construction products (EEC, 1989) the working group (WG) 7 of TC 264 "air quality" at the European standardizing organisation CEN defines general standards for environmental test chambers for the determination of VOC emissions from building products (prENV 13419-1, 1998). As a member of the WG 7 the, author (Wensing, 1996) contributes to this standardization, for which important preparatory work has been published (Tichenor, 1989; ASTM 1990; Nordtest, 1990; EC, 1991; EC, 1993; EC, 1995). The future EN is going to require a certain quality standard of important parameters of a VOC emission examination of building products with environmental test chambers. Table 2.1-3 gives an overview.

2.1.3.1 Suitable Materials for Constructing a VOC Environmental Test Chamber

All the components of the chamber that are in direct contact with the inlet and outlet air and the chamber atmosphere have to be constructed in such a way that they emit no relevant amounts of VOC under test conditions and show only a very small sink effect (see 2.1.3.2). Suitable materials for the walls of a chamber are electro-polished stainless steel and glass. Seals are especially critical components. As little of their surface as possible should come into contact with the chamber atmosphere. Additional appliances such as fans are also problematical. The TVOC background concentration should be lower than 10 μg/m³; the background concentration of single compounds should be lower than 2 μg/m³.

2.1.3.2 Sink Effects and Recovery

Adsorption of the emissions released inside the test chamber, e.g. on the chamber's walls, is called the "sink effect". Because of this effect the concentration can be incorrectly determined at the exit of the chamber, leading to faulty results and an incorrect calculation of the SER. In principle a (slight) sink effect can be observed in every test chamber. However, the amount of adsorption (or the extent of the recovery) can vary greatly for different VOCs in one single chamber. They depend inter alia on the degree of volatility of the particular VOC. It can be stated simplistically that the higher the boiling point of the VOC the more the adsorption is promoted inside the test chamber. This was discovered, e.g., by Levsen and Sollinger (1993) on the basis of the homologous series of *n*-alkanes. A comparatively higher polarity of a compound can also favor adsorption on the chamber's walls (Meyer et al., 1994). Sollinger and Levsen (1993) also proved that the sink effect can be increased markedly by introducing a sample with an adsorbing surface into the test chamber. Colombo et al. (1993) also described how the sample to be examined added to the sink effect.

This sink effect caused by the sample is always of importance in any examination in a chamber, and it cannot be avoided. Thus, it is of fundamental significance that an empty chamber guarantees as good a recovery as possible. For selected VOCs in a concentration range which is typical for emission examinations, the recovery is determined by one of the following procedures. Test gases with a known concentration of VOCs are introduced into the chamber, or small permeation or diffusion tubes are placed in the test cell. At the air outlet of the empty chamber the VOC concentration is measured in each case and it is then compared with the theoretical nominal value. The recovery for target VOCs should be $\geq 80\,\%$.

In connection with the sink effect, the development of the so-called "memory effect" needs to be taken into account in the following analyses. The effect should be avoided by adequately purifying the chamber. It has proved to be advantageous if the test chamber can be heated up to higher temperatures for reasons of purification (Meyer et al., 1994).

2.1.3.3 Quality of the Inlet Air and Regulation of the Air Exchange

By a continuous air current streaming in from outside, a defined air exchange rate is adjusted in the test chamber. The inlet air can be regulated most accurately by electronic mass flow controls. As inlet air, ultrapure synthetic air from pressure vessels or purified air from the surroundings can be used. The requirements of the purity of the inlet air follow from the permissible background concentration.

The area-specific air flow rate with the units $m^3/(hm^2)$ results from a combination of air exchange rate and product loading factor. It is recommended to test building products at area-specific air flow rates comparable to the relevant ventilation rates in buildings. The relevant emission rates may be underestimated if the ventilation rates during testing are lower (Gunnarsen, 1997). For examinations of floor coverings, for instance, a value of 1.25 $m^3/(hm^2)$ is suggested (EC 1997). However, it has to be taken into account that varying concentrations can occur in the non-steady state even though combinations of different air exchange rates and product loading factors lead to the same value for the area-specific air flow rate (Guo, 1993).

2.1.3.4 Regulation of the Climatic Parameters Temperature and Humidity

Suitable examinations have shown that emissions from building products strongly depend on the temperature (e.g. van der Wal, 1997). For the purposes of standardized examinations it is thus important to have the possibility to regulate the temperature inside the test chamber to, e.g., 23 °C, with a maximum variation of ± 1 K. This can be achieved by using the chamber inside an air-conditioned room (ambient air temperature = chamber air temperature), by controlling the temperature of the chamber walls or by installing a heating appliance inside the chamber.

The humidity inside the test chamber can be adjusted to a defined state of, e.g., 50 ± 5 % by moistening the purified, dry inlet air to a defined degree:

- continuous addition of water vapor into the inlet air current by means of a small pump,
- moistening at dew point. The inlet air is led through a container filled with distilled water of a defined temperature. The degree of humidity of the inlet air corresponds to the saturation moisture at the given temperature.

The variation of temperature and humidity with time should be continuously monitored during the examination. For this purpose, sensors and measuring instruments

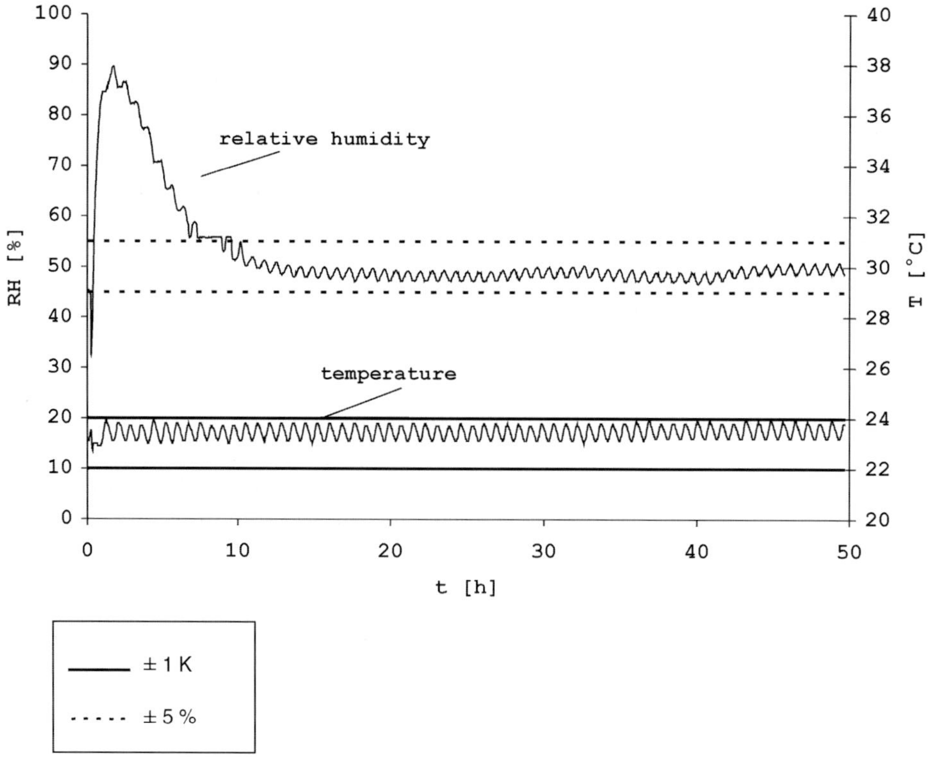

Figure 2.1-2. Changes in temperature and humidity during the examination of a water-based dispersion adhesive.

should be installed inside the chamber. This procedure is especially appropriate for documenting the actual climatic test conditions for samples with a high moisture content or for samples that cannot be conditioned to climatic standard conditions. Figure 2.1-2 shows the changes in temperature and humidity during the examination of a water-based dispersion adhesive. Because of the high water content of the adhesive, the relative humidity is well above 50 ± 5 % during the first hours of the experiment. In contrast to this, the temperature is in the range 23 ± 1 °C from the beginning.

2.1.3.5 Air Tightness

An environmental test chamber has to be hermetically sealed off from the outside atmosphere. The tightness of the chamber is expressed as the leakage rate, which should have a numerical value of < 1 %. A test chamber is usually operated at a slightly positive pressure of a few mbar. By measuring the half-life period during which the positive pressure after the closure of the air inlet diminishes by half, the leakage rate can be fixed (EC, 1997). A leakage rate of < 1 % is also achieved if the leakage per minute amounts to less than 0.001 % of the chamber's volume at a positive pressure of 1000 Pa (Nordtest, 1990). The amount of air that is let in and let out can easily be measured by air flow meters installed ahead of and after the test chamber. In the ideal case these flow meters are installed permanently. The tightness of the system can then be continuously controlled during an emission examination in progress.

2.1.3.6 Internal Mixing

The atmosphere of the test chamber has to be well mixed. This can be achieved by the following means:

- Fan(s)
- Multi-port inlet and outlet diffusers
- Baffle plates
- Perforated plates

The installation of a fan coupled to a magnet is technically complicated (Meyer et al., 1994).

The mixing behavior inside the test chamber can easily be checked by the "tracer gas method", in which an inert tracer gas (e.g. SF_6) is continuously added to the inlet air stream of the test chamber by means of the mass flow control. After a certain period of time the tracer gas supply is cut off. At the exit of the chamber the concentration of tracer gas is determined continuously. The mixing should be checked inside the gas chamber including a sample to be tested. Figure 2.1-3 shows a typical concentration/time profile for this test set-up. Should the measured values of the increasing or decreasing tracer gas concentration differ more than ± 10 % from the theoretical curve, the mixing of the chamber atmosphere is unsatisfactory. This happens, e.g., in the case of a technical ventilation problem, when the air inlet and the air outlet of the test chamber are "short-circuited".

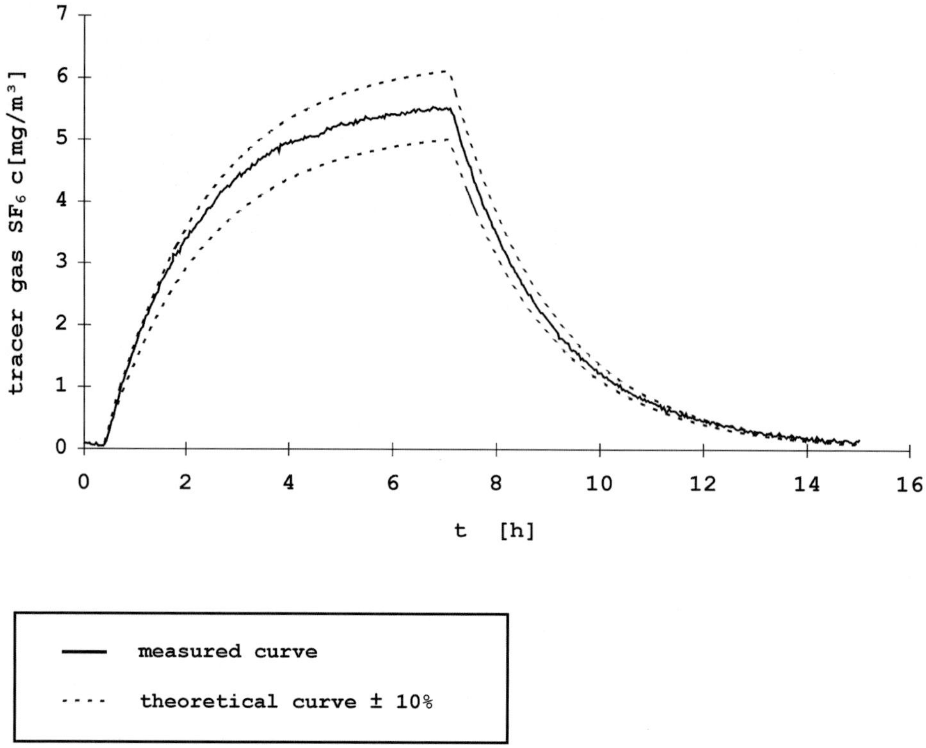

Figure 2.1-3. Typical concentration/time profile during checking of the mixing of the chamber atmosphere.

2.1.3.7 Air Velocity

The internal mixing of the chamber atmosphere results in air turbulence effects above the surface of the sample to be examined. When the fan is in operation these effects are independent of the air exchange rate and only depend on the speed of the fan. The question arises whether and how the SER is influenced by the air velocity above the sample surface.

A fundamental distinction can be made between evaporation-controlled and diffusion-controlled emission processes. In between these states there is a smooth transition for liquid materials such as paints. When liquid materials are applied to a solid ground, a transition from evaporation-controlled emission to diffusion-controlled emission takes place during the drying process (Clausen, 1993). To simplify matters, it can be said that the height of the SER is directly influenced by the air velocity in the case of evaporation-controlled emission (Clausen, 1993). For diffusion-controlled emissions this connection has not yet been proved and seems to be non-existent. To achieve realistic air stream conditions in the test chamber compared to normal interiors, a surface air velocity of 0.1–0.3 m/s is recommended (EC, 1997).

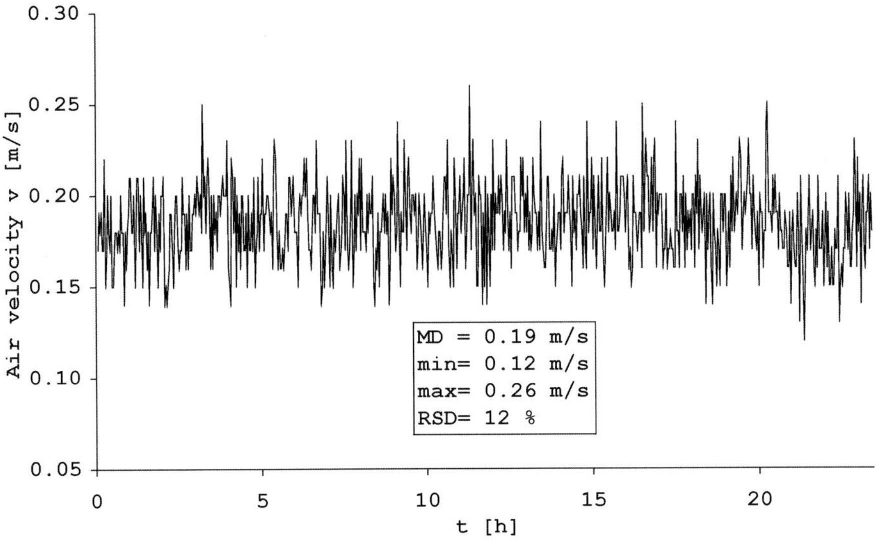

Figure 2.1-4. Variation of the air velocity with time.

The air velocity inside the sealed chamber can be measured by a hot wire anemometer above the sample. Figure 2.1-4 represents the variation of the velocity with time. It was continuously measured. The diagram shows that the velocity is not constant, but that it fluctuates and is subject to obvious variations at different points of time.

2.1.4 Environmental Test Chambers for Special Emission Measurement Methods

The "classical" chemical-analytical VOC examination of building products is usually carried out at 23 °C and 50 % RH. Some of the numerous possible applications of the environmental test chamber besides the "classical" examination are presented in the following.

2.1.4.1 Measurement of the Perceived Air Quality

Emissions from building products influence the perceived air quality, which can be determined e.g. with the aid of a CLIMPAQ (chamber for laboratory investigations of materials, pollution and air quality) and with sensory measurements (Gunnarsen et al., 1994). The CLIMPAQ has a volume of 50.9 l and is made of panes of window glass, stainless steel and anodically oxidized aluminium. It can be operated at an air exchange rate of 0.02 1/h to 140 1/h. A speciality of the CLIMPAQ is its funnel-shaped exit for the outlet air, at which point the smell of the examined product is analyzed by means of sensory measurements.

2.1.4.2 Examination of Elements of Cars' Interiors

Similarly to building products, the elements of the interior furnishings of a vehicle can emit a multitude of VOCs and SVOCs. In connection with an examination of whole vehicles (see Bauhof and Wensing, Chapter 1.8) a special VOC/SVOC environmental test chamber was developed and used for the examination of single components (assignment of sources). The test chamber is 1 m^3 in size and it consists of stainless steel. Examinations at temperatures of 65 °C and 100 °C are possible (Meyer et al., 1994; Bauhof et al., 1996). Recently, standardized test series have been developed for routine examinations of elements of cars' interiors. The test series have been put into practice in laboratories, and they have been validated by means of comparative internal and external measurements (Möhle and Wensing, 1999).

2.1.4.3 Examination of Electronic Devices

In contrast to building products, electronic devices are so-called "active products" which can emit heat and a multitude of VOCs and SVOCs into the surrounding air when they are in operation (Wensing, 1999a, 1999b, 1999c). Possible emissions range from solvents to fire-proofing agents. The latter belong to the so-called SVOCs. In an examination of emissions in a 1-m^3 stainless steel environmental test chamber, they can be enriched and determined by taking samples of "fogging". The fact that SVOCs tend to precipitate on surfaces inside the chamber due to a sink effect is taken advantage of.

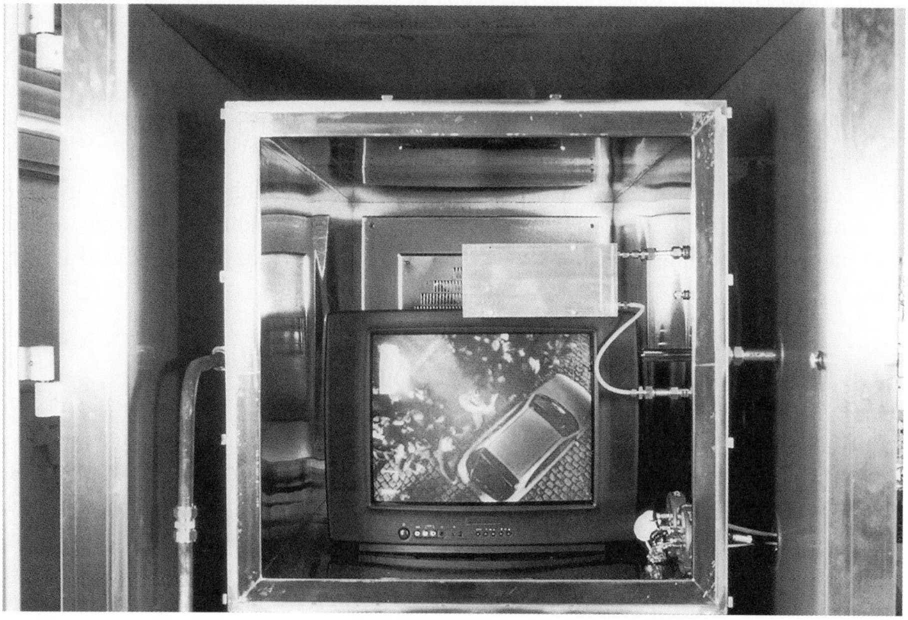

Figure 2.1-5. View into an open test chamber containing a test object (television set) and apparatus for the condensation of SVOCs ("fogging").

Inside the test chamber a cooled glass surface is installed for sample taking. The SVOCs disengage on the surface. Figure 2.1-5 gives a view into an open test chamber with an object to be tested (television set) and an apparatus for the condensation of SVOCs ("fogging"). In Fig. 2.1-6 an example of a SVOC GC/MS screening chromatogram of a television set is shown with some of the identified substances.

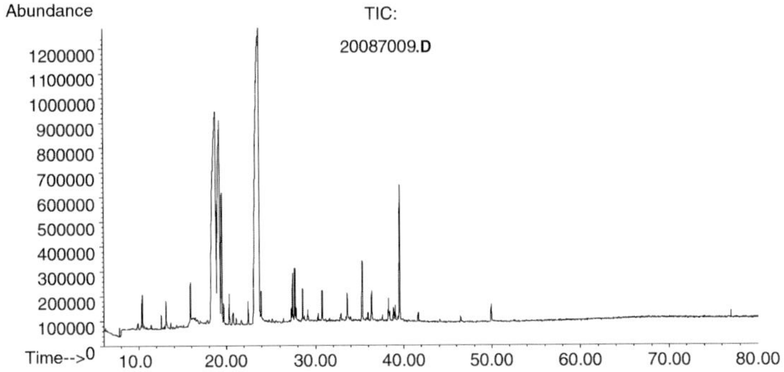

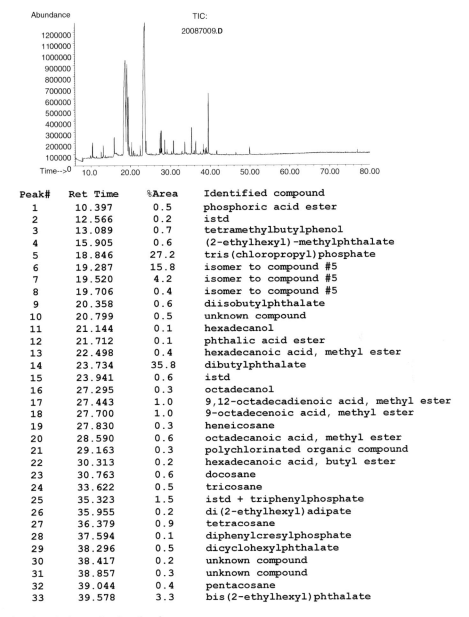

Peak#	Ret Time	%Area	Identified compound
1	10.397	0.5	phosphoric acid ester
2	12.566	0.2	istd
3	13.089	0.7	tetramethylbutylphenol
4	15.905	0.6	(2-ethylhexyl)-methylphthalate
5	18.846	27.2	tris(chloropropyl)phosphate
6	19.287	15.8	isomer to compound #5
7	19.520	4.2	isomer to compound #5
8	19.706	0.4	isomer to compound #5
9	20.358	0.6	diisobutylphthalate
10	20.799	0.5	unknown compound
11	21.144	0.1	hexadecanol
12	21.712	0.1	phthalic acid ester
13	22.498	0.4	hexadecanoic acid, methyl ester
14	23.734	35.8	dibutylphthalate
15	23.941	0.6	istd
16	27.295	0.3	octadecanol
17	27.443	1.0	9,12-octadecadienoic acid, methyl ester
18	27.700	1.0	9-octadecenoic acid, methyl ester
19	27.830	0.3	heneicosane
20	28.590	0.6	octadecanoic acid, methyl ester
21	29.163	0.3	polychlorinated organic compound
22	30.313	0.2	hexadecanoic acid, butyl ester
23	30.763	0.6	docosane
24	33.622	0.5	tricosane
25	35.323	1.5	istd + triphenylphosphate
26	35.955	0.2	di(2-ethylhexyl)adipate
27	36.379	0.9	tetracosane
28	37.594	0.1	diphenylcresylphosphate
29	38.296	0.5	dicyclohexylphthalate
30	38.417	0.2	unknown compound
31	38.857	0.3	unknown compound
32	39.044	0.4	pentacosane
33	39.578	3.3	bis(2-ethylhexyl)phthalate

istd = internal standard

Figure 2.1-6. SVOC GC/MS screening chromatogram of a television set.

2.1.5 Outlook

The environmental test chamber method is a powerful tool for the comparative characterization of various products and materials under test conditions with regard to their emissions. When choosing the test parameters and when interpreting the results it has to be taken into account that it is a question of a conventional procedure, and that it is impossible to simulate all real conditions indoors. In the future, it will be possible to make frequent use of the results to develop low-emitting products. By selecting such products with the aid of a source control, the environmental test chamber can make an essential contribution to an improved IAQ.

References

ASTM–American Society for Testing and Materials (1990): Standard Guide for Small-scale Environmental Chamber Determinations of Organic Emissions from Indoor Materials/Products. ASTM designation D 5116–90. American Society for Testing and Materials, Philadelphia, PA.

Bauhof H., Wensing M., Zietlow J. and Möhle K. (1996): Prüfstandsmethoden zur Bestimmung organisch-chemischer Emissionen des Pkw-Innenraums. ATZ Sonderheft 25 Jahre FAT, 37–42.

CEC–Commission of the European Communities (1993): Council Directive 89/106/EEC on Construction Products: Essential Requirement No. 3 "Hygiene, Health and the Environment". Interpretative Document. Directorate-General for the Internal Market and Industrial Affairs, Brussels.

Clausen P.A. (1993): Emission of volatile and semivolatile organic compounds from waterborne paints–the effect of the film thickness. Indoor Air, *3*, 269–275.

Colombo A., De Bortoli M., Knöppel H., Pecchio E. and Vissers H. (1993): Adsorption of selected volatile organic compounds on indoor surfaces. Indoor Air, *3*, 276–282.

Colombo A., Jann O. and Marutzky R. (1994): The estimate of the steady state formaldehyde concentration in large chamber tests. Staub Reinh. Luft, *54*, 143–146.

EC–European Commission (1989): Formaldehyde Emission from Wood Based Materials: Guideline for the Determination of Steady State Concentrations in Test Chambers. Indoor Air Quality and its Impact on Man. Report No. 2, Luxembourg.

EC–European Commission (1991): Guideline for the Characterization of Volatile Organic Compounds Emitted from Indoor Materials and Products Using Small Test Chambers. Indoor Air Quality and its Impact on Man, Report No. 8, Luxembourg.

EC–European Commission (1993): Determination of VOCs Emitted from Indoor Materials and Products. Interlaboratory Comparison of Small Chamber Measurements. Indoor Air Quality and its Impact on Man, Report No. 13, Luxembourg.

EC–European Commission (1995): Determination of VOCs Emitted from Indoor Materials and Products. Second Interlaboratory Comparison of Small Chamber Measurements. Indoor Air Quality and its Impact on Man, Report No. 16, Luxembourg.

EC–European Commission (1997): Evaluation of VOC Emissions from Building Products. Indoor Air Quality and its Impact on Man, Report No. 18, Luxembourg.

EEC–European Economic Community (1989): Council Directive 89/106/EEC of 21 December 1988 on the Approximation of Laws, Regulations and Administrative Provisions of the Member States Relating to Construction Products. O.J.No L 40, 11.02.98, pp 12–25.

EN 717–1 (1996): Bestimmung der Formaldehydabgabe, Teil 1: Formaldehydabgabe nach der Prüfkammer-Methode. Beuth, Berlin.

Gunnarsen L. (1997): The influence of area-specific ventilation rate on the emissions from construction products. Indoor Air, *7*, 116–120.

Gunnarsen L., Nielsen P.A. and Wolkoff P. (1994): Design and characterization of the CLIMPAQ, chamber for laboratory investigations of materials, pollution and air quality. Indoor Air, *4*, 56–62.

Guo Z. (1993): On Validation of Source and Sink models: Problems and Possible Solutions. In: Nagda, N.L., Modeling of Indoor Air Quality and Exposure, ASTM STP 1205.

Gustafsson H. (1992): Building materials identified as major sources for indoor air pollutants; a critical review of case studies. Swedish Council for Building Research, Document No. D10 : 1992, Stockholm.
Horn W., Ullrich D. and Seifert B. (1998): VOC emissions from cork products for indoor use. Indoor Air, *8*, 39–46.
Levsen K. and Sollinger S. (1993): Textile floor coverings as sinks for indoor air pollutants. Proceedings of Indoor Air '93, Helsinki, Vol. 2, 395–400.
Meyer U., Möhle K., Eyerer P. and Maresch L. (1994): Entwicklung, Bau und Inbetriebnahme einer 1 m^3 Bauteilkammer zur Bestimmung von Emissionen aus Endprodukten. Staub Reinh. Luft, *54*, 137–142.
Möhle K. and Wensing M. (1999): Ein Standardprüfverfahren zur Ermittlung des Emissionsverhaltens von Kfz-Innenraumbauteilen mit einer $1m^3$-Emissionsprüfkammer. In: Neuere Entwicklungen bei der Messung und Beurteilung der Luftqualität, VDI-Bericht 1443, VDI, Düsseldorf, in press.
Nordtest (1990): Building Materials: Emission of Volatile Compounds, Chamber Method. NT Build 358. Nordtest, Postbox, Esbo, Finland.
prENV 13419–1 (1998): Building products–Determination of the emission of volatile organic compounds–Part 1: Emission test chamber method, Final Draft, December 1998.
Salthammer T. (1997): Emission of volatile organic compounds from furniture coatings. Indoor Air, *7*, 189–197.
Sollinger S. and Levsen K. (1993): Parameter zur Charakterisierung von Adsorptionserscheinungen bei dynamischen Emissionsprüfungen nach der Prüfkammermethode. Staub Reinh. Luft, 53, 153–156.
Tichenor B.A. (1989): Indoor air sources; using small environmental test chambers to characterize organic emissions from indoor materials and products. U.S. Environmental Protection Agency (EPA), Report No. EPA-600/8–89-074.
Van der Wal J.F., Hoogeveen A.W. and Wouda P. (1997): The influence of temperature on the emission of volatile organic compounds from PVC flooring, carpet and paint. Indoor Air, *7*, 215–221.
Wensing M. (1996): Bestimmung gasförmiger Emissionen von Bauprodukten–Anforderungen an die Prüfkammermethode und exemplarische Untersuchungsergebnisse. In: Aktuelle Aufgaben der Meßtechnik in der Luftreinhaltung, VDI-Bericht 1257, VDI, Düsseldorf, 405–438.
Wensing M. (1999a): Abschlußbericht zum BMBF-Vorhaben 07INR35: Bestimmung organisch-chemischer Emissionen als Anstoß für zukünftige Minderungsmaßnahmen. TÜV Nord (1996–1998).
Wensing M. (1999b): Emissionen elektronischer Geräte. In: Neuere Entwicklungen bei der Messung und Beurteilung der Luftqualität, VDI-Bericht 1443, VDI, Düsseldorf, in press.
Wensing M. (1999c): Determination of organic chemical emissions from electronic devices. In: Proceedings of the 8th International Conference on Indoor Air Quality and Climate, Indoor Air 1999, Edinburgh, in press.

2.2 The Field and Laboratory Emission Cell – FLEC

Hans Gustafsson

2.2.1 Background

The field and laboratory emission cell (FLEC) is a microchamber designed for emission testing of volatile organic compounds (VOCs) from, e.g., building materials. In contrast to traditional climate chambers, in the FLEC and other emission cells the test material becomes part of the cell itself.

The FLEC is mainly used for the determination of area-specific emission rate, (SER_A) at constant temperature, relative humidity and air exchange rate.

The FLEC was developed to meet the need for a portable testing device in research and product development of low-emitting materials and for identifying emission sources on-site in buildings (Wolkoff et al., 1991).

Prior to the design of the FLEC, a large number of patents, standard methods and devices for emission measurements with industrial applications were evaluated (Gustafsson and Jonsson, 1991). Devices for emission determination are used in various applications where low-emitting products are required e.g. building materials, automobile interior trim (fogging), and material in submarines and space cabins. Such chamber devices are also applied in the determination of emissions from packaging materials, solid waste, petroleum products, pesticides, fragrant compounds in the perfume industry, paints and other materials.

Many devices for industrial use are limited to very specific applications, e.g. UF-bonded products, and the testing is often performed at elevated temperatures.

2.2.2 Design

The FLEC (Fig. 2.2-1) differs considerably from the traditional types of chambers. The cell is of the shape of a lid which is positioned on the emission source to be tested.

It is circular (I.D. 150 mm), with a maximum test material area of 0.0177 m^2 and a volume of 0.035 l. The construction of the cell permits an efficient air flow over the test specimen, with the exception of a small area in the center. Contrary to traditional chambers the air is mixed without any fans due to the cylindrical cross-sectional area which is constant from the perimeter.

The cell is made of stainless steel (DIN 17 440, x5CrMo 17333, AISI 316). The inner surface is lathe made and hand polished. All tubes and couplings are made of high-quality stainless steel.

An O-ring seals the interface between the cell and the material specimen. The sealing material is emission-free and tolerates 200 °C.

The air is introduced into two diagonally positioned inlets at the perimeter of the cell and then distributed by a channel following the perimeter, from where the air is distributed over the surface of the test specimen. The air leaves the cell at the top of its center.

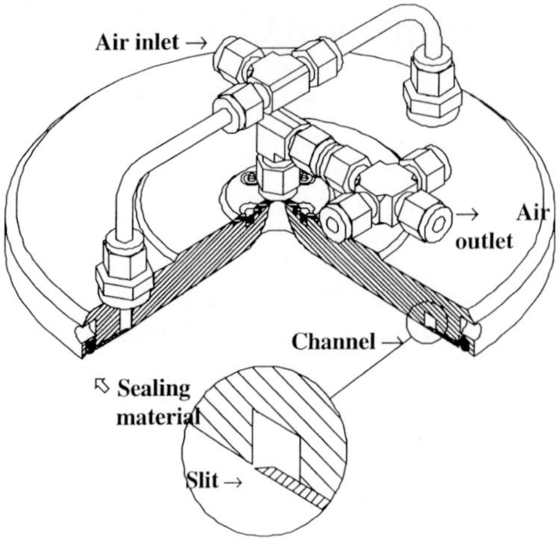

Figure 2.2-1. Schematic drawing of field and laboratory emission cell.

The area of the test object is large relative to the 35-ml cell air volume to provide for sufficient analytical sensitivity. The maximum loading factor is 507 m^2/m^3. The high loading factor is compensated by a high ventilation rate required to operate the cell. With an air flow of 100 ml/min the air exchange rate is 171 h^{-1}.

The air velocity over the test specimen is dependent on the air flow and is normally somewhat lower than in a traditional chamber.

Clean air is introduced into the FLEC with (PTFE) tubing from an air supply unit as described by, e.g., Wolkoff et al. (1995). The air flow into the cell is controlled by needle valves. The outlet is a union cross with an overflow tube to avoid false air intake during

Table 2.2-1. Technical specifications of the FLEC (Wolkoff, 1996a).

Volume in m^3	3.5×10^{-5}			
Max. exposed test surface area in m^2	0.0177			
Inlet slit in mm	1.0			
Diameter in mm	150			
Height (centre) in mm	18			
Maximum material loading in m^2/m^3	507			
Airflow rate in l/min	0.100	0.300	1.400	2.800
Air exchange rate n in h^{-1}	171	514	2400	4800
Air velocity (slit) in m/sec	0.0035	0.01	~0.05	~0.1
Area specific airflow rate[a] in $m^3/(sm^2)$	0.1	1	5	9
Reynold's number	1			10
Wall surface micro structure[b] R_a	<0.1 μ			

[a] Total exposed area of the test surface.
[b] Inner surface is hand polished.

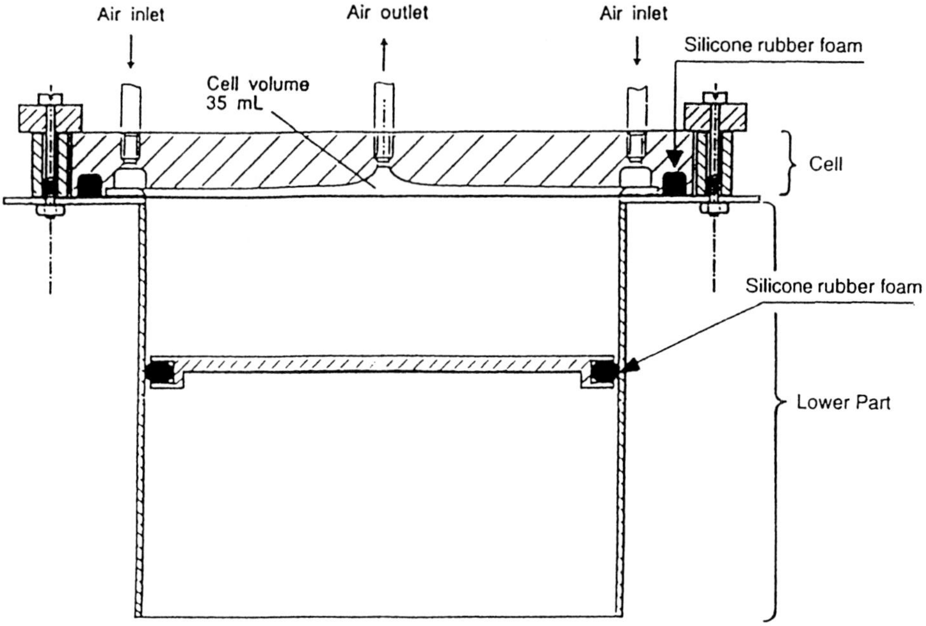

Figure 2.2-2. The cell and an example of a lower part with an adjustable bottom to adapt to the test material, so that the internal volume of the apparatus is unchanged when the test specimen is positioned in the lower part.

air sampling. During standby the left and right positions of the union cross are closed with metal rods. During active sampling of the outlet air from these positions, calibrated low-flow sampling pumps are used. Sampling and analysis of the emitted VOCs are mainly conducted using adsorbent tubes. The cell may also be used for emission measurements of compounds detectable with other analytical techniques.

The cell is intended to be positioned on the surface of the material being measured. However, if the surface of the material is textured, the sample is cut to a disc and placed in a lower part (Fig. 2.2-2). Non-setting or rugged materials as sealants, fillers for walls etc. can be accommodated in the lower part, e.g., in a Petri dish of the same size as the internal diameter of the lower part and the cell. Technical specifications of the FLEC are summarized in Table 2.2-1 (see Wolkoff, 1996a).

2.2.3 General Maintenance and Testing Procedures

Initially, Wolkoff et al. (1991) developed a testing protocol for the FLEC cell. Later, a Nordtest method for laboratory measurements was approved (Nordtest, 1995). This Nordtest method mainly deals with the FLEC as a testing device for the determination of VOCs at standard climate and using adsorbent tubes intended for analysis by GC. Trade standards describing materials sampling etc. have been developed for specific products like paints (SVEFF, 1997) and flooring materials (GBR 1992a, GBR

1992b). FLEC testing is also a part of labeling systems for building materials (Wolkoff and Nielsen, 1996; Larsen et al., 1997; FiSIAQ, 1995; FiSIAQ, 1996).

Test procedures for building materials covering the FLEC are also developed within the European standardization organization (CEN, 1999).

A procedure for the determination of area-specific emission rates (SER_A) from sources on site in buildings has also been developed (Nordtest, 1998). During such measurements the air supply may consist of pure synthetic air from, e.g., a mobile gas bottle.

2.2.4 Examples of Applications with the FLEC

The use of FLEC and traditional climate chambers has been demonstrated for several materials and applications by various investigators. SER_A from materials tested in the laboratory can be:

- used in the development of low-emitting materials and in factory production control
- used in labeling schemes
- transformed to room air concentrations.

During recent years the interest in emission processes has focused on building materials. Materials in building structures can be considered as major emission sources in buildings and especially those applied to surfaces in large quantities permanently exposed to indoor air. A review of case studies (Gustafsson, 1992) identifies some main categories of sources for VOCs in buildings:

- polymer materials as vinyl floor covering
- water-borne or solvent-based dried paints
- products based on linseed oil as alkyd paint and linoleum
- various materials decomposed by moisture.

Most of these long-term emissions are due to internal diffusion within the building material itself. In such cases, with solid and dried materials, SER_A is not affected to any great extent by the air velocity over the material surface. An overview of parameters affecting SER_A of VOCs from building products is given by Wolkoff (1995) and experimentally evaluated (Wolkoff, 1996b).

In order to transfer SER_A to indoor conditions, information about emission processes over a long period of time is required. The FLEC has therefore been used in the study of time-versus-emission data for vinyl flooring (Gustafsson et al., 1995), carpet, sealant, paint and varnish (Wolkoff, 1996b), and, with advanced mathematical modeling, for carpet, linoleum, paint and sealant (Wolkoff et al., 1993a).

In contrast to traditional chambers, the FLEC is also used for identification of emission sources on site in buildings, as outlined in Section 2.2.4.4.

2.2.4.1 Resilient Flooring

Linoleum often has a characteristic odor due to emission of VOCs.

Jensen et al. (1995a) investigated the emissions from 12 different linoleums and identified, e.g., aldehydes, fatty acids, glycol ethers and terpenes. The emission from some of the test pieces and compounds were modeled using an exponential diffusion model. In a further investigation (Jensen et al., 1995b), the emitted compounds were correlated with odor intensity.

With two FLECs positioned on two similar pieces of linoleum and supplied with air or nitrogen, multivariate analysis revealed that an increased emission of aldehydes was affected by oxidative reactions at the linoleum surface exposed to air (Jensen et al., 1996a).

FLEC measurements of a complaint case of malodorous linoleum showed that wetting of the material increased the emission of, e.g., odorous fatty acids. (Wolkoff et al., 1995). The impact of humidity has also been investigated for other building products (Wolkoff, 1996b). In this article, the impact of temperature, nitrogen instead of air and air velocity in the FLEC have also been investigated for VOCs with low human odor thresholds.

FLEC testing on a number of newly produced vinyl floorings has shown that SER_A decreases by about 2/3 during the following six months (Gustafsson and Jonsson, 1993). A model for emission controlled by internal diffusion in vinyl flooring has been developed by Clausen et al. (1993).

Vinyl flooring contains plasticizers. When the flooring material is laid on damp concrete, the plasticizers can give rise to emission of odorous alcohols due to moisture-induced ester hydrolysis of the plasticizer. The released octanols may then migrate into the subfloor. FLEC measurements have revealed that screeds (self-levelling compounds) laid over concrete floors can be contaminated with octanols. Due to their volatility, such octanols are often present in the room air in buildings with moisture problems (Bornehag, 1996). The origin of these problems – moistened concrete combined with vinyl floorings – has been confirmed in a number of laboratory investigations with the FLEC positioned over the test specimen with concrete cast in cups in a lower part (Gustafsson, 1996).

The technique with a lower part has also been applied in investigations of materials with a fleecy surface such as carpets (Zellweger et al., 1997a).

2.2.4.2 Paints and Other Surface Treatments

The SER_A values from various paints and varnishes with a dry film thickness of 40–50 μm have been investigated during 26 weeks (Jonsson, 1995).

Water-based single-component parquet sealing has been shown to release, e.g., aromatic compounds and butylglycol (Zellweger et al., 1997b). The effect of the film thickness on the SER_A from acrylic paint was investigated by Zellweger et al. (1997c).

Emissions of VOCs from wooden floors where the surface has been treated with oils have been investigated with the FLEC (Englund et al., 1996).

2.2.4.3 Other Products

Residual solvent in rotogravure-printed brochures has given rise to complaints from postal workers (Jensen et al., 1996b). The emission decay [SER_A (t)] of toluene was measured, e.g. using a photoionization detector (PID) in combination with a FLEC positioned on the front page of the brochures. These measurements correlated well with personal exposure measurements in a 41-m^3 test chamber during simulated mail handling.

VOCs as aromatic hydrocarbons and plastic monomer residues emitted from copied, laser-printed and matrix printed paper have been measured by the FLEC (Wolkoff et al., 1993b). Several of these compounds have also been identified in the corresponding tone powders. From a freshly photocopied paper, the initial SER_A of, e.g., styrene was 5 μg/(m^2h). When handling 200 such copies in an office, the styrene concentration in the room air could be calculated to reach 12 μg/m^3.

Evaporation of VOCs from 10 cleaning agents with different properties has been measured (Vejrup and Wolkoff, 1994). The nonpolar VOCs gave rise to peak emissions immediately after the application of the products. Cleaning agents have also been investigated with a photoacoustic detector (PAD) in combination with a FLEC (Vejrup and Wolkoff, 1995).

The emission from wood-based materials such as a sealed table top (massive wood construction) and a varnished cupboard door (pinewood) has been investigated (Zellweger et al., 1997d).

2.2.4.4 On-Site Measurements

The FLEC provides the possibility of non-destructive emission testing on site in buildings.

Ekberg et al. (1995) investigated one office in each of three buildings where FLEC was mounted on the wall, floor and desk surface. The emission contribution to the room air concentration was calculated using the SER_A, the total area of the surface materials and the supply air rate in the room. The majority of the VOCs identified in the office air were also present in the FLEC samples. In one of the three offices, the calculated values corresponded quite well to the measured room air concentration. In one of the other two offices the calculated room air concentration exceeded the measured concentration, while the opposite was found in the third office. This lack of correlation was assumed to be mainly due to sorption effects. Generally, sinks are more pronounced in real rooms than in all types of chambers (Tichenor, 1996).

In another case study, Zellweger et al., (1997e) studied the emission of hexanal from linoleum in an office. SER_A measured with the FLEC (both on-site and in-lab) correlated well with the area-specific emission rates (SER_A) calculated on ventilation rate and room air concentrations. The FLEC measurements also revealed that low-volatility compounds such as glycol ethers from the room air had absorbed on the linoleum, although the linoleum initially did not emit any glycols. Relatively often, low-volatility VOCs are subject to sink effects.

A compound like hexanal has many indoor sources besides linoleum. This complicates the possibility of connecting a compound identified in the room air to a specific indoor source. However, for materials with an emission of specific compounds, SER_A measured with FLEC on site can be quantitatively matched with the room air concentration. Relatively good correlation has been shown for Texanol from parquet sealing in a dining room (Zellweger et al., 1997f) and butoxyethoxyethanol from a vinyl flooring in an apartment (Gustafsson and Rosell, 1995).

Based on testing with FLEC, indoor air levels of material-specific compounds can be estimated in a room assuming no sinks. A „worst case" approach with a 17-m^3 standard room (INF, 1994) is applied for the purpose of health and comfort evaluation of building products and materials in the Danish Labeling Scheme (Wolkoff and Nielsen, 1996; Larsen 1997).

2.2.5 Evaluation of the FLEC

The repeatability and reproducibility in FLEC testing have been demonstrated in the measurement of various objects such as liquid wax, latex paint (Roache et al., 1996), vinyl flooring (Gustafsson et al., 1995), sealant and linoleum (Gunnarsen et al., 1993), carpet, vinyl flooring, sealant, paint and varnish (Wolkoff, 1996b).

The performance of the FLEC has been evaluated in a recovery study with a polar and an non-polar VOC by comparison of measured concentrations with the gravimetric loss from the sources (Wolkoff et al., 1995). In this experiment, an artificial source positioned halfway the perimeter and the center was used. No significant differences in the measured concentrations were observed when operating the FLEC symmetrically and asymmetrically to the source.

In contrast, Uhde et al. (1997, 1998) have found distinct deviations with three different point sources with constant SER_A positioned on a plate at different distances from the center of the FLEC.

On stepwise rotation of the FLEC by 90° on the plate, a dependence of SER_A on the location of the source in the cell was observed. These findings are in agreement with a mathematical approach of flow theory, air velocity measurements and visualization of the flow field given in the same article. With a total flow of 250 ml/min, air velocities ranged from < 0.1 to 0.9 cm/s. As predicted by the theory, the highest air velocities were measured at the axis through the air inlets. The authors conclude that the sources used did not represent common emission characteristics and that the results can be applicable to inhomogeneous materials such as wooden materials with knot holes or encapsulated solvent residues. If the homogeneity of the sample is doubtful, it would be a reasonable procedure to perform the emission measurement twice with a 90° turn between the measurements. For a test specimen with homogeneous surfaces the effect will average out.

The FLEC is mainly applied in the laboratory for emission testing of materials with homogeneous surfaces. For this kind of material, satisfactory results have been achieved when comparing the FLEC with other chambers (EC–European Commission, 1993; EC–European Commission, 1995; Gunnarsen et al., 1994; Roache et al., 1996; Wolkoff et al., 1995).

2.2.6 Conclusion

Emission cells such as the FLEC may be useful in many applications, e.g. for routine testing in manufacturers product development and production control.

Despite the portable size of the FLEC and the relatively high air exchange rate, the experimental conditions seem to be compatible with larger chambers. A disadvantage is the limited sample size and thus the possibility of sample inhomogeneity.

The FLEC can also be used on site in buildings on planar material surfaces for the identification of emission sources.

References

Bornehag C.G.(1996): Evaluation of a method for dealing with humidity and odor problems in concrete floors. Proceedings of Indoor Air '96, Nagoya, Japan, Vol. 4, 325–330.

CEN 1999: European Prestandard, prENV 13419–2, Building products–Determination of the emission of volatile organic compounds, Part 2: Emission test cell method.

Clausen P., Laursen B., Wolkoff P., Rasmusen E. and Nielsen P.A. (1993): Emission of volatile organic compounds from a vinyl floor covering, modeling of indoor air quality and exposure. In: Nagda N. L. (ed) ASTM STP 1205, American Society for Testing and Materials, Philadephia, 1–13.

EC–European Commission (1993): Determination of VOCs emitted from Indoor Materials and Products–Interlaboratory Comparisons of Small Chamber Measurements. Joint Research Centre, Report No. 13, Luxembourg.

EC–European Commission (1995): Determination of VOCs emitted from Indoor Materials and Products–Interlaboratory Comparisons of Small Chamber Measurements, Joint Research Centre, Report No. 16, Luxembourg

Ekberg L.E., Gunnarsen L., Knutti R., Haas J. and Mogl S. (1995): Identification of air pollution sources in office buildings using the Field and Laboratory Emission Cell. Proceedings of Healthy Buildings '95, Milan, Italy, Vol. 3, 1353–1358.

Englund F., Larsen A., Winther Funch L., Saarela K., Tirkkonen T. (1996): Emissions of VOC from wooden floors, surface treated with oils and waxes. Proceedings of Indoor Air '96, Nagoya, Japan, Vol. 3, 89–94.

FiSIAQ (1995): Finnish Society of Indoor Air Quality and Climate–Classification of Indoor Climate, Construction and Finishing Materials, June 15, 1995.

FiSIAQ (1996): Finnish Society of Indoor Air Quality and Climate–Classification of Finishing Materials, General Instructions. Building Information Institute, Committe TK 185, Helsinki, February 1996.

GBR (1992a): Trade standard–Measurement of Chemical Emission from Flooring Materials, Approved by The Swedish Flooring Trades Association and The Swedish National Testing and Research Institute, 1992.

GBR (1992b): Trade standard: Measurement of Chemical Emission from Smoothing Compounds for Floors, Approved by The Swedish Flooring Trades Association and The Swedish National Testing and Research Institute, 1992.

Gunnarsen L., Nielsen P.A. and Wolkoff P. (1994): Design and characterization of the CLIMPAC (chamber for laboratory investigations of materials, pollution and air quality). Indoor Air, Vol. 4, 56–62.

Gunnarsen L., Nielsen P.A., Nielsen J.B., Wolkoff P., Knudsen H. and Thøgersen K. (1993): The influence of specific ventilation rate on the emissions from construction products. Proceedings of Indoor Air Quality and Climate 93, Helsinki, Finland, Vol. 2, 501–506

Gustafsson H.(1992): Building materials identified as major emission sources for indoor air pollutants–a critical review of case studies. Document D10 : 1992, ISBN 91–540-5471–0, Swedish Council for Building Research.

Gustafsson H. (1996): Moisture-induced chemical degradation and emission from flooring materials on damp concrete–a critical review of investigations (in German), SP-report 96 : 25t. Swedish National Testing and Research Institute.

Gustafsson H. and Jonsson B. (1991): Review of small-scale devices for measuring chemical emission from materials. Report 1991 : 25, Swedish National Testing and Research Institute.
Gustafsson H. and Jonsson B. (1993): Trade standards for testing chemical emission from building materials, Part 1: Measurements of Flooring Materials. Proceedings of Indoor Air'93, Helsinki, Finland, Vol.2, 437–442.
Gustafsson H. and Rosell L. (1995): Kemisk emission från golvmaterial – undersökning av i vilken utsträckning golvmaterial avger kemiska ämnen på plats i byggnader och under laboratorieförhållanden, SP:AR 1995 : 29 (in Swedish). Swedish National Testing and Research Institute.
Gustafsson H., Jonsson B. and Lundgren B. (1995): Chemical emission from paints and other surface materials – measurements with Field and Laboratory Emission Cell (FLEC) and other climate chambers. In: Bagda E. (ed) Emissionen aus Beschichtungsstoffen, Chapter 8, Stand der Technik, Analyse der Emissionen und deren Einfluss auf die Innenraumluft, 95–105. Expert Verlag.
INF (1994): Danish Standard/INF 90; Directions for the determination and evaluation of the emission from building products/Anvisning for bestemmelse og vurdering af afgasning fra byggevarer.
Jensen B., Wolkoff P., Wilkins C.K., Clausen P.A. (1995a): Characterization of linoleum, Part 1: Measurement of volatile organic compounds by use of the field and laboratory emission cell „FLEC". Indoor Air, 5, 38–43.
Jensen B., Wolkoff P. and Wilkins C.K. (1995b): Characterization of linoleum, Part 2: Preliminary odour evaluation. Indoor Air, 5, 44–49.
Jensen B., Wolkoff P. and Wilkins C.K. (1996a): Characterization of Linoleum: Identification of oxidative emission processes, characterizing sources of indoor air pollution and related sink effects, ASTM STP 1287, Bruce Tichenor (ed). American Society for Testing and Materials, 145–152.
Jensen B., Olsen E. and Wolkoff P. (1996b): Toluene in rotogravure printed brochures: high speed emission testing and comparision with exposure data. Appl. Occup. Environ. Hyg. 11(8), August 1996.
Jonsson B. (1995): Results from emission testing of paints. Proceedings of Indoor Air Quality in Practice '95, Oslo, Norway, 181–183.
Larsen A. (1997): Introduction to the principles behind the Danish indoor climate labeling. Danish Indoor Climate Labeling Association, February 1997.
Nordtest (1995): Nordtest method; Building Materials: Emission of volatile organic compounds – Field and Laboratory Emission Cell (FLEC), NT BUILD 438.
Nordtest (1998): Building Materials: Emission of volatile compounds – on site measurements with Field and Laboratory Emission Cell (FLEC), NT BUILD 484.
Roache N.F., Guo Z., Fortman R., and Tichenor B.A. (1996): Comparing the Field and Laboratory Emission Cell (FLEC) with traditional emissions testing chambers. Characterizing sources of indoor air pollution and related sink effects. ASTM STP 1287, Tichenor B. (ed) American Society for Testing and Materials, 98–111.
SVEFF (1997): Trade standard: Measurement of Chemical Emission from Paint and Varnish, Approved by The Swedish Paint and Printing Ink Makers Association.
Tichenor B., (ed). (1996): Characterizing Sources of Indoor Air Pollution and Related Sink Effects, ASTM STP 1287, American Society for Testing and Materials.
Uhde E., Borgschulte A. and Salthammer T. (1997): Characterization of the Field and Laboratory Emission Cell (FLEC): Impact of Air Velocities on VOC Emission Rates. Proceedings of Healthy Buildings/ IAQ '97, Washington DC, Vol. 3, 503–508.
Uhde E., Borgschulte A. and Salthammer T. (1998): Characterization of the Field and Laboratory Emission Cell (FLEC): Flow field and air velocities Atmos. Environ., 32, 773–781.
Vejrup K. and Wolkoff P. (1994): VOC Emission from some floor cleaning agents. Proceedings of Healthy Buildings '94, Budapest, Hungary, Vol. 1, 247–252.
Vejrup K. and Wolkoff P. (1995): High-speed emission testing of cleaning agents with the FLEC. Part I: Method development. Proceedings of Healthy Buildings '95, Milan, Italy, Vol. 2, 983–988.
Wolkoff P. (1995): Volatile Organic Compounds – sources, measurements, emissions, and the impact on indoor air quality. Indoor Air, Suppl. No. 3, 1–73.
Wolkoff P. (1996a): An emission cell for measurement of volatile organic compounds emitted from building materials for indoor use – the Field and Laboratory Emission Cell (FLEC). Gefahrst. Reinhalt. Luft, *56*, 151–157.
Wolkoff P. (1996b): Characterization of emissions from building products: long-term chemical evaluation. The impact of air velocity temperature, humidity, oxygen and batch/repeatability in the FLEC. Proceedings of Indoor Air '96, Nagoya, Japan, Vol. 1, 579–584.

Wolkoff P. and Nielsen P.A. (1996): A new approach for indoor climate labeling of building materials–Emission testing, modeling and comfort evaluation. Atmos. Environ., *30*, 2679–2689.

Wolkoff P., Clausen P.A., Nielsen P.A., Gustafsson H., Jonsson B., Rasmusen E. (1991): Field and Laboratory Emission Cell (FLEC), Conf. Proceedings of Healthy Buildings 91, Washington D.C., USA, Sept 4–8, 160–165.

Wolkoff P., Clausen P.A., Nielsen P.A. and Gunnarsen L. (1993a): Documentation of Field and Laboratory Emission Cell (FLEC): Identification of emission processes from carpet, paint and sealant by modeling. Indoor Air, 3, 291–297.

Wolkoff P., Wilkins C.K., Clausen P.A. and Larsen K. (1993b): Comparison of Volatile Organic Compounds from processed paper and toners from office copiers and printers: methods, emission rates, and modeled concentrations. Indoor Air, *3*, 113–123.

Wolkoff P., Clausen P.A., Nielsen P.A. (1995): Application of Field and Laboratory Emission Cell "FLEC"-Performance study, intercomparison study, and case study of damaged linoleum in an office. Indoor Air, *5*, 196–203.

Zellweger C., Hill M., Gehrig R. and Hofer P. (1997a): Emissions of Volatile Organic Compounds (VOC) from building materials–methods and results, A38, A46, A47, A48 and A49. Bundesamt für Energiewirtschaft.

Zellweger C., Hill M., Gehrig R. and Hofer P.(1997b): Emissions of Volatile Organic Compounds (VOC) from building materials–methods and results, A30–A32. Bundesamt für Energiewirtschaft.

Zellweger C., Hill M., Gehrig R. and Hofer P.(1997c): Emissions of Volatile Organic Compounds (VOC) from building materials–methods and results, 66 and A34–A35. Bundesamt für Energiewirtschaft.

Zellweger C., Hill M., Gehrig R. and Hofer P.(1997d): Emissions of Volatile Organic Compounds (VOC) from building materials–methods and results, A63 and A70. Bundesamt für Energiewirtschaft.

Zellweger C., Hill M., Gehrig R. and Hofer P.(1997e): Emissions of Volatile Organic Compounds (VOC) from building materials–methods and results, 74–80 and A42. Bundesamt für Energiewirtschaft.

Zellweger C., Hill M., Gehrig R. and Hofer P.(1997f): Emissions of Volatile Organic Compounds (VOC) from building materials–methods and results, 84. Bundesamt für Energiewirtschaft.

2.3 Mathematical Modeling of Test Chamber Kinetics

Stylianos Kephalopoulos

2.3.1 Introduction

The prediction of pollutant concentrations in indoor environments using mathematical models has received increased interest. Mathematical modeling is commonly used to improve the understanding of chemical and physical processes that affect indoor pollutant concentrations and to predict indoor concentrations and human exposures under different situations. Mathematical modeling is of particular importance for evaluating test chamber experiments widely used to characterize the emission of volatile organic compounds (VOCs) from indoor materials and products, their ad/absorption on/by indoor materials and their re-emission (secondary emission) from them. One of the primary advantages of mathematical modeling is that predictions of human exposure to various indoor pollutants can be made without any chemical sampling. For example, prediction of exposure to VOC emissions from building materials based on emission data and ventilation rates can be made before construction. Models can be used to predict future concentrations and to extrapolate present concentrations when the factors affecting exposure are being altered. Other applications of mathematical models include prediction and optimization of the effectiveness of various control strategies, such as ventilation and air cleaning, in reducing concentrations of indoor pollutants. After validation, a model can have considerable power, but conditions that limit the validity of the model should always be acknowledged and respected.

Several models have been developed in the last decade aimed at describing the emission of volatile organic compounds (VOCs) from indoor materials. These models may be broadly distinguished with respect to their conceptual background (physical-mass transfer models and/or empirical-statistical models) as well as their ability to describe different emission profiles. Physical models are models based on principles of physics and chemistry, whereas the empirical models do not necessarily require fundamental knowledge of the underlying physical, chemical and/or biological mechanisms. Many models used in the indoor air quality field in practice are "hybrid" models, in which aspects of both physical and empirical approaches are combined.

In this Chapter, both physical and empirical models will be discussed in an attempt to put into evidence several aspects of the model building and validation procedures as they apply to the indoor pollution field. There will be thorough discussion of how these procedures can be applied to model the test chamber kinetics, this being the most common approach to characterizing emissions from indoor materials and products, including the interaction of these emissions with indoor sinks. This will provide the reader with the necessary elements to develop and apply modeling techniques as part of his/her own indoor pollution practices. To this end, examples will be given to cover aspects of both model building and model validation processes, some of them arising from the author's own practice and some others from reviewing other researchers' work. Research over the last decade has shown that emissions from indoor sources dominate indoor air concentrations and exposures much more so than factors such as ventilation. Desorption of adsorbed compounds from an indoor surface is suspected to contribute significantly to indoor VOC exposures. Understanding of complexities such as re-emitting sinks is just beginning. As sources and sinks are basic components of indoor air quality (IAQ) modeling and considerations of sources and sinks are often linked to one another, the discussion included in this Chapter will mostly focus on these two components of IAQ modeling. The issue of how these components are combined to model the test chamber kinetics will be properly addressed.

2.3.2 Model Building Process

In practical terms, an indoor air quality model should provide a reasonable description of the mass balance of the test chamber experiments, trying to address factors such as material emissions, airflows into and out of the chamber and chemical/physical decay or other removal and/or transformation processes of the VOCs. VOC concentrations are increased by emissions within the defined volume of the chamber and by infiltration from external air to the chamber. Similarly, concentrations are decreased by transport via exiting chamber air, by removal to chemical and physical sinks within the chamber air, or by transformation of a VOC to other chemical forms. A general mass balance equation concerning the concentration of a VOC in a test chamber can be written in the form of one or more differential equations representing the rate of accumulation and the VOC gain and loss. This concept for a VOC concentration C (mass units/m^3) in a chamber of volume V (m^3) is translated into the following differential equation:

$$V \cdot \frac{\mathrm{d}C}{\mathrm{d}t} = ER - F - Q_{\mathrm{out}} \cdot C + Q_{\mathrm{in}} \cdot C_{\mathrm{ext}} - A + D + V \cdot X \tag{1}$$

where ER = emission rate of a source in the chamber (mass units/h); F = removal rate by an air cleaning device inside the chamber (mass units/h); Q_{out} = air flow rate from the chamber to the air space externally to the chamber (m^3/h); Q_{in} = air flow rate from outdoors to indoors (m^3/h); C_{ext} = outdoor VOC concentration; A = rate of adsorption by interior surfaces of the chamber (mass units/h); D = rate of desorption from interior surfaces of the chamber (mass units/h); X = rate of generation through a gas-phase chemical reaction (mass units/m^3/h). X is negative if the reaction consumes the VOC of interest.

The accuracy of any model to predict VOC concentrations in test chamber experiments mostly depends on the accuracy of the source (term ER in Eq. 1) and sink submodels (terms A and D in Eq. 1) incorporated into the IAQ model. In other words, a realistic estimate of human exposure to VOCs emitted from indoor materials and products requires knowledge not only of their emission rates but also of their adsorption/desorption capacity or the "buffer" effect on VOC concentrations in indoor air. Small environmental test chambers are increasingly used in order to characterize the emission of VOCs (i.e. the source term in Eq. 1) from materials and products present indoors, whether they are used to realize the building environment, maintenance work (including cleaning) or other activities (e.g. do-it-yourself work). More recently, small test chambers have also been used for investigating the adsorption of pollutants onto indoor surfaces (i.e. the terms *A* and *D* in Eq. 1). Several models have been proposed for describing adsorption and desorption of VOCs on/from indoor surface materials (Dunn and Tichenor, 1988; Tichenor et al., 1991; Colombo et al., 1993; De Bortoli et al., 1996). Most work performed so far uses the model proposed by Tichenor et al. (1991), which assumes a Langmuir isotherm for adsorption and a low coverage of the adsorption sites. De Bortoli et al. (1996) proposed a "two-sink" model because their experimental data could not be satisfactorily fitted by the Tichenor model. Sorption models have also been proposed by Axley (1991), that might be useful in extrapolating results from small chambers to the situation in full scale. These models are based on mass transfer and boundary layer theory and take into account the effect of the air velocity past the sorbent.

In the following, a real case study will be given to practically demonstrate how the chamber kinetics of VOCs can be modelled by applying mathematical theory to handle experimental data resulting from experiments performed in test chambers. Through this case study, the different steps of model building and model validation processes will be clearly identified and thoroughly discussed.

The case study concerns experiments designed and a model developed by M. De Bortoli et al. (1996), who attempted to characterize the sink effects in small test chambers. This was motivated by the fact that small environmental test chambers may present wall adsorption for VOCs, which must be known for correct emission or adsorption rate determinations on indoor materials. Experiments have been performed in a 0.28-m^3 stainless steel chamber in the dynamic (air flow through the chamber) and static mode (no flow), with *n*-decane and *n*-dodecane. Further experiments with a 0.45-m^3 glass chamber, more compounds and typical indoor sink materials (i.e., carpet, vinyl wall covering, gypsum board) were also performed. Details concerning the properties of the chambers used and the chemical analysis of the vapor concentration in the chamber air are given in the paper of De Bortoli et al. (1996). The model developed accounts for both fast and slow sinks. Let us see in the following how this model has been developed, going through the different steps of the model's building and validation processes.

The first step in the model building process is the setup of an appropriate experimental design.

In our example, M. De Bortoli et al. (1996) carried out three types of experiments: (a) dynamic adsorption experiments with air flow through the chamber (air exchange rate k_2) and simultaneous introduction of vapor(s) at a constant known rate (source strength k_1) leading to concentration build-up in the chamber; (b) static adsorption experiments, i.e. injection of vapor(s) into the chamber and follow-up of the concentration decay in the sealed chamber (no air flow); (c) dynamic desorption experiments, i.e purging the chamber with a clean air flow after a (dynamic or static) adsorption experiment and follow-up of the concentration decay. The static and the dynamic approach for adsorption measurements correspond to two different pollution situations in indoor environments: the static approach resembles too short-term emissions as they are often caused by activity-related sources such as cleaning or smoking, whereas the dynamic approach reflects the impact of long-term emissions from building materials or of outdoor pollutants or pollutants injected into the room by recirculated air on indoor surfaces.

The second step in the model building process is the model conceptualization.

Through this step and based on experimental evidence we try to develop the appropriate model to describe the test chamber kinetics. As was anticipated in the introduction of this Chapter, from a conceptual point of view, two broad categories of models can be developed: empirical-statistical and physical-based mass transfer models. It should be emphasized that, in several cases, even the fundamentally based mass transfer models are indistinguishable from the empirical ones. This happens because the mass transfer models are generally very complex in both the physical concept involved and the mathematical treatment required. This often leads the modelers to introduce approximations, making the mass transfer models not completely distinguishable from some empirical models in terms of both functional formulations and descriptive capabilities. Considering the current status of models which have been developed to describe VOC emissions (and/or sink processes), we could define the mass transfer models as "hybrid-empirical" models.

An example of the above consideration is the equivalence of the power model proposed by Colombo et al. (1994) and Clausen's model (1993) for emission controlled by internal diffusion in the source. The assumptions made in the latter – more physically based – model lead to a final description of the emission rate at the surface of the source which is equivalent to the description of the former – purely empirical – model. The equivalence of the models is valid when the parameter C of the empirical model takes the value 1. This can readily be seen if we compare the mathematical equations of the two models:

$$C(t) = A/(F_0 \cdot V \cdot k_2)/[1 + (F_0 \cdot k/L) \cdot t] \qquad C(t) = A/(1 + B \cdot t^C) \quad (2)$$

Clausen's diffusion-controlled emission model — *Colombo's power model*

A second example refers to the comparison between the vapor pressure model proposed by Dunn and Tichenor (1988) and Tichenor's mass transfer model (1993). The analytical formulas of both models which describe the concentration in the chamber as a function of time are completely equivalent. They only reinterpret the k_1, k_2, k_5 coefficients of the vapor pressure model in terms of N, D_f, C_v, δ, M_0 and L of the mass transfer model, i.e.

$$k_2 \rightarrow N, \qquad k_1 \rightarrow \frac{D_f \cdot C_v}{\delta \cdot M_0}, \qquad k_5 \rightarrow \frac{L \cdot D_f}{\delta} \qquad (3)$$

Consequently, it would be expected that the predictive capability of the two models will also be equivalent, and any limitations in terms of prediction of one model will reflect analogous limitations of the other model. Indeed, if the vapor pressure model fails to describe the tailing of the concentration versus time curve – which is an indication of sink effects – the mass transfer model is also unable to describe the same part of the experimental data (Tichenor et al., 1993). It should be noted, however, that the parameters of the mass transfer model have well-defined physical meanings [e.g., vapor pressure (C_v), molecular diffusivity (D_f), boundary layer thickness (δ)], and the parameter estimation does not rely heavily on curve fitting. The parameter estimation is the first step in the model performance and validation process, as we will see later in this Chapter.

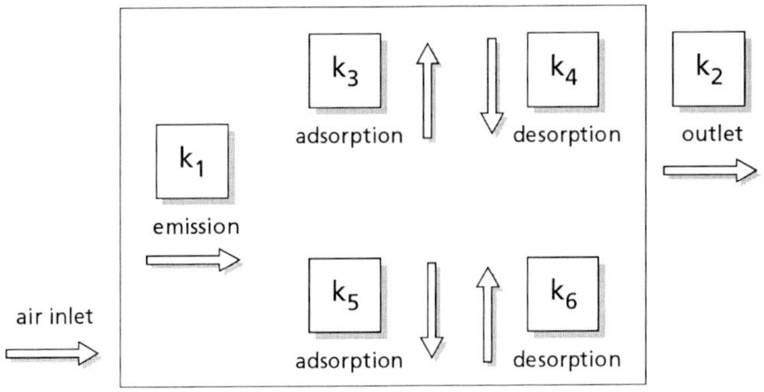

Figure 2.3-1. Schematic representation of the "two-sink" model.

In our case study, the experimental observations (i.e. concentration versus time data) were used as input to the conceptualization phase of a mathematical model with two sink compartments (the so-called "two-sink" model). Following the above discussion, this model (schematically represented in Fig. 2.3-1) can be considered as a "hybrid-empirical" model. Conceptually, this model describes the test chamber kinetics of a VOC for the three types of experiments which have been carried out. The adsorption-desorption kinetics is described by the rate constants k_3, k_4, k_5, k_6. Given that the conceptualization phase of any model is a progressive procedure which combines over the time past experience with new experimental evidence, it is interesting to see how this has happened for the "two-sink" model.

Adsorption of vapors on test chamber walls has been previously described by means of models including two or three rate constants for adsorption/desorption processes in the case of dynamic experiments (Dunn et al., 1988; Colombo et al., 1993) and with three adsorption/desorption constants in the case of static experiments (Colombo et al., 1993). Two rate constants describe a reversible sink whereas three rate constants describe a reversible and an "irreversible" (i.e. leak type) sink. However, these models did not adequately describe the sorption process(es), especially in the case of long-term tests, as resulted from two observations (Colombo et al., 1993): (a) the model with three sorption rate constants (reversible + irreversible sink) provided a better description of the experimental data than the one-sink model and (b) desorption experiments following adsorption gave strong indications that the "irreversible" sink was in fact slowly rever-

Table 2.3-1. Comparison between Dunn and Tichenor (1988) models and De Bortoli et al. (1996) model.

Model	Parameters	Ref.
Dilution	$k_3, k_4, k_5, k_6 = 0$	Dunn and Tichenor (1988)
One-sink	$k_5, k_6 = 0; k_3, k_4 \neq 0$	Dunn and Tichenor (1988)
Vapor Pressure	$k_3, k_4, k_6 = 0, k_5 \neq 0$	Dunn and Tichenor (1988)
Full	$k_6 = 0$	Dunn and Tichenor (1988)
Two-sink	k_3, k$_4$, $k_5, k_6 \neq 0$	De Bortoli et al. (1996)

sible. These observations led to the development of the "two-sink" model including four sorption rate constants or two reversible sinks. This model represents an extension of Dunn and Tichenor's models (Dunn et al., 1988) to establish a mass balance of the vapors introduced into the chamber, also accounting for the deposition into the chamber sink.

The system of differential equations which describe the mass balance of the adsorption-desorption experiments of our case study are the following:

$$V \cdot \frac{\mathrm{d}C}{\mathrm{d}t} = k_1 - (k_2 + k_3 + k_5) \cdot V \cdot C + k_4 \cdot W_1 + k_6 \cdot W_2 \tag{4}$$

$$\frac{\mathrm{d}W_1}{\mathrm{d}t} = -k_4 \cdot W_1 + k_3 \cdot V \cdot C \tag{5}$$

$$\frac{\mathrm{d}W_2}{\mathrm{d}t} = -k_6 \cdot W_2 + k_5 \cdot V \cdot C \tag{6}$$

where C = the concentration in the chamber air (mass units $\cdot$ m^{-3}); W_1 = the mass in the 1st sink (mass units); W_2 = the mass in the 2nd sink (mass units); k_1 = the source strength (mass units $\cdot$ h^{-1}); k_2 = the air exchange rate (h^{-1}); k_3 (h^{-1}), k_4 (h^{-1}) and k_5 (h^{-1}), k_6 (h^{-1}) are the adsorption and desorption rate constants for the 1st and 2nd sink respectively and V = the chamber volume (m^3).

By comparing Eq. (1) and Eq. (4) it can readily be seen that, in our case study, $R = k_1$ (or 0 for the desorption phase of dynamic experiments and for the static experiments), $F = 0$, $Q_{out} = k_2 \cdot V$, $Q_{in} = 0$, $A = -k_3 \cdot V \cdot C - k_5 \cdot V \cdot C$, $D = k_4 \cdot W_1 + k_6 \cdot W_2$, and $X = 0$.

For the sake of comparison, the parameters concerning the models developed by Dunn and Tichenor (1988) and De Bortoli et al. (1996) are summarized in Table 2.3-1. The dilution model represents the simplest situation and results if the sink effects and the vapor pressure effect (k_5) are neglected.

As alternative expression to be used for the terms R, A, D and X, (Guo, 1996) provides a detailed description of different source and sink sub-models in the User's Guide of his Z-30 IAQ simulator. This very useful compilation of existing widely used source and sink sub-models includes 25 source and 6 sink sub-models. The 25 source sub-models include 3 for constant source, 7 for a first-order (exponential) decay source, 3 for a higher order decaying source, 1 for an instant source, 1 for a time-varying source, 2 for ambient air as an indoor source and 8 mass transfer sub-models. The 6 sink sub-models include one irreversible and five reversible sub-models. Among the five reversible sink sub-models two are equilibrium models and the remaining three compute simultaneously both adsorption and desorption rates.

In order to resolve the above system, we need a set of initial conditions (i.c.) at time zero for C, W_1 and W_2. These are different for the 3 types of experiments performed. For the dynamic adsorption experiments the vector of i.c. is: [$C(0) = 0$, $W_1(0) = 0$, $W_2(0) = 0$]. This means that, at time zero, the concentration in the chamber air and the masses in the two sinks are all zero. For the static adsorption experiments the vector of i.c. is: [$C(0) = C_0$, $W_1(0) = 0$, $W_2(0) = 0$]. In this case, the initial concentration is equal to the concentration C_0 of the vapors injected into the chamber, and the initial masses in the two sinks are still zero. Given that the static experiments are performed in a sealed chamber (no air flow), the value of k_2 in Eq. (4) is zero. Finally, for the dynamic desorption experiments which follow the dynamic adsorption or static experiments the corresponding i.c. vector is: $[C(0) = C_0, W_1(0) = W_{10}, W_2(0) = W_{20}]$. Here, the initial concentration C_0 is the experimental value of the concentration in the chamber air and the calculated values W_{10} and W_{20} of the masses deposited in the two sinks at the end of the adsorption experiment. The solution of Eqs. (4), (5) and (6) for all three initial conditions has a closed analytical – though complicated – form (De Bortoli et al., 1996), which can be explicitly used for the estimation of the model parameters. For diffusion-controlled sources and sinks the solution of the corresponding models in most cases can be obtained only numerically (if no model approximations are made).

The next step is to run the model and estimate its parameters. As was anticipated earlier in this Chapter, the estimation of the model parameters is the *first step* of the model performance and validation process.

2.3.3 Model Performance and Validation Process

After building the model, it is necessary to test its performance and validate it with independent measurement data. Basically, there are two reasons why measured data might be out of line with model predictions: measurement or sampling errors and model failure. If sampling and measurement errors can be excluded, the model needs adjustment and modification. An adequate model is seldom obtained at the first attempt. In general, an iterative procedure is needed, where improvements are continuously made until an adequate model is achieved and measurement results are within the confidence limits of the model.

In our case study, to test the performance of the "two-sink" model we need to determine the unknown model parameters (k_3, k_4, k_5, k_6) that could give the best agreement between the model prediction and the experimental observations. These parameters were estimated from the experimental data by a non-linear least squares regression (curve fitting) algorithm described elsewhere (Colombo et al., 1993). No matter what data are used, such model development depends heavily on statistical estimations. This methodology is applied to the empirical or hybrid-empirical models. In contrast, the parameters of a mass transfer model often have well-defined physical meanings such as vapor pressure, diffusivity, adsorption energy, molecular weight and boundary layer thickness. Most of these parameters are obtained from the literature or from well-established models. Therefore, parameter estimation for mass transfer models does not rely heavily on regression. Prior to discussing the estimation of the "two-sink" model parameters relative to our case study, some mathematical aspects concerning the parameter estimation step by applying regression techniques will be briefly outlined. This is necessary to emphasize the importance of thoroughly examining the parameter values that result from the parameter estimation step before trying to interpret them in physical terms.

Some basic features of regression analysis. When a model is grossly mis-specified or the estimation procedure gets "hung up" in a local minimum, the standard errors for the parameter estimates can become very large. This means that regardless of how the parameters were moved around the final values, the resulting residual sum of squares function *S* did not change much. Also, the correlations between parameters may become very large, indicating that parameters are redundant, or, in other terms, when the estimation algorithm moved one parameter away from the final value the increase in the loss function could be almost entirely compensated for by moving another parameter. Thus, the effect of these two parameters on the residual sum of squares was very redundant.

Significance of the parameter estimates. Convergence to significantly different values of the residual sum of squares function *S* and of the parameters estimates for different initial guesses suggests local minima. Convergence to identical values of *S* and parameters for any reasonable initial guesses indicates that the global minimum has been found.

Final parameter values are given with their standard errors; small standard errors are good indicators of a satisfactory fit and give an indication of the significance of the model parameters. If the values of the standard errors are much greater than the corresponding values of the parameter estimates, for these parameters little confidence exists about their importance even if the global fitting result is satisfactory.

The fine-tune of the model parameters is of particular interest for the inter-laboratory exercises in which it is desired to maximize the reproducibility and comparability of the results.

Over-parameterization. This is one of the two aspects of what is defined in regression analysis as an "ill-conditioned" problem. When a problem is ill-conditioned the parameter vector which minimizes the residual sum of squares function *S* may be difficult to obtain computationally. Ill-conditioning could indicate a model that is over-parameterized, that is, a model that has more parameters than are needed, or could reveal the existence of inadequate data (e.g. too few data), which will not allow us to estimate the parameters postulated. Because these are two sides of the same coin, the choice of whether one or the other is the culprit depends on a priori knowledge about the practical problem and one's point of view (Seber and Wild, 1989; Salthammer, 1996).

Situations in which there are more parameters in the model than are needed to represent the data (i.e., over-parameterization) generally show up in the pattern of the correlation matrix of the estimated parameters. When some of these correlations are large, this indicates that one or more parameters may not be useful or, more accurately, that a re-parameterized model involving fewer parameters might fit the experimental data almost as well. Note that this does not necessarily mean that the original model is inappropriate for the physical situation under study. It may simply be an indication that the data in hand are not adequate to the task of estimating all the original parameters. Correlation is usually not a serious problem if off-diagonal matrix elements are smaller than about 0.98. A higher value reveals an interdependence of the parameters, i.e., their final values will depend on one another. If the correlation is total, the parameter estimates will depend entirely on the initial guesses (i.e., no unique estimates for each parameter will be found).

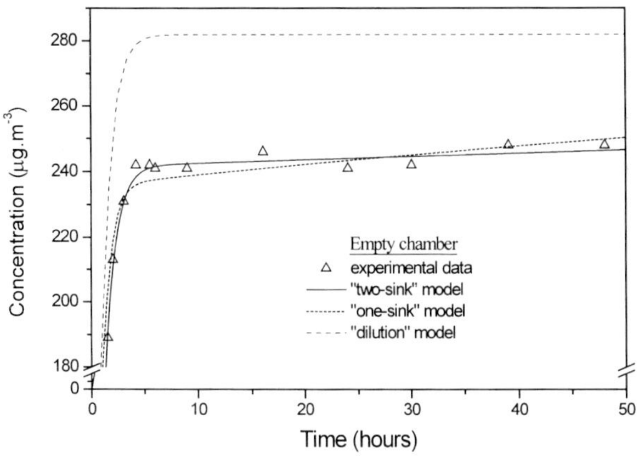

Figure 2.3-2. Concentration of *n*-decane versus time in the test chamber and curve fitting with the "one-sink" and "two-sink" models (experiment No. 1, dynamic mode).

In our case study, regression of the experimental data with the "two-sink" model yielded a "fast" and a "slow" sink, i.e. the rate constants of one sink were in general significantly greater than those of the other sink. Resulting from Fig. 2.3-1 and from the model equations, this model is symmetric with respect to the two sinks, i.e. the model equations are symmetric with respect to the two pairs of adsorption/desorption constants. Therefore the model cannot distinguish between the two sinks, and it depends on the initial guesses of the model parameters which of the two pairs of constants describes the "fast" and which the "slow" sink. Because of the greater number of regressed parameters, the "two-sink" model is more easily subject to "over-parameterization" (i.e. the parameters are not fully independent of each other) than the simpler "one-sink" model. This difficulty can be overcome by an appropriate selection of the number of air samples to be collected and the points in time at which they are collected.

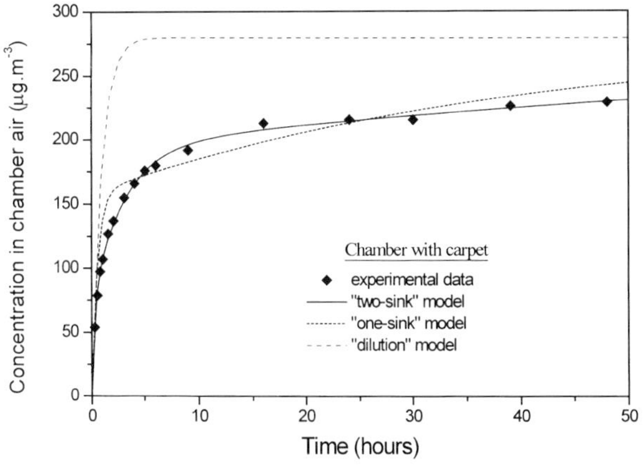

Figure 2.3-3. Concentration of *n*-decane versus time in the test chamber with carpet and curve fitting with the "one-sink" and "two-sink" models (experiment No. 2, dynamic mode).

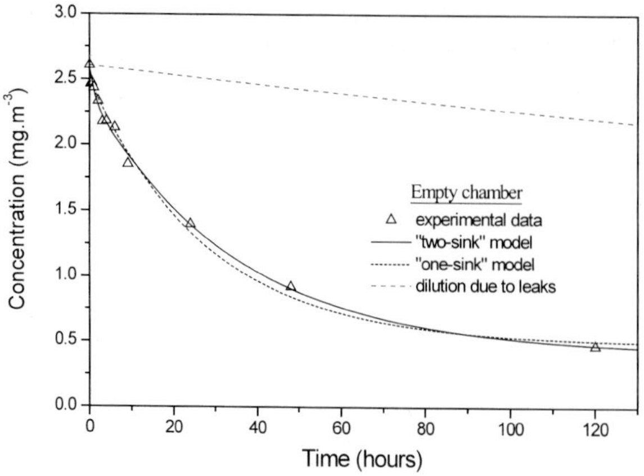

Figure 2.3-4. Concentration of *n*-decane versus time in the test chamber and curve fitting with the "one-sink" and "two-sink" models (experiment No. 3, static mode).

Dynamic adsorption experiments with *n*-decane in the empty 0.45-m^3 glass chamber and in the chamber with a carpet sample have been used to test the performance of the "two-sink" model and compare it with the "one sink" model. The "two-sink" model provided a much better regression of data points than the previous "one sink" model. Figures 2.3-2 and 2.3-3 show the improved fitting of the "two-sink" model to the concentration data in the chamber without and with carpet. Contrary to the "two-sink" model, the "one-sink" model is not able to follow the slow increase of the *n*-decane concentration in the adsorption phase. The performance of the "two-sink" model has been also tested in the case of static adsorption experiments by injecting *n*-decane into a sealed stainless steel chamber (0.28 m^3). Also in this case, the "two-sink" model adheres significantly better to the experimental data than the "one-sink" model as shown in Fig. 2.3-4.

Table 2.3-2 summarizes the main parameters and model estimates for two dynamic experiments (No. 1 and No. 2) and one static experiment (No. 3). It should be emphasized that the discrimination between the two models was based not only on graphic comparisons – which of course is absolutely necessary – but also on the output of statistical tests. These two different ways of discriminating among models are briefly reviewed below.

Comparison of different models. In addition to estimating parameters, another major use of regression analysis is to compare the performance of several possible models that could explain the observed data. The usual goodness-of-fit criteria – deviation plots and summary statistics – are used for this task. A small standard deviation with respect to the size of the experimental values along with a random deviation plot can usually be taken as good evidence of an acceptable fit. To distinguish a series of models, the data can be regressed onto each model in succession. If the number of parameters in each model is the same, the best standard deviation consistent with an uncorrelated model and a random deviation plot can be used as the criterion for choosing the best model. The use of summary statistics alone for such comparisons is not recommended because they give little indication of systematic deviations. Deviation plots are often better indicators of good-

Table 2.3-2. Dynamic experiments (No.1 and No.2) and static experiment (No.3) with *n*-decane in a 0.45 m^3 glass chamber and in a 0.28 m^3 stainless steel chamber respectively.

Dynamic experiments	Experiment No.1 (empty chamber)	Experiment No. 2 (chamber with carpet)	Static experiment	Experiment No. 3 (empty chamber)
Air exchange rate k_2 (h^{-1})	1.06	1.07	Concentration at start ($mg.m^{-3}$)[a]	2.61
Source strength k_1 ($mg\ h^{-1}$)	0.135	0.135		
Duration (h)	48	48	Duration (h)	120
Concentration at end ($mg\ m^{-3}$)	0.247	0.231	Concentration at end ($mg\ m^{-3}$) MVPE[b] (mg)	0.48 0.13
Mass in 1st sink at end (mg)	0.026	0.219	Mass in 1st sink (mg)	0.011
Mass in 2nd sink at end (mg)	0.851	1.395	Mass in 2nd sink (mg)	0.58
Rate constants of 1st sink k_3 (h^{-1}) k_4 (h^{-1})	 930.0 39374	1.548 0.730	Rate constants of 1st sink k_3 (h^{-1}) k_4 (h^{-1}) k_3/k_4 Mass in 1stsink/MVPE	 0.064610 0.765620 0.084 0.082
Rate constants of 2nd sink k_5 (h^{-1}) k_6 (h^{-1})	0.178 0.0032	0.442 0.0164	Rate constants of 2nd sink k_5 (h^{-1}) k_6 (h^{-1}) k_5 / k_6 Mass in 2nd sink/MVPE	 0.025871 0.005001 5.2 4.3
Steady state concentration ($mg\ m^{-3}$)	0.282	0.280	Apparent percentage of saturation of 2nd sink[c]	83
Estimated time to reach 99 % of steady state concentration (h)	975	298		

[a] Mass injected divided by the chamber volume of 0.28 m^3.
[b] MVPE = mass in vapor phase at end of experiment.
[c] Mass in 2nd sink/MVPE divided by k_5/k_6 times 100.

ness of fit than summary statistics because each data point is checked separately for its adherence to the regression model.

If the models being compared have different numbers of regression parameters, a slightly different approach is needed to compare goodness of fit. Here, the individual regression analyses have different numbers of degrees of freedom, defined as the number of data points (n) less the number of parameters (p). The difference in degrees of freedom must be accounted for when using a summary statistic to compare the applicability of the two models. A statistical test that can be used in such situations employs the extra sum of squares F-test (Haswell, 1992). This involves

$$F(p_2 - p_1, n - p_2) = \frac{(S_1 - S_2)/(p_2 - p_1)}{S_2/(n - p_2)} \tag{7}$$

where S_1 and S_2 are the residual error sums from the regression analyses of the same data onto the models 1 and 2, p_1 and p_2 now represent the respective numbers of parameters, and the subscripts refer to the specific models. To use Eq. (7), model 2 should be a generalization of model 1. To distinguish between the two models we should compare the value of the $F(n, p)$ ratio in Eq. (7) with the corresponding entry (n, p) in the F tables at a certain percentage value of confidence level α (e.g., 95 %). If this latter value is much smaller than the former (e.g., the experimental value), then the second model gives the best fit at the α level of confidence. If the contrary happens, this means that model 2 does not reduce the variance significantly, and consequently we can still apply model 1 for the description of the emission data. As this model has fewer parameters, over-parameterization is also less probable.

In our case study, the "two-sink" model fitted most experimental data very efficiently. However, it is not supposed to provide a complete physical description of the real adsorption process(es). Increasing the number of regression parameters in any equation improves its adherence to the data. However, the model is supposed to be a better approximation to the more complex reality. This reality may include adsorption sites with a distribution of adsorption enthalpies instead of simply two distinct values, and may also include diffusion processes within the sorbing material. A further development of the "two-sink" model could take into consideration the transfer of molecules from the "faster" (more superficial) to the "slower" (deeper) sink sites. The model in its present form can be used to estimate four unknown parameters, i.e. the adsorption and desorption rate constants for each of the two sinks.

Desorption experiments are crucial for the validation of adsorption models and in particular for confirming the estimates of the adsorbed masses. Two desorption experiments have been carried out in order to test the reliability of the "two-sink" model for estimating the masses in the sinks. One of the desorption tests followed dynamic adsorption of *n*-dodecane, the other the static adsorption of *n*-decane. The results showed that: (1) the masses deposited into the sinks are released very slowly; (2) the "two-sink" model estimates the masses in the sinks reasonably well; therefore, this model may be considered a satisfactory tool for estimating chamber (or test material) sinks. It coherently confirmed the existence of two sinks with different saturation rates, also giving reasonable estimates of the relative parameters.

Overall, the "two-sink" model provided a reasonable description of the mass balance of the adsorption-desorption experiments performed and reproducible parameter estimates. However, the reported results have given evidence that the reproducibility of parameter estimates for adsorption of a compound on chamber walls may critically depend on residues of the compound in the sinks from earlier experiments. It highlighted the very long time needed to attain equilibrium between the vapor phase and the adsorbed phase and vice versa. This effect was particularly pronounced for adsorption on chamber walls. The chamber sink, compared with the adsorption on three test materials, showed that, in general, the former will introduce only a minor or negligible bias in the measurement of the latter, provided that a sufficiently large surface of the test material is used.

In view of the largely differing time constants of adsorption/desorption processes on/from indoor materials and the very long test times required for their complete characterization, the question which adsorption/desorption data (which test mode, which test concentrations, which test times) will be most useful for assessing the impact of these processes on human exposure needs more careful consideration.

However, the "two-sink" model as well as other existing adsorption (sink) models do not seem to be able to describe the strong asymmetry between the adsorption/desorption of VOCs on/from indoor surface materials (the desorption process is much slower than the adsorption process). Diffusion combined with internal adsorption is assumed to be capable of explaining the observed asymmetry. Diffusion mechanisms have been considered to play a role in interactions of VOCs with indoor sinks. Dunn and Chen (1993) proposed and tested three unified, diffusion-limited mathematical models to account for such interactions. The phrase "unified" relates to the ability of the model to predict both the ad/absorption and desorption phases. This is a very important aspect of modeling test chamber kinetics because in actual applications of chamber studies to indoor air quality (IAQ), we will never be able to predict when we will be in an accumulation or decay phase, so that the same model must apply to both. Development of such models is underway by different research groups. An excellent reference, in which the theoretical bases of most of the recently developed sorption models are reviewed, is the paper by Axley and Lorenzetti (1993). The authors proposed four generic families of models formulated as mass transport modules that can be combined with existing IAQ models. These models include processes such as equilibrium adsorption, boundary layer diffusion, porous adsorbent diffusion transport, and convection-diffusion transport. In their paper, the authors present applications of these models and propose criteria for selection of models that are based on the boundary layer/conduction heat transfer problem.

If a model is statistically based (such as the "two-sink" model), sensitivity analysis of the model with respect to its parameters is of crucial importance in the model validation procedure. Sensitivity analysis is aimed at the identification of the most influential system parameters, which could be proven very useful, particularly, if the model is to be used as a part of an IAQ model. This is because any extrapolation errors generated by the model can be propagated during an IAQ simulation and exposure modelling. For illustrative purposes, the sensitivity of the "two-sink" model is tested with respect to the source strength k_1. For this test case the concentration data concerning the emission of hexanedioic acid bis(2-methyl propyl) ester from gypsum board over a 60-h time period have been used (Colombo et al., 1989). A small variation of k_1 in the range from 0.1 to 0.148 mg/h causes a nearly 68 % increase in the value of the estimated concentration at steady state. This means that the extrapolation by the model beyond the 60-h data range reflects the sensitivity of the model to its parameter k_1. In other words, if we are uncertain about the value of k_1, even a small degree of variation in k_1 will be translated into a significant variation of the extrapolated concentrations.

In the context of the model validation process, another important point to address is how two models with very similar or even identical predictive capabilities in a certain range of experimental data are differentiated with respect to their extrapolation power beyond this data range. For the experimental data of the gypsum board (Colombo et al., 1989), this can be demonstrated by comparing the concentrations at steady state as they are estimated by Dunn's "full" model (Dunn and Tichenor, 1988) and the "two-sink" model. These two models behave identically in the data range of the given experiment. The full agreement is valid also for the rate constants. The parameter vector

(k_1, k_3, k_5, k_6) of the "two-sink" model corresponds very well to the parameter vector (k_1, k_5, k_3, k_4) of the "full" model for the experimental data range of 60 h. Nevertheless, a substantial discrepancy is observed for the extrapolated concentrations at steady state. These values for the "two-sink" and the "full" model are 1.2789 mg m^{-3} and 0.24873 mg m^{-3}, respectively. The difference between the extrapolated concentrations is attributed to the fifth parameter of the "two-sink" model which cannot be accounted for by the "full" model.

A further step in the model validation process is to assess the stability of the model parameter estimates over the sample space. If the model has been constructed using observations taken over a long time span, one can test the stability of the estimated parameters by fitting the model to data concerning shorter time spans and determining the pattern of successive estimates of the parameters. If the estimated model parameters were not stable and, for example, showed trend-like patterns, use of the parameter estimates obtained from fitting all the data for predicting concentrations would be unwise.

An example is given below concerning the stability test of parameter estimates over the sample space for the "two-sink" model by employing the "half and half" approach. This example refers to a static experiment designed to study the adsorption of *n*-decane in a 0.28-m^3 stainless steel chamber over a 770-h period. It was observed that, if we perform estimations starting with employing the full data set and repeating the procedure by leaving out successively 7, 12, 17 or 18 observations (half data of observations), we obtain a largely fluctuating pattern for the equilibrium concentration. The equilibrium concentration values varied from 1.236 µg m^{-3} to 239.07 µg m^{-3}.

From the above it becomes evident that for any model validation study a validity range and validity conditions should be specified, otherwise the statement that it is "validated" can be very confusing. Without specifying the validity range of a model, the whole validation process becomes meaningless. This should be kept in mind for those wishing to study the long-term effects of indoor sources and sinks by developing and validating models using test chamber data over a limited time period.

Guo (1993), in his state-of-the-art paper on validation of source and sink models, observed that few source and sink models have been validated. He outlined five major problem areas and suggested three levels of model validation: (a) to check the agreement between the model and a single set of observations, (b) to check the agreement between the model and multiple set of observations, and (c) to perform the so-called "scale-up" verification of the model. The first level of the model validation process follows the end of the model building process and is required to test the model performance over just a single experimental data set. If the model survives this step then its performance should be further tested over some more experimental data sets (2nd level of validation). This step, apart from being decisive for the model's survival, also offers the possibility either to fine-tune the parameters or to find correlations between the values of the model parameters and the test conditions. Whereas the first two levels of the model validation process are quite obvious and almost straightforward, the third one, concerning the scale-up problem, is the most troublesome in the validation of source and sink models. This because the scale-up verification requires data from either real buildings or large test chambers to effectively test the applicability of test chamber results to real building conditions. Concerning scale-up verification, while the empirical models are useful, they do not provide information necessary to scale the chamber data to buildings nor do they provide information necessary to understand the processes controlling emissions. Mass-transfer-based source emission models can overcome many of the disadvantages of the empirical models. Mass-transfer-based models allow separation of the source and environmental factors. This allows the model parameters developed from chamber experiments to be scaled to buildings. Sparks et al. (1996) developed a mass transfer model for gas-phase-limited mass transfer. Their research program aims at developing a generalized mass-transfer model and includes the development of a suitable theory, small chamber experiments to develop a model parameter, test house experiments to provide scaling factors, and test house experiments to verify the theory.

Model validation is perhaps the weakest aspect of indoor air quality model development. Recognizing a critical need, ASTM has published a Standard Guide for Statistical Evaluation of Indoor Air Quality Models (ASTM D 5157) that provides quantitative tools for evaluation of indoor air quality models. These tools include statistical formulae for assessing the general agreement between predicted and measured values as well as for evaluating bias. The guide also proposes specific ranges of values for various statistical indicators that can be used in judging model performance.

The case study included in this Chapter provided the reader with a real working example concerning the modeling of test chamber kinetics. The issues of both model building and model validation were tackled and thoroughly discussed. As sources and sinks are basic components of indoor air quality (IAQ) modeling, and considerations of sources and sinks are often linked, the discussion included in this Chapter mostly focused on how these components are combined to model the test chamber kinetics. Concerning model validation, in this case study only the first two levels of Guo's model validation process have been applied. However, the importance of also performing the third level of model validation was recognized, this being absolutely necessary if one wishes to extend the model's results obtained from test chamber experiments to real buildings.

References

Axley J. W. (1991): Adsorption modelling for building contaminant dispersal analysis. Indoor Air, *2*, 147–171.

Axley J.W. and Lorenzetti D. (1993): Sorption Transport Models for Indoor Air Quality Analysis. Proceedings of Modeling of Indoor Air Quality and Exposure, ASTM STP 1205, N. L. Nagda (ed), 105–127.

Clausen P.A. (1993): Emission of volatile and semivolatile organic compounds from waterborne paints – The effect of the film thickness. Indoor Air, *3*, 269–275.

Colombo A., De Bortoli M., Pecchio E., Schauenburg H., Schlitt H. and Vissers H. (1989): Assessing the emission of organic compounds from building and furnishing materials via small test chambers. EC - European Commission, EUR Report 12220 EN, Luxembourg.

Colombo A., De Bortoli M, Knöppel H., Pecchio E. and Vissers H. (1993): Adsorption of selected volatile organic compounds on a carpet, a wall coating, and gypsum board in a test chamber. Indoor Air, *3*, 276–282.

Colombo A., Jann O. and Marutzky R. (1994): The estimate of the steady state formaldehyde concentration in large chamber tests. Staub-Reinhaltung der Luft, Springer-Verlag, Vol. 54, 143–146.

De Bortoli M, Knöppel H., Colombo A. and Kephalopoulos S. (1996), Attempting to characterize the sink effect in a small stainless steel test chamber. Proceedings of Characterizing Sources of Indoor Air

Pollution and Related Sink Effects, Pittsburgh, PA, USA, ASTM STP 1287, B. A. Tichenor (ed), 305–318.

Dunn J.E. and Tichenor B. A. (1988): Compensating for sink effects in emission test chambers by mathematical modeling. Atmos. Environ., 22, 885–894.

Dunn J.E. and Chen T. (1993): Critical evaluation of the diffusion hypothesis in the theory of porous media volatile organic compounds (VOC) sources and sinks. In: Nagda N.L.(ed) Proceedings of Modeling of Indoor Air Quality and Exposure, ASTM STP 1205, 64–80.

Guo Z. (1993): On validation of source and sink models: problems and possible solutions. Proceedings of Modeling of Indoor Air Quality and Exposure, ASTM STP 1205, Nagda N.L. (ed), 131–144.

Guo Z. (1996): Z-30 Indoor Air Quality Simulator, User's Guide. Acurex Environmental Corporation, Research Triangle Park, NC (USA).

Haswell S.J. (1992): Practical Guide to Chemometrics. Dekker, New York.

Salthammer T. (1996): Calculation of kinetic parameters from chambes tests using nonlinear regression. Atmos. Environ., 30, 161–171.

Seber G.A.F. and Wild C.J. (1989): Nonlinear Regression, Wiley, New York.

Sparks L.E., Tichenor B.A., Chang J. and Guo Z. (1996): Gas-phase mass transfer model for predicting volatile organic compound (VOC) emission rates from indoor pollutant sources. Indoor Air, *6*, 31–40.

Tichenor B.A., Guo Z., Dunn J.E., Sparks L.E. and Mason, M.A. (1991): The interaction of vapor phase organic compounds with indoor sinks. Indoor Air, *1*, 23–35.

Tichenor B.A., Guo Z. and Sparks L.E. (1993): Fundamental Mass Transfer Models for Indoor Air Pollution Sources. Indoor Air, *3*, 263–268.

3 Release of Organic Compounds from Indoor Materials

3.1 Occurrence of Volatile Organic Compounds in Indoor Air

Stephen K. Brown

3.1.1 Introduction

The types of VOCs and their concentrations in indoor air were presented previously (Brown et al., 1994). This information was gathered by stratifying measurements according to the class of building (dwelling, office, school, hospital), building age (new, established) or presence of occupant complaint. Measurements were pooled to derive VOC concentration distributions for each stratum, and mean and 90th percentile concentrations were estimated. This led to a number of findings which will be briefly reiterated here. However, the prime focus will be to identify and present new findings on VOCs occurring in indoor air and the major factors associated with their presence and occupant exposure. Particular attention will be given to the presence in indoor air of:

- VOCs from building occupants,
- VOCs from microbial sources,
- VOCs from automobile exhausts,
- VOCs from new construction and renovations, and
- major VOCs and total VOC (TVOC) levels from large studies of established buildings.

Other sections of this book will discuss VOC emissions from specific sources, some common to the above, and, to avoid duplication, the present discussion will have as its focus the occurrence of these VOCs in indoor air.

3.1.2 Indoor Air VOCs

The previous review found that most measurements had been made in established dwellings, with quantitative information available for 80 compounds. Nineteen of these VOCs exhibited weight-averaged (weighted by number of measurements) geometric mean (GM) concentrations above 5 $\mu g/m^3$, and these were considered the predominant indoor air VOCs (Table 3.1-1). With the exception of ethanol, none exceeded a GM concentration of 50 $\mu g/m^3$. Similar VOCs occurred in public buildings, and comparative concentrations of VOCs in dwellings and public buildings existed for 36 compounds; for most of these, the GM concentrations in dwellings significantly exceeded those in offices by a factor of 2 or more. This was also found for TVOC concentrations. These concentrations were derived without considering sampling and analytical differences from different studies when estimating averages. These differences must be significant to the measurements made individually (Hodgson, 1995), but it was assumed that data pooling would average the differences and allow strata comparison. Results in Table 3.1-2 are presented with this limitation and show that TVOC concentrations in established dwellings were much higher than those in established public buildings. This might reflect lower ventilation rates in the former or different sampling strategies (e.g. personal versus area sampling). A specific explanation was lacking.

Table 3.1-2 also shows that TVOC concentrations were much greater in new dwellings and offices (less than three months old) than established buildings. Similarly, concentrations of individual VOCs were often an order of magnitude greater in new buildings,

Table 3.1-1. Derived concentrations (weight averaged geometric means and 90 percentiles) for predominant VOCs in the indoor air of established buildings (Brown et al., 1994).

Compound	Building		Derived concentrations (μg/m^3)		
	No.	Type[a]	No. of meas.	WAGM	90 %ile
n-Decane	852	D	1085	5	20
Ethylbenzene	1225	D	1867	5	22
Diethyl ketone	1	S	12	6	26
o-Xylene	891	D	1518	6	25
1,2,4-Trimethylbenzene	619	D	619	6	23
Nonanal	15	D	15	7	27
Tetrachloroethylene	1276	D	1919	7	28
m-Methylethylbenzene	3	O	168	8	33
p-Dichlorobenzene	1256	D	1881	8	32
n-Nonane	586	D	592	5	19
Ethyl acetate	248	D	302	8	31
Ethyl acetate	1	S	12	10	42
Benzene	1388	D	2171	8	34
1,2,3-Trimethylbenzene	3	O	152	9	36
Phenol	1	S	5	9	36
Chloroform	1	O	20	10	40
1,2-Dichloroethylene	35	D	35	11	47
n-Hexane	8	O	26	12	47
n-Hexane	602	D	656	5	19
Acetic acid	1	S	5	12	50
Camphene	44	MD	44	14	55
Dichloromethane	41	D	101	17	70
m-& *p*-Xylene	945	D	1587	18	73
Limonene	584	D	584	21	85
Methylethylketone (MEK)	262	D	316	21	85
1,1,1-Trichloroethane	937	D	1580	24	96
Butyric acid	1	S	5	25	100
Acetone	32	D	86	32	130
Methanol	1	S	11	29	118
Toluene	712	D	792	37	150
Ethanol	39	D	39	120	490

[a] D = dwelling; O = office; S = school; H = hospital (includes nursing homes); MD = mobile dwelling.

Table 3.1-2. TVOC concentrations derived for different building types (Brown et al., 1994).

Building type	No. of buildings	No. of meas.	Derived TVOC concentration (μg/m³)	
			WAGM	90 %ile
Dwelling				
established	1081	2478	1130	4630
new	33	33	4500	18400
complaint	14	28	520	2130
Office				
established	60	384	180	740
new	1	100	4150	17000
complaint	51	331	490	2000
School				
established	21	27	70	290
complaint	9	56	480	1970
Hospital				
established	2	8	410	1680
complaint	6	51	2080	8500

though they were generally less than 100 μg/m³. Most VOCs quantified in new buildings also occurred in established buildings, but additional VOCs in new buildings were 2-ethoxyethylacetate, *n*-butanol, α-pinene, undecane, dodecane, tridecane, tetradecane and 2-propanol.

VOC and TVOC (Table 3.1-2) concentrations had been measured in relatively few buildings with occupant complaints of poor indoor air quality (IAQ), limiting their interpretation. Higher concentrations were observed in some complaint buildings, but not others. In view of the potentially multi-factorial nature of these complaints, better controlled studies (and TVOC definitions) are necessary to test the relationship between VOC concentrations and occupant complaints (Norbäck et al., 1990; 1995).

3.1.3 VOCs from Building Occupants

VOCs from occupants include any compounds emitted metabolically (bioeffluents) or any compounds emitted from their persons (personal products) or clothing. These VOCs are expected to bear a close relationship to occupant density, while building-related pollutants will not.

Shields et al. (1996) observed that concentrations of two VOCs – decamethylcyclotetrasiloxane (D4) and decamethylcyclopentasiloxane (D5) – were significantly greater in office buildings with higher occupant densities, with GM concentrations of 9–10 and 26–40 μg/m³, respectively. Similar compounds may be seen in GC column phase artefacts, often ignored by analysts. However, as common constituents of personal deodorant products, they are occupant-related VOCs. Higher (though low) levels of C12 to C16 *n*-alkanes were also observed for higher occupant densities, but these have many potential sources.

Wang (1975) determined indoor air concentrations of several bioeffluents in a mechanically ventilated 2400-m^3 lecture theatre containing 225–389 occupants. These were [with range of average concentrations (μg/m^3)]: ethanol (43–84), acetone (49–70), methanol (37–72), butyric acid (42–54), acetic acid (21–24), phenol (15–18), amyl alcohol (13–27), diethyl ketone (6–20), ethyl acetate (9–31), toluene (7–36), acetaldehyde (2–8) and allyl alcohol (4–9). Many of these were considered to originate from metabolic breakdown of foodstuffs or from food components.

Exhaled breath is probably the major contributor to VOC bioeffluents. A recent review of previous studies (Fenske and Paulson, 1998) reported the following average exhaled breath concentrations (μg/m^3) at 23 °C and 760 mmHg: acetone (2400), ethanol (1590), isoprene (614), methanol (485), 2-propanol (370), methylethyl ketone (47), hexanal (45) and *n*-pentane (36). However, these might originate from inside the body as endogenous products of metabolism or from outside the body from exogenous sources including the air inhaled. The former will be important to indoor air as it relates to the occupants as VOC sources, although indoor air VOCs could also be catabolized to different compounds which are then exhaled. Exact origins of exhaled VOCs are still largely unknown, although Phillips et al. (1994) assigned origin with 12 normal volunteers based on mean alveolar gradient: the concentration in breath minus the concentration in inspired air. Strong negative gradients indicated ingestion of air pollutants, and these were found for isoprene, tetrachloroethylene, 1,1,1-trichloroethane, benzene and *n*-pentane.

An exhaled breath VOC of concern is benzene. The exhaled breath of smokers exhibited mean benzene concentrations of 14–50 μg/m^3, compared to 1–8 μg/m^3 for non-smokers (Brugnone et al., 1989; Wallace et al., 1987; Reidel et al., 1996), and immediate smoking can elevate breath benzene concentrations to 90–220 μg/m^3 (Wallace, 1989; Brown and Knott, 1997). Assuming a breathing rate of 6 litres/minute and a worst case exhaled breath benzene concentration of 50 μg/m^3, a smoker may exhale 18 μg/h of benzene into building air even where smoking does not occur; for a 20-m^3 room with a ventilation rate of 0.35 h^{-1} and no pollutant sinks, this would lead to an equilibrium benzene concentration of 2.6 μg/m^3. Similar calculations could be made for many of the exhaled VOCs described earlier provided it is known that the compounds are endogenous, e.g., for isoprene at an exhaled breath concentration of 600 μg/m^3, indoor air con-

Table 3.1-3. List of MVOCs used to investigate Swedish problem buildings (Wessén and Schoeps, 1996a).

3-Methylfuran	3-Methylbutan-1-ol	3-Methylbutan-2-ol
2-Pentan-ol	2-Hexanone	2-Heptanone
3-Octanone	3-Octanol	1-Octen-3-ol
2-Octen-1-ol	Geosmine	Dimethyl disulfide
1-Butanol	2-Methylpropan-1-ol	Ethyl isobutyrate
Thujopsene	Karveol	Terpineol
Endoborneol	Fenchone	2-Nonanone
2-Pentylfuran		4-Methylheptan-3-one

centrations up to 30 μg/m^3 are expected; however, with a boiling point of 34 °C this is classified as a very volatile organic compound and may not be generally sampled. This link between exhaled breath VOCs and indoor air exposure has been demonstrated for tetrachloroethylene (TCEE) exhaled by dry-cleaning workers. Aggazzotti et al. (1994) found the GM exhaled TCEE concentrations for the workers and their family members were 5140 and 225 μg/m^3, and control subjects 3 μg/m^3. Corresponding mean indoor air concentrations in homes were found to be 265 and 2 μg/m^3 respectively.

3.1.4 VOCs from Microbial Sources

This is a complex subject which will receive detailed discussion in Chapter 3.7. The purpose here is to consider some microbial VOCs (MVOCs) that have been identified and quantified in buildings.

MVOCs may be present at much lower concentrations than other VOCs and require specific analytical methods (e.g. mass spectrometry in SIM mode) for their measurement. Analytically, this means a list of MVOCs needs to be assembled and specifically sought in indoor air investigations. Wessén and Schoeps (1996a) utilized the list presented in Table 3.1-3 for Swedish "problem" buildings and found total MVOC concentrations of 0.1 – 17.4 μg/m^3 (mean 1.0 μg/m^3) in these buildings. Additional useful MVOC tracers may be 2-ethyl-2-hexenal (Whillans and Lamont, 1995), and Wessén and Schoeps (1996b) nominated three MVOCs and the ratios of their concentrations as being excellent indicators of micro-organisms: 3-methylfuran, dimethyldisulfide and 1-octen-3-ol.

Norbäck et al. (1993) measured 1-week concentrations of 1-octen-3-ol in the bedroom air and personal air of 39 Swedish subjects. The subjects were selected from a random population, but their willingness to participate may have been influenced by their perceptions of sick building syndrome symptoms and poor IAQ at their workplaces. Concentrations of 1-octen-3-ol in bedroom air and personal air were 2 and 6 μg/m^3 respectively (arithmetic means) with concentration ranges of <1–50 and <1 – 140 μg/m^3 respectively. A significant correlation ($p < 0.05$) was observed between the occurrence of facial itching/rashes and the concentration of 1-octen-3-ol.

Fungal colonization of fibreglass duct lining is an issue of concern for IAQ, but complexities in fungal species and nutrient contribution led to difficulties in assigning MVOCs in indoor air to this source. Hexanol, isooctanol and 6-methyl-1-heptanol (Ahearn et al., 1996), and 2-ethyl – 1 – hexanol (Ezeonu et al., 1994) have been identified as major VOCs from this source.

3.1.5 VOCs from Automobile Exhausts

Indoor air exposure to automobile exhausts can arise in three ways:

- from contamination of air used to ventilate a building,
- from automobile garages within or attached to a building, and
- from occupancy within an automobile.

This last factor requires that automobile interiors be considered indoor air environments, which is not unreasonable since urban populations can spend an average of 7 % of their time (1.7 hours per day) in this environment (Jenkins et al., 1992). Also, it is a relevant component to consider in an individual's total exposure to air pollutants that has been overlooked in some studies (Fishbein, 1992).

Many indoor air VOCs can arise from both automobile exhaust and indoor materials, making it difficult to evaluate the contribution from the former. VOC concentration patterns have been found useful for this purpose. Lagoudi et al. (1996) evaluated four offices from urban areas and two offices from rural areas. VOCs for which the indoor/outdoor concentration ratio was less than 1.3 were classified as outdoor air pollutants; the pattern of concentrations of these VOCs showed good correlation between indoor and outdoor air for urban areas but not for rural areas. For these compounds (mainly branched alkanes, *o*-xylene and trimethylbenzenes), they estimated that outdoor air contributed to 25–68 % of indoor air pollution for urban buildings and 17–29 % for rural buildings. In a study of two urban office buildings, Ekberg (1993) estimated that outdoor air contributed a similar VOC load to that from construction materials or office equipment. Perry and Gee (1994) reviewed changes in fuel components and the effect this would have on outdoor levels of air toxins, especially benzene and 1,3-butadiene. They predicted increases in outdoor air pollutants with a resultant increase in indoor air VOC pollution, especially in naturally ventilated buildings.

Several studies have demonstrated that ingress of petrol fumes from attached garages led to higher indoor air levels of these VOCs. Gammage et al. (1988) found that petrol VOC concentrations were four times higher in eight dwellings with attached garages than

Table 3.1-4. Benzene concentrations measured in automobile interiors during transit.

Country	Benzene concentration (μg/m^3)	Sample description	Ref.
USA	14	Median for car in mixed traffic conditions	Lawryk et al. (1995)
	45	Mean for car with faulty carburettor	
Germany	50	Average for car in mixed traffic conditions	Fromme (1995)
Korea	31	Mean for car in urban traffic	Jo & Choi (1996)
	18	Mean for car in suburban traffic	
	20	Mean for bus in urban traffic	
	12	Mean for bus in suburban traffic	
Taiwan	146	Median for bus in urban traffic	Chan et al. (1993)
	335	Median for motorcycle in urban traffic	
Australia		Mean for catalyst-equipped car	Duffy & Nelson (1996)
	153	– urban traffic	
	8	– freeway traffic	
		Mean for pre-catalyst car	
	153	– urban traffic	
	20	– freeway traffic	

in dwellings without attached garages. Cases confirming this contribution have been reported (Brown, 1996), and have demonstrated the importance of the presence of a car and fuel storage containers in the garages (Levsen et al., 1996) as well as the presence of a door between the spaces. Design factors to minimize the effect appear to have received little or no study, although greater ventilation of garage spaces to outdoors has been suggested (Furtaw et al., 1993).

Significantly higher concentrations of benzene and other aromatic hydrocarbons have been found in enclosed automobiles in city traffic than in building air. Table 3.1-4 presents a summary of results for benzene from several studies. Measured benzene concentrations vary widely due to:

- the level of ambient pollution, probably as affected by traffic flow,
- the type of vehicle (e.g. car compared to bus), and
- mechanical faults leading to petrol fumes in vehicle interior.

The high benzene concentrations observed for bus travel in Taiwan reflected a high benzene content in petrol and congested, slow moving traffic. For several of the other studies, the proportion of daily benzene exposure from car or bus travel in the absence of industrial exposure was estimated to be approximately 10–30 % (Adlkoffer et al., 1993). The exposure of individuals to petrol-derived VOCs during automobile travel in the United States was estimated to range from 15 to 60 % of daily exposure (Weisel et al., 1992).

3.1.6 VOCs from New Construction or Renovation

This area was discussed previously (Brown et al., 1994), an arbitrary period of three months after construction being adopted for classifying buildings as new. Using this classification, new buildings were found to exhibit significantly higher TVOC (Table 3.1-2) and VOC concentrations than established buildings. Also, similar compounds were found in new and established buildings. The period after construction or renovation over which concentrations of VOCs remain elevated is an important factor for exposure of occupants and for pollutant control strategies. Discussion will focus on this aspect.

The rate of decay of VOC concentrations in new or renovated buildings will be a difficult factor to generalize about since it will depend on:

- physical properties of the VOCs (volatility, molecular volume, polarity),
- the types and amounts of materials used,
- physical conditions in the building, and
- introduction of VOC sources by occupants (Farant et al., 1992).

However, many case studies have shown the decay can occur over periods up to several months to years for some VOCs (Gustafsson, 1992). One study of six low-energy houses, which remained unoccupied and unfurnished, found TVOC concentrations halved over periods of 2–6 months.

Recent comprehensive studies of buildings have confirmed that VOC and TVOC concentrations decay for periods of several months or longer. Englund and Harderup

(1996) carried out measurements over 10 months after commissioning of a new three-storey apartment building. While TVOC concentrations reduced by approximately tenfold, most of this reduction occurred in the first three months. Fewer VOCs were quantifiable by three months, but some VOCs persisted with only small reductions in concentrations over three months. These were aldehydes (particularly hexanal) and terpenes; sources of these VOCs were not identified.

Crump et al. (1997) determined VOC concentrations in four unoccupied test houses during the period between two months and two years after completion of construction and furnishing. TVOC concentrations declined rapidly in the first six months (from 5000–9000 μg/m^3 to approximately 500 μg/m^3), followed by a more gradual decline to approximately 200 μg/m^3 after 16 months. Many materials from the houses were examined in emission test chambers. While most materials were found to be potential sources of VOCs in the buildings, it was difficult to link the chamber emission decay behaviors to those observed in the buildings. For example, most products such as paints and adhesives exhibited the highest emission rates, but these declined very rapidly in the 24 h after application, a much faster rate than observed in the buildings. The potential for sink adsorption/re-emission from surfaces and the concurrent emissions from several sources were considered to be confounding factors.

Overall, it is apparent that VOC emissions following new construction or renovation should be considered to occur predominantly in the first six months but to persist for longer periods (perhaps years) depending on the VOC and the physical properties of its source material.

3.1.7 Major VOCs and TVOC Levels from Extensive Studies of Buildings

Tables 3.1-1 and 3.1-2 present the VOCs and TVOC levels found in established buildings from a previous review of published studies. The finding that TVOC levels in established dwellings were higher than those in public buildings was unexpected and not readily explained. Several extensive studies of VOCs in buildings have been carried out in recent years. These will be reviewed for their consistency with earlier findings.

Brown et al. (1994) listed 80 VOCs for which quantitative data were available for established dwellings (Table 3.1-5); the major 29 of these compounds (from all types of established buildings) are presented in Table 3.1-1. Similar VOCs have been found in dwelling surveys from Canada (Otson et al., 1994), England (Brown and Crump, 1996), Denmark (Harving et al., 1992), United States (Lindstrom et al., 1995), Japan (Park et al., 1996), China (Li et al., 1996) and Australia (Brown et al., 1996). Established office surveys have also found similar VOCs to those shown in Table 3.1-5, although with some extra compounds related to building materials and occupant activities:

Daisey et al. (1994)	2-butoxyethanol, 2-propanol
Ekberg (1994)	trimethyl-1,3-pentanediol diisobutyrate (TXIB), dibutyl phthalate, phenoxyethanol, Texanol®, di-*tert*-butylcresol

Table 3.1-5. Summary of WAGM concentrations (μg/m^3) derived for VOCs in established dwellings (Brown et al., 1994).

<1	1–<5	5–<10	10–<20	20–<50	>50
i-Amyl alcohol	Butanal	Benzene	Camphene	Acetone	Ethanol
Bromodichloromethane	2-Butanone	*n*-Decane	1,2-Dichloroethylene	Limonene	
1-Butanol	2-Butanol	*p*-Dichlorobenzene	Dichloromethane	Toluene	
n-Butylbenzene	*n*-Butyl acetate	Ethyl acetate	*m*-& *p*-Xylene	1,1,1-Trichloroethane	
Chlorobenzene	Carbon tetrachloride	Ethylbenzene	(*o*-& *m*-& *p*-Xylene)		
o-Dichlorobenzene	Chloroform	Nonanal			
m-Dichlorobenzene	Cyclohexane	Tetrachloroethylene			
1,2-Dichloroethane	1,1-Dichloroethene	1,2,4-Trimethylbenzene			
1,2-Dichloropropane	*n*-Dodecane	*o*-Xylene			
Diethyl ether	Ethylene dibromide				
Dimethylcyclopentane	2-Ethyl-1-hexanol				
1,4-Dioxane	*n*-Heptane				
n-Hexadecane	Hexanal				
1-Methylnaphthalene	*n*-Hexane				
p-Methyl-*i*-propylbenzene	Methylcyclohexane				
Naphthalene	Methylcyclopentane				
N-Nitrosodiethylamine	*o*-& *m*-& *p*-Methyl ethyl benzene				
N-Nitrosodimethylamine	2-& 3-Methyl hexane				
N-Nitrosopyrrolidine	3-Methyl pentane				
β-Pinene	*n*-Nonane				
1,2,3,5-Tetramethylbenzene	*n*-Octane				
1,2,4,5-Tetramethylbenzene	*n*-Pentadecane				
1,2,3-Trichlorobenzene	α-Pinene				
1,2,4-Trichlorobenzene	*n*-& *i*-Propylbenzene				
1,3,5-Trichlorobenzene	Styrene				
1,2,3-Trimethylbenzene	Tetradecane				
4-Methyl-2-pentanone	Trichloroethylene				
	1,3,5-Trimethylbenzene				
	n-Undecane				
	α-Terpinene				

Lagoudi et al. (1996)	2-methylbutane, 2-methyl-1,3-butadiene, methylpentane (outdoor air sources probable)
Bluyssen et al. (1996)	1-ethoxy-2-propanol, 2-butoxyethanol, 2-phenoxyethanol, 2-(2-ethoxyethoxy) ethanol, cyclohexanone, benzaldehyde, decanal, acetic acid butyl (or ethyl) ester, 2-(2-butoxy-ethoxy) ethyl acetate, benzoic acid, dodecanoic acid

Many of these additional VOCs are compounds identified from emission tests of paints and other building materials; it is assumed their appearance in recent surveys has been the result of greater analytical focus on such compounds rather than a change in the VOC mixture present in established buildings. While these surveys had been carried out in established buildings (usually some years old), the finding of paint-related VOCs may be related to a regular presence of new paints in such buildings. Brown and Crump (1996) observed that 20 % of a dwelling population sampled monthly for two years had some painting activity in any month; this activity more than doubled the monthly average TVOC concentration, with the elevation persisting for 1–2 months.

There is some utility in having a definitive list of VOCs relevant to indoor air exposure. For example, analytical parameters can be tuned to suit defined compounds and a database of toxicological effects can be developed. The European Commission (1994) has recommended such a list (Table 3.1-6) based on field data, emission studies and frequency of presence in indoor air. This includes many of the VOCs from Table 3.1-5 and several of the extra compounds described above.

Table 3.1-6. List of VOCs frequently present in indoor air (European Commission, 1994).

Chemical compound	CAS-no.	Frequency[a]
Acetaldehyde	75–07–0	5
Acetone	67–64–1	6
Benzaldehyde	100–52–7	6
Benzene	1076–43–3	7
1-Butanol	71–36–3	7
2-(2-Butoxyethoxy)ethanol	112–34–5	4
Butyl acetate	123–86–4	5
Butylglycol	111–76–2	5
Δ^3-Carene	13466–78–9	5
Cumene	98–82–8	4
n-Decane	124–18–5	7
Dibutylphthalate	84–74–2	6
p-Dichlorobenzene	106–46–7	3
n-Dodecane	112–40–3	7
2-Ethyl-1-hexanol	104–76–7	5
Ethyl acetate	141–78–6	8
Ethylbenzene	100–41–4	7
Formaldehyde	50–00–0	6
Heptanal	111–71–7	5
n-Heptane	142–82–5	7
Hexanal	66–25–1	7

Table 3.1-6. (Continued)

Chemical compound	CAS-no.	Frequency[a]
n-Hexane	110–54–3	7
iso-Butyl acetate	110–19–0	6
Limonene (+)	138–86–3	6
α-Methylstyrene	98–83–9	5
Nonanal	124–19–6	5
n-Nonane	111–84–2	7
n-Octane	111–65–9	7
1-Octanol	111–87–5	5
n-Pentadecane	629–62–9	6
1-Pentanol	71–41–0	5
Phenol	108–95–2	4
α-Pinene	80–56–8	7
1,2-Propanediol	57–55–6	5
2-Propanol	67–63–0	4
Propylbenzene	103–65–1	6
Styrene	100–42–5	7
Tetrachloroethylene	127–18–4	7
n-Tetradecane	629–59–4	6
Toluene	108–88–3	7
Trichloroethylene	79–01–6	3
1,1,1-Trichloroethane	71–55–6	4
n-Tridecane	629–50–5	7
1,3,5-Trimethylbenzene	108–67–8	7
1,2,4-Trimethylbenzene	95–63–6	7
1,2,3-Trimethylbenzene	526–73–8	7
TXIB	6846–50–0	5
n-Undecane	1120–21–4	7
Pentanal	110–62–3	5
o-Xylene	95–47–6	7

[a] Numerical indicator of the frequency of presence in indoor air.

A summary of studies of TVOC concentrations in different types of buildings, predominantly established dwellings and offices, is presented in Table 3.1-7. Statistical comparison of this data is not possible since differences in analytical and sampling procedures must have an impact on the measured concentrations. Generally, less difference is observed than found previously between TVOC concentrations in established dwellings and offices, mean concentrations tending to be less than 500 $\mu g/m^3$ for both building classes.

3.1.8 Conclusions

It is clear that VOCs have a multitude of sources in buildings and that VOC concentrations will vary according to building characteristics (e.g. age, urban location, moisture control), occupant factors (e.g. occupancy loading, occupant-related sources, occupant

Table 3.1-7. Total VOC (TVOC) concentrations in different types of buildings.

Type of building	No. of meas.	TVOC[a] concentration (μg/m^3)		Ref.
		Ave. method	Result	
Established dwellings	36	median	700	Harving et al. (1992)
Established dwellings	3000	mean	400	Brown & Crump (1996)
Established dwellings	142	mean	220	Brown & Crump (1996)
Complaint dwellings	43	mean	1040	Brown et al. (1996)
Established dwellings	16	range	190–640	Park et al. (1996)
Established dwellings	20	GM	210	Brown (1996)
Complaint dwellings	10	GM	200	Brown (1996)
Complaint offices	29	mean	70	Sundell et al. (1993)
Established offices	~ 25	GM	510	Daisey et al. (1994)
Established offices	5	range	160–350	Ekberg (1994)
Established offices from nine European countries	270	GM	228	Bluyssen et al. (1996)

[a] For most of these studies TVOC was the GC-FID sum of all chromatogram peaks in toluene equivalents; approaches where only a limited number of compound concentrations have been summed (i.e. ΣVOC) have not been included.

activities) and many other factors (e.g. oxidant reactions). Identification of major VOC sources in new buildings and emission control are the clearest strategies for reducing VOCs from new construction and renovation in the first year after source introduction. Reduction of VOC levels in established buildings needs to consider more diverse sources and compounds of more specific concern. This increased focus on specific compounds is becoming apparent in recent building surveys.

References

Adlkofer F., Angerer J., Ruppert T., Scherer G. and Tricker A.R. (1993): Determination of benzene exposure from occupational and environmental sources. In: "Volatile Organic Compounds in the Environment", G. Leslie and R. Perry (eds), Indoor Air International, London, 511–518.

Aggazzotti G., Fantuzzi G., Predieri G., Righi E. and Moscardelli S. (1994): Indoor exposure to perchloroethylene (PCE) in individuals living with dry-cleaning workers. Sci. Total Envir., *156*, 133–137.

Ahearn D.G., Crow S.A., Simmons R.B., Price D.L., Noble J.A., Mishra S.K. and Pierson D.L. (1996): Fungal colonization of fiberglass insulation in the air distribution system of a multi-story office building: VOC production and possible relationship to a sick building syndrome. J. Ind. Microbiol., *16*, 280–285.

Bluyssen P.M., Fernandes E.O., Groes L., Clausen G., Fanger P.O., Valbjorn O., Bernhard C.A. and Roulet C.A. (1996): European indoor air quality audit project in 56 office buildings. Indoor Air, *6*, 221–238.

Brown S.K. (1996): Assessment and control of volatile organic compounds and house dust mites in Australian buildings. Proceedings of Indoor Air '96, Nagoya, Japan, Vol. 2, 97–102.

Brown S.K. and Knott P. (1997): Benzene concentrations in the exhaled breath of pot liners at an aluminium refinery. Proceedings of 16th Annual Conference of Australian Institute of Occupational Hygenists, Albury, AIOH.

Brown S.K., Sim M.R., Abramson M.J. and Gray C.N. (1994): Concentrations of volatile organic compounds in indoor air–a review. Indoor Air, *4*, 123–134.
Brown V.M. and Crump D.R. (1996): Volatile organic compounds. In: "Indoor Air Quality in Homes: Part 1–The Building Research Establishment Indoor Environment Study", Berry R.W. et al. (eds) Building Research Establishment, Watford, 38–66.
Brown V.M., Cockram A.H., Crump D.R. and Mann H.S. (1996): The use of diffusive samplers to investigate occupant complaints about poor indoor air quality. Proceedings of Indoor Air '96, Nagoya, Japan, Vol. 2, 115–120.
Brugnone F., Perbellini L., Faccini G.B. and Pasini F. (1989): Benzene in the breath and blood of normal people and occupationally exposed workers. Amer. J. Ind. Med., *16*, 385–399.
Chan C.-C., Lin S.-H., Her G.-R. (1993): Students' exposure to volatile organic compounds while commuting by motorcycle and bus in Taipei city. J. Air Waste Manag. Assoc., *43*, 1231–1238.
Crump D.R., Squire R.W. and Yu C.W.F. (1997): Sources and concentrations of formaldehyde and other volatile organic compounds in the indoor air of four newly built unoccupied test houses. Indoor Built Envir., *6*, 45–55.
Daisey J.M., Hodgson A.T., Fisk W.J., Mendell M.J. and Brinke J.T. (1994): Volatile organic compounds in twelve California office buildings: classes, concentrations and sources. Atmos. Envir., *28*, 3557–3562.
Duffy B.L. and Nelson P.F. (1996): Exposure to benzene, 1,3-butaidiene and carbon monoxide in the cabins of moving motor vehicles. Proceedings of 13th International Clean Air and Environment Conference, Adelaide, Australia, Clean Air Society of Australia and New Zealand, 195–200.
Ekberg L.E. (1993): Sources of volatile organic compounds in the indoor environment. In: "Volatile Organic Compounds in the Environment", G. Leslie and R. Perry (eds) Indoor Air International, London 573–579.
Ekberg L.E. (1994): Volatile organic compounds in office buildings. Atmos. Envir., *28*, 3571–3575.
Englund F. and Harderup L.E. (1996): Indoor air VOC levels during the first year of a new three-storey building with wooden frame. Proceedings of Indoor Air '96, Nagoya, Japan, Vol. 3, 47–52.
EC–European Commission (1994): European database on indoor air pollutant sources in buildings. Newsletter, December.
Ezeonu I.M., Price D.L., Simmons R.B., Crow S.A. and Ahearn D.G. (1994): Fungal production of volatiles during growth on fiberglass. Appl. Envir. Microbiol., *60*, 11, 4172–4173.
Farant J.P., Baldwin M., de Repentigny F. and Robb R. (1992): Environmental conditions in a recently constructed office building before and after the implementation of energy conservation measures. Appl. Occ. Envir. Hyg., *7*, 2, 93–100.
Fenske J.D. and Paulson S. (1998): Human emissions of volatile organic compounds. J. Air Waste Manag. Assoc. (in press).
Fishbein L. (1992): Exposure from occupational versus other sources. Scand. J. Work Envir. Health, *18*, Suppl. 1, 5–16.
Fromme H. (1995): Significance for the health of the general population of exposure to benzene in traffic. Medline PMID: 7542450 (abstract only).
Furtaw E.J., Pandian M.D. and Behar J.V. (1993): Human exposure in residences to benzene vapors from attached garages. Proceedings of Indoor Air '93, Helsinki, Finland, Vol. 5, 521–526.
Gammage R.B., Higgins C.E., Dreibebbis W.G., Guerin M.R., Buchanan M.V., White D.A., Olerich G. and Hawthorne A.R. (1988): Measurement of VOCs in Eight East Tennessee Homes. Oak Ridge National Laboratory, Tennessee, Report ORNL-6286.
Gustafsson H. (1992): Building materials identified as major sources for indoor air pollutants: a critical review of case studies. Swedish Council for Building Research, Stockholm, Document D10: 1992.
Harving H., Dahl R., Korsgaard J. and Linde S.A. (1992): The indoor environment in dwellings: a study of air-exchange, humidity and pollutants in 115 Danish residences. Indoor Air, *2*, 121–126.
Hodgson A.T. (1995): A review and a limited comparison of methods for measuring total volatile organic compounds in indoor air. Indoor Air, *5*, 247–257.
Jenkins P.L., Phillips T.J., Mulberg E.J. and Hui S.P. (1992): Activity patterns of Californians: use of and proximity to indoor pollutant sources. Atmos. Envir., *26A*, 12, 2141–2148.
Jo W.-K. and Choi S.-J. (1996): Vehicle occupants' exposure to aromatic volatile organic compounds while commuting on an urban-suburban route in Korea. J. Air Waste Manage. Assoc., *46*, 749–754.
Lagoudi A., Loizidon M. and Asimakapoulos D. (1996): Volatile organic compounds in office buildings: 2–Identification of pollutant sources in indoor air. Indoor Built. Envir., *5*, 348–354.

Lawryk N.J., Lioy P.J. and Weisel C.P. (1995): Exposure to volatile organic compounds in the passenger compartment of automobiles during periods of normal and malfunctioning operation. J. Expos. Anal. Environ. Epidem., *5*, 4, 511–531.

Levsen K., Schimming E., Angerer J., Wichmann H.E. and Heinrich J. (1996): Exposure to benzene and other aromatic hydrocarbons: indoor and outdoor sources. Proceedings of Indoor Air '96, Nagoya, Japan, Vol. 1, 1061–1066.

Li Y., Hu J., Liu G. and Mu W. (1996): Determination of volatile organic compounds in residential buildings. Proceedings of Indoor Air '96, Nagoya, Japan, Vol. 3, 601–605.

Lindstrom A.B., Proffitt D. and Fortune C.R. (1995): Effects of modified residential construction on indoor air quality. Indoor Air, *5*, 258–269.

Norbäck D., Torgen M. and Edling C. (1990): Volatile organic compounds, respirable dust, and personal factors related to the prevalence of sick building syndrome in primary schools. Brit. J. Ind. Med., *47*, 733–741.

Norbäck D., Edling C., Wieslander G. and Ramadhan S. (1993): Exposure to volatile organic compounds (VOC) in the general Swedish population and its relation to perceived air quality and sick building syndrome. Proceedings of Indoor Air '93, Helsinki, Finland, Vol. 1, 573–578.

Norbäck D., Björnsson E., Janson C., Widström J. and Boman G. (1995): Asthmatic symptoms and volatile organic compounds, formaldehyde, and carbon dioxide in dwellings. Occup. Env. Med., *52*, 388–395.

Otson R., Fellin P. and Tran Q. (1994): VOCs in representative Canadian residences. Atmos. Envir., *28*, 3563–3569.

Park J.S., Fujii S., Yuasa K., Kagi N., Toyozumi A. and Tomura H. (1996): Characteristics of volatile organic compounds in residences. Proceedings of Indoor Air ''96, Nagoya, Japan, Vol. 3, 579–584.

Perry R. and Gee I.L. (1994): Vehicle emissions and effects on air quality: indoors and outdoors. Indoor Envir., *3*, 224–236.

Phillips M., Greenberg J. and Awad J. (1994): Metabolic and environmental origins of volatile organic compounds in breath. J. Clin. Pathol., *47*, 1052–3.

Reidel K., Ruppert T., Conze C., Scherer G. and Adlkofer F. (1996): Determination of benzene and alkylated benzenes in ambient and exhaled air by microwave desorption coupled with gas chromatography-mass spectrometry. J. Chrom. A., *719*, 383–389.

Shields H.C., Fleischer D.M. and Weschler C.J. (1996): Comparisons among VOCs measured in three types of US commercial buildings with different occupant densities. Indoor Air, *6*, 2–17.

Sundell J., Andersson B., Andersson K. and Lindvall T. (1993): TVOC and formaldehyde, as risk indicators of SBS. Proceedings of Indoor Air '93, Helsinki, Finland, Vol. 2, 579–584.

Wallace L. (1989): The exposure of the general population to benzene. Cell Biol. Toxicol., *5*(3), 297–314.

Wallace L., Pellizxzari E., Hartwell T.D., Perritt R. and Ziegenfus R. (1987): Exposures to benzene and other volatile organic compounds from active and passive smoking. Arch. Envir. Health, *42*(5), 272–279.

Wang T.C. (1975): A study of bioeffluents in a college classroom. ASHRAE Trans., *81*, 1, 32–44.

Weisel C.P., Lawryk N.J. and Lioy P.J. (1992): Exposure to emissions from gasoline within automobile cabins. J. Expos. Anal. Envir. Epidem., *2*, 1, 79–96.

Wessén B. and Schoeps K.-O. (1996a): Microbial volatile organic compounds–what substances can be found in sick buildings? Analyst, *121*, 1203–1205.

Wessén B. and Schoeps K.-O. (1996b): MVOC ratios–an aid for remediation of sick buildings. Proceedings of Indoor Air ''96, Nagoya, Japan, Vol. 3, 557–561.

Whillans F.D. and Lamont G.S. (1995): Fungal volatile metabolites released into indoor air environments: variation with fungal species and growth media. In "Indoor Air: An Integrated Approach", Morawska L., Bofinger N. D. and Maroni M. (eds) Elsevier Science, Oxford, 47–50.

3.2 Emission from Floor Coverings

Kristina Saarela

Emissions from building materials have a profound effect on indoor air quality; more than half of the VOCs in indoor air are derived from materials. Of all the building materials, flooring materials have been of special interest because they form large, thick surfaces which can contribute to high and long-lasting emissions, and these emissions have been the subject of complaints.

3.2.1 Flooring Material Types

Natural wood-based flooring materials range from plain wood boards and parquets to finished, assembled plank parquet composites of various wood species. Cork materials are manufactured from the outer layer of the tree, mainly oak.

Typical synthetic flooring materials are the vinyl floor coverings consisting of qualities varying from hard, wear-resistant PVC tiles and semi-hard calendered products used in public spaces to soft, cushion vinyl floor coverings with good sound-damping properties, used in dwellings. Linoleum materials are traditional, and represent one of the oldest synthetic materials manufactured from linseed oil or equivalent wood-based oil. Linoleum floorings are also wear-resistant and are used both in public premises and dwellings. Wall-to-wall textile carpets have been very popular in public spaces and homes because of their warm looks and good sound-absorbing properties. Rubber flooring materials are extremely wear-resistant, have good antislip properties and are usually used in highly loaded public premises such as traffic terminals.

Flooring materials are usually fixed on the slab by gluing so that they form a composite with the bearing structure, levelling agent, adhesive, and the covering itself. Thus, a flooring material should never in practice be treated as a single emission source, since the emissions depend on all components used in the structure and on the chemical interaction of the whole floor system. Additionally, factors such as humidity in the floor structures and temperature have an impact on the emission behavior of the flooring system and consequently on the flooring material itself. However, this factor is difficult to take into account, e.g. in labelling systems, because of its complex nature involving several manufacturers. Nearly all published emission results only deal with plain flooring material emissions either as factory new or aged for a certain time described in emission testing or evaluation methods (EC – European Commission, 1997; GBR, Trade Standard, 1992; Produktstandard, 1998a; Produktstandard, 1998b; FiSIAQ, 1995). In modelling of indoor air quality and especially in source characterization of problem buildings, the complete system should, however, be considered.

3.2.2 Sources of Chemical Emissions in Flooring Materials

The different types of synthetic materials are produced using manufacturing processes and raw materials specific to each product. Materials consist of, e.g., thermoplastic polymers compounded with process aids, such as stabilizers, viscosity modifiers and cosolvents. Some materials are produced from monomers, oligomers or other low-molecular-weight reactive organic compounds through curing processes such as crosslinking, vulcanization or oxidation. Most materials also require additives such as plasticizers, antioxidants and surface-treating agents and colorants to reach the desired mechanical properties, to be attractive and to have a long performance life. Even natural materials such as wood and cork contain volatile compounds. Typical of these are the etheric oils and other extractives of wood. The natural flooring materials are very seldom used as such, but include surface-treatment agents such as varnishes oils or waxes in order to obtain their looks and performance properties. The prefabricated parquet planks from different wood species are glued together. Synthetic coatings are frequently used in cork products.

Emissions are also generated as a result of decomposition of polymer chain is caused by temperature during the manufacturing process. In products applied on site, the humidity can retard curing reactions, and this is usually followed by the emission of unreacted monomers.

In summary, chemical emissions from newly produced flooring materials are caused by:

- Plasticizers,
- Cosolvents or viscosity modifiers,
- Unreacted monomers,
- Additives such as antioxidants, stabilizers, antistats, and flame retardants,
- Finishing agents, colorants and surface coatings,
- Extractives from wood-based materials,
- Solvents used in paints, varnishes, oils and waxes.

Additives such as antioxidants and stabilizers are technical mixtures of numerous chemical compounds, which in addition to the effective substance also bring along solvents and sometimes processing residues, all of which are potential emission sources in the flooring material. In the late 1980s PVC flooring materials contained plasticizers in which the proportion of VOCs was up to 29 %, viscosity modifiers (up to 100 % VOCs), and stabilizers (up to 54 %) (Saarela, 1989).

Thermoplastic materials and in some cases even other products are exposed to high temperatures during processing, and, depending on the effectiveness of the heat stabilization, some oxidation always occurs, which brings along a potential emission hazard through heat degradation of the material or its components. Materials based on natural oils and manufactured through oxidation/polymerization reactions used as e.g. binding resin in linoleums, may also contain organic decomposition products such as lower-molecular-weight mono- and dicarboxylic acids, ketones, and saturated and unsaturated aldehydes. The process also includes the use of accelerators and inhibitors, which may contain volatile substances as phenol and amines.

On-site construction processes, such as the use of leveling agents, may cause emissions, which degrade flooring materials. The use of curing injection resin in flooring systems may cause emissions when the curing reaction is retarded.

3.2.3 Effect of Material Structure on Emissions

The emission process from a flooring material has been characterized by two principal mechanisms; a diffusion-controlled mass transfer in the material and an evaporation-controlled emission mechanism from the surface. These mechanisms, together with the air velocity and concentration effects over the evaporative surface, are considered the principal mechanisms governing the physical factors affecting the emissions, and models have been developed to describe the effect on indoor air quality (Gunnarsen, 1991; Matthews et al., 1987; Neretnieks et al., 1993; Sparks, 1996).

3.2.4 Sorption

In predicting the impact of flooring materials on indoor air, the adsorption-desorption properties of materials (sink effects) should also be considered, as it has a major effect on the development of the final quality of indoor air (Kirchner et al., 1995; Kjaer and Nielsen, 1991; Bluyssen and van der Wal, 1994; Tirkkonen and Saarela, 1997; Jörgensen et al., 1993). The magnitude of the adsorption/desorption effect is controlled by van der Waals forces. The impact of the effect on indoor air quality is affected by the strength of the interaction between adsorbed compounds and the adsorbing area. In addition to the above physical adsorption effect, chemisorption processes may occur especially in soft flooring materials (Colombo et al., 1993). An example of chemisorption is e.g. the effect on foamed rubber backings in textile carpets (Levsen et al., 1993). The chemisorption especially promotes the sorption of aromatic and chlorinated compounds into the material as they are solvents for rubbers. One example of the sink effects caused by both of the above-mentioned sorption mechanisms is the wall-to-wall textile carpets with foamed rubber backings. When new, they contain plenty of emittable compounds, especially in the foamed rubber backing because of the manufacturing process of rubbers. The emissions are governed by diffusion processes in the rubber backing, but are additionally affected by the chemisorption in the system and the van der Waals forces on the textile fibres, and consequently the offgassing of new materials is delayed. Installed, they collect additional emissions from the surroundings, particularly those caused by human activity. Therefore they tend to become not only continuous but growing emission sources. A typical example of the type of emissions they fix is tobacco smoke. Flooring materials with high sink effects adsorb tobacco smoke from the air and after getting saturated begin to re-emit the adsorbed compounds and their oxidation products. In spaces where people smoke and odorous cleaning agents are used, only smooth, easily cleaned surfaces would therefore be preferable.

The above-described physical characteristics of materials illustrate that the structures of materials have practical consequences with respect to emission properties, and a proper selection of materials requires not only that the amount of potential emissions contained in them is known, but also that the physical factors affecting emission behavior should be taken into account.

3.2.5 Moisture

During construction or while renovating a building, the structures of the building, e.g. concrete slabs, are wet, dampness may accumulate within the building or the building structures may become wet through leaks in pipelines or the building envelope. The amount of moisture will partly depend on the method of construction and the weather construction or conditions during renovation. The dampness or moisture should be removed by heating and drying prior to finishing building interiors. For instance, in floor covering work, the set limits for concrete moisture content vary in the range 60–90 % RH, depending on the type of covering material to be applied (Saarela, 1992).

The effect of humidity and moisture on building materials is most often based on hydrolytic decomposition reactions and degradation products resulting from such reactions. The harm caused by moisture and humidity are amplified in such polycondensate products as thermosetting plastics, whose polymerizing reaction is reversible. A good example of this phenomenon is embodied in urea-formaldehyde resin, perhaps best known as a binder in chipboard and also extensively used in acid-catalyzed lacquers and parquet glues. A strong and continuous formaldehyde emission results from this degradation reaction.

Materials produced from natural oils by oxidative polymerization reactions, e.g. binding agents in linoleum mats, will under the effects of moisture produce and emit degradation products such as fatty acids, higher aldehydes, and alcohols.

Water vapor emitted from moist concrete structures is alkaline and promotes hydrolysis reactions in many polymeric materials.

Casein and other natural compounds have been used in building materials for decades. These compounds appear in glues, adhesives, levelling compounds and natural organic fibres (e.g. jute). Interestingly, the proteins (casein) contained in levelling agents, e.g. in floor structures, has been found to degrade in moist, alkaline conditions commonly appearing in modern buildings. Among the emitted products are ammonia, amines, sulfur compounds and possibly alcohols. These emissions have been found to be detrimental to the indoor air quality, and in Finland the Ministry of Social and Health Care has set a guide value of 40 $\mu g/m^3$ for ammonia in indoor air. The above-mentioned degradation reactions due to moisture can often be recognized by their pungent, musty odors.

PVC flooring materials, especially the cushion vinyls, can degrade in the presence of moisture under alkaline conditions. The effect of moisture-triggered degradation reactions, in particular of alkaline moisture containing ammonia, is shown as color changes in materials, e.g. darkening of oak and color changes in PVC mats. As a result of the degradation reactions of the polymer chain in the material, aggressive chemical degradation products are formed and emitted. Some of these degradation products even trigger the decomposition reaction of additives such as phthalate plasticizers. As phthalate-based plasticizers degrade, they produce such organic degradation products as 2-ethyl-1-

Table 3.2-1. Examples of materials susceptible to damage by humidity and alkaline humidity, and resulting emissions.

Material	RH (%)	Type of damage caused and emitting compounds
Casein-based levelling compounds	75–85	Hydrolysis, ammonia, amines, organic sulfur compounds
PVC mats and covers in damp structures	>95	Staining, decomposition reactions, 2-ethyl-1-hexanol
Water-based glues in damp structures	>85–95	Saponification, material-specific products of hydrolysis
Urea-formaldehyde resins: as binding agents in thermal insulation and textile treatment, as glue in chip board, wood floorings, and lacquers	>60–70	Degradation, formaldehyde
Two component paints, injection resins plastic fillers	Material-specific	Inhibition of curing reactions, monomer emissions
Micro-organisms	>85	Bioaerosols

hexanol, a compound with a distinct, bitter-sweet odor and a suspected cause of many indoor air problems (Gustafsson, 1990). Sjöberg (1997) has reported on the effect of moisture and alkali on floor structures. The TVOC emission from PVC mats on moist structures is more than double that of the original PVC mat, the main individual VOCs being 2-ethyl-1-hexanol and butanol.

Examples of the effect of moisture of materials used in floor systems are given in Table 3.2-1.

3.2.6 Temperature

Temperature affects the material emissions in many ways depending on the offgassing mechanisms taking place in the material. These mechanisms may accelerate offgassing of low-molecular-weight compounds, give rise to new emissions of higher-molecular-weight compounds, and accelerate chemical reactions due to e.g. oxidation or accelerate hydrolytic reactions if humidity is present in the system. The mechanism and magnitude of the temperature effect have been studied by several researchers (Seifert et al., 1989; van der Wal, 1997; Bluyssen et al., 1996). According to the results, a simple dependence of offgassing on temperature does not exist, which can be expected since offgassing is controlled by different types of both chemical and physical interactions in the material. Therefore the proposal of offgassing new spaces using elevated temperatures does not always have the desired result and is not acceptable from a chemical and physical chemical point of view.

A slow decomposition of building materials caused by heat can start at temperatures as low as 100 °C or even below if the material is susceptible to oxidation reactions, is poorly stabilized, or is aged to the point where the stabilizer system has been worn out (Saarela,

1982). Such heat-sensitive materials include, for example, polyurethane, PVC, linoleum and foamed rubbers. The effect of light, especially UV light, promotes the degradation reactions.

At temperatures above 200 °C, oxidation and decomposition of nearly all polymeric materials used in construction takes place. The oxidization products include aldehydes and ketones. Certain ester-based materials such as acrylates emit corresponding monomers as decomposition products. Such products of thermal decomposition are often extremely irritating compounds. Other typical volatile heat-degradation products of materials are aldehydes, ketones and sometimes organic acids.

3.2.7 Oxidation

Oxidation is a consequence of the ageing of the material. As has been mentioned in the preceding Chapters, the polymeric and organic compounds contained in the flooring materials are subject to oxidative reactions during their whole life cycle starting from the processing. Heat, light, humidity, and foreign chemicals trigger the oxidative reactions. The ability of the materials to withstand these reactions depends on the chemical structure of the material and the success of the stabilizers and antioxidants in resisting the deterioration caused by these reactions (Saarela, 1982). During their long service life, the flooring materials are also subject to other ageing effects, as discussed in the above Chapters. Of these, the attack of light and aggressive chemicals such as ozone, $NO_{x,}$ sulfuric acid and other acidic compounds in ambient and sometimes even in indoor air is a potential oxidation threat to susceptible materials, as it has been shown that these compounds may accumulate on building surfaces (Weschler and Shields, 1996).

Light- and heat-induced degradation of polymers produces, as intermediate degradation compounds, free radicals, which further catalyze the degradation of the polymer itself and react with the other released degradation products of the material, thus producing many types of VOCs of which the aldehydes, organic acids, and alcohols are considered to be irritants.

3.2.8 Analytical Techniques

The analytical procedure for VOCs from flooring materials consists of some or all of the following: sampling, transport, storage before testing, preparation of the test specimen, starting of the testing during which the test specimen is kept in the test chamber or outside the chamber in similarly defined conditions, sampling of the VOCs, analytical procedures, and reporting (Nordtest, 1990; EC – European Commission, 1991; Nordtest, 1995; Nordic Wood, 1997; Saarela et al.,1996). In CEN TC/264 WG7 a method using the experience gained from, e.g., the quoted works has been drafted.

For sampling of VOCs from material emissions the Tenax® sorbent tubes have shown to be the most versatile and widely used in Europe (Tirkkonen et al., 1995). Methods based on using Tenax® sorbent in collecting VOCs from material emission test chambers and indoor air are drafted in ISO TC 146 SC/6.

3.2.9 Emissions from Flooring Materials.

In order to give an overview of emissions from flooring materials, data is extracted from the material emission data base, DAME, of VTT, Chemical Technology (Saarela et al., 1994) containing emission data of nearly three hundred materials. Emission measurements have been performed according to methods described in the above Chapter using emission chambers and the FLEC. In the Tables, mainly data consisting of three-day and four-week measurements of the materials are used since these are the principal ages of materials for emission evaluation in the ECA guideline (EC, 1997). When the data in VTT's database has not been sufficient, e.g. regarding the age of the material, less complete data is used and augmented with data from other sources. The Tables included in this article give guidance to emissions from flooring materials during a period between about ten years ago and now and thus cannot be considered as a basis for material selection, because the synthetic materials have developed immensely since the early days, although even today high-emitting materials still exist.

From Table 3.2-2 it can been seen that the emissions from kiln-dried wood species are small. The critical point for wooden floorings is the surface treatment. From Tables 3.2-3 to 3.2-5 it can be seen that emissions, especially from factory-varnished and oil-treated parquets, are low, but wax treatment may bring an emission problem when solvents are used in applying the product. The solvents are often turpentine- or mineral oil-based. When using heat application the problem should be negligible.

Table 3.2-2. VOC emissions, SER_A [$\mu g/m^2$ h], from plain wood boards three days and four weeks old.

		Plain oak		Plain pine		Plain birch	
Compound	CAS	3 days	28 days	3 days	28 days	3 days	28 days
α-Pinene	000080–56–8	28	10	130	17		
Acetic acid	000064–19–7	23	14				
β-Pinene	000127–91–3	10	4	47	7		
2,2,4,6,6-Pentamethylheptane	013475–82–6	6	0	16			
Hexanal	000066–25–1	5	8	34	6	23	20
Limonene	000138–86–3	5	2	48	10		
Benzoic acid	000065–85–0	5	2				
Pentanal	000110–62–3	3	1	10	2	8	6
Undecane	001120–21–4	2	1				
2-Furancarboxaldehyde	000098–01–1	2	1				
Δ^3-Carene	013466–78–9	1	1	24	6		
1-Pentanol	000071–41–0					9	6
Styrene	000100–42–5					9	3
Toluene	000108–88–3					3	1
Hexane	000110–54–3			3	2	3	1

Table 3.2-3. VOC emissions, SER_A [μg/m^2 h], from six varnished parquets three days and four weeks old.

Compound	CAS	Max	Average	Occur./6	Max	Average	Occur. /6
		3 days	3 days	3 days	28 days	28 days	28 days
Hexanal	000066–25–1	35	16	4	24	13	5
Benzaldehyde	000100–52–7	28	11	6	20	8	6
Methanone, diphenyl-	000119–61–9	27	16	5	83	27	5
α-Pinene	000080–56–8	25	15	6	19	8	6
Cyclohexanone	000108–94–1	23	8	6	21	8	5
β-Pinene	000127–91–3	13	8	5	10	5	4
Benzoic acid, methyl ester	000093–58–3	12	9	4	14	10	4
2-Propenoic acid, 2-methyl-, 2-hydroxypropyl ester	000923–26–2	11	11	1	12	12	1
Hexane, 3-methyl-	000589–34–4	10	10	1			
Decanal	000112–31–2	10	8	4	6	5	4
Toluene	000108–88–3	9	7	3	13	10	5
Phenol	000108–95–2	8	6	5	7	5	4
Nonanal	000124–19–6	7	5	4	6	5	4
Ethanol, 2-butoxy-	000111–76–2	6	6	1	5	5	1
Acetic acid, butyl ester	000123–86–4	6	5	2	6	5	2
1-Hexanol, 2-ethyl-	000104–76–7	5	4	3	5	4	3
Cyclohexane, methyl-	000108–87–2	4	4	1	8	8	1
Limonene	000138–86–3	4	3	3	8	4	3
Ethanol, 2-(2-ethoxyethoxy)-	000111–90–0	3	3	2	4	3	2
Δ^3-Carene	013466–78–9	3	2	2	4	3	2
1-Pentanol	000071–41–0	3	3	1	4	4	1
Styrene	000100–42–5	3	3	1			
Octanal	000124–13–0	3	3	1			
Hexane	000110–54–3	2	2	1			
Ethanol, 2-[2-(2-ethoxy-ethoxy)ethoxy]-	000112–50–5	2	2	1	1	1	1
Ethanone, 1-phenyl-	000098–86–2	1	1	1			
Benzene, ethyl-	000100–41–4	1	1	1	1	1	1
Benzene, 1,4-dimethyl-	000106–42–3	1	1	1	1	1	1
Decane	000124–18–5				1	1	1
Hexanoic acid	000142–62–1				9	6	2
Heptane	000142–82–5				32	18	3
Hexane, 2-methyl-	000591–76–4				7	7	1
Xylenes	001330–20–7				10	7	4

Table 3.2-4. VOC emission, SER_A [µg/m² h], from two oiled parquets.

Compound	CAS	Max	Ave	Occuren./2	Max	Ave	Occur./2
		3 days	3 days	3 days	28 days	28 days	28 days
Hexanoic acid	000142–62–1	99	67	2	8	6	2
Hexanal	000066–25–1	68	66	2	30	23	2
1-Pentanol	000071–41–0	22	13	2	2	1	2
Pentanal	000110–62–3	18	16	2	8	6	2
α-Pinene	000080–56–8	17	14	2	12	10	2
Heptanoic acid	000111–14–8	13	13	1	0	0	1
Octanal	000124–13–0	12	10	2	4	4	2
β-Pinene	000127–91–3	10	7	2	7	6	2
Heptanal	000111–71–7	6	6	2	2	2	2
Limonene	000138–86–3	5	5	1	8	7	2
Pentanoic acid	000109–52–4	3	2	2			
Δ^3-Carene	013466–78–9	3	3	1	4	4	2

Table 3.2-5. VOC emission, SER_A [µg/m² h] from five waxed parquets as three and four weeks old.

Compound	CAS	Max 3 days	Ave 3 days	Occur./5 3 days	Max 28 days	Ave 28 days	Occur./5 28 days
Undecane	001120–21–4	7195	3028	5	392	209	5
Dodecane	000112–40–3	3937	1484	5	188	112	5
Undecane, 4-methyl-	002980–69–0	2322	583	5	86	29	5
Decane, 4-methyl-	002847–72–5	1755	440	5	14	9	4
α-Pinene	000080–56–8	1499	341	5	1371	286	5
Decane, 3-methyl-	013151–34–3	749	279	4	26	16	4
Hexanal	000066–25–1	540	313	5	54	33	5
Undecane, 2-methyl	007045–71–8	521	521	1	21	13	4
Decane	000124–18–5	510	229	5	19	11	5
Hexanoic acid	000142–62–1	441	198	3	32	20	4
Undecane, 5-methyl-	001632–70–8	399	208	4	34	20	4
Δ^3-Carene	013466–78–9	388	388	1	375	189	2
Decane, 2-methyl-	006975–98–0	348	178	4	20	13	4
Undecane, 2,6-dimethyl-	017301–23–4	299	198	3	38	20	4
Undecane, 3-methyl-	001002–43–3	239	128	4	21	13	4
Decane, 5-methyl-	013151–35–4	233	108	4	13	8	4
Nonane, 2,6-dimethyl-	017302–28–2	214	81	4	8	6	3
Tridecane	000629–50–5	207	74	5	6	4	4
Undecane, 2-methyl-	007045–71–8	202	116	3	28	17	4
Cyclohexane, 1,4-dimethyl-	000589–90–2	192	145	2	4	4	1

Table 3.2-5. (Continued)

Compound	CAS	Max 3 days	Ave 3 days	Occur./5 3 days	Max 28 days	Ave 28 days	Occur./5 28 days
Cyclohexane, pentyl-	004292–92–6	178	178	1			
Cyclohexane, (2-methylpropyl)-	001678–98–4	169	102	2	6	6	1
Cyclohexane, 1,1,2-trimethyl-	007094–26–0	157	157	1			
Cyclohexanone	000108–94–1	136	64	3	11	5	3
β-Pinene	000127–91–3	132	132	1	139	139	1
Pentanal	000110–62–3	112	73	5	21	13	5
Toluene	000108–88–3	85	43	2	20	12	2
Octanal	000124–13–0	73	73	1	7	6	2
Limonene	000138–86–3	67	67	1	75	40	2
Cyclohexane,butyl-	001678–93–9	61	61	1	6	6	1
Xylenes+pentanoic acid		58	58	1	13	13	1
Nonane, 3,7-dimethyl-	017302–32–8	43	42	2	4	4	1
Xylenes	001330–20–7	40	39	2	4	4	1
Hexane	000110–54–3	38	26	3	12	7	3
Heptanal	000111–71–7	34	18	3	5	4	4
Heptane	000142–82–5	33	21	4	4	3	
2-Heptenal	018829–55–5	30	23	3	1	1	1
Octane	000111–65–9	27	17	4	2	2	4
2(3h)-Furanone, 5-ethyldihydro-	000695–06–7	26	26	1	2	2	1
Camphene	000079–92–5	25	25	1	27	27	1
Cyclohexane, 1-methyl-3-propyl-	004291–80–9	23	23	1			
1-Pentanol	000071–41–0	22	14	4	4	3	4
Cyclohexane, methyl-	000108–87–2	17	15	3	3	2	3
β-Myrcene	000123–35–3	16	16	1	17	17	1
Nonanal	000124–19–6	9	9	1	8	8	1
Terpinolene	000586–62–9	9	9	1	6	6	1
Cyclopentane, 1,2-dimethyl-, *trans*-	000822–50–4	8	8	1			
Decanal	000112–31–2	6	6	1	5	5	1
Benzene, 1-methyl-4-(1-methylethyl)-	000099–87–6	6	6	1	7	7	1
γ-Terpinene	000099–85–4	5	5	1	3	3	1
Benzene, 1-methyl-3-(1-methylethyl)-	000535–77–3	5	5	1	6	6	1
3,6-Dimethyldecane	017312–53–7	4	4	1	16	9	2
Benzene, methyl (1-methylethenyl)-	026444–18–8	2	2	1	2	2	1

Table 3.2-5. (Continued)

Compound	CAS	Max 3 days	Ave 3 days	Occur./5 3 days	Max 28 days	Ave 28 days	Occur./5 28 days
Sabinene	003387–41–5	2	2	1	2	2	1
1-Heptene	000592–76–7	2	2	1	1	1	1
Pentanoic acid	000109–52–4	2	2	1	1	1	1
2-Heptanone	000110–43–0	1	1	1	1	1	1
Acetic acid, butyl ester	000123–86–4	1	1	1	1	1	1
Benzene, ethyl-	000100–41–4	1	1	1	0	0	1
Benzene, 1,3,5-trimethyl-	000108–67–8	0	0	1	1	1	1
Benzene, 1,3-dimethyl-	000108–38–3				1	1	1

New linoleums are often odorous materials because of emitted hexanal, propanoic acid and nonanal (Jensen et al., 1995), but emissions from linoleums in use for several years still cause odor problems. The emissions of linoleums in chemical terms are given in Table 3.2-6. It seems, as expected, that the main emissions of both new and used linoleums contain volatile substances which are similar in new and used products, degradation reactions occur in the material during use, and some emissions are common for both new and old materials.

Table 3.2-6. VOC emission from new and old linoleum floorings.

Compound	CAS	New material FID-area [%][1)]	Used material SER_A[μg/m^2 h][2)]
Hexanoic acid	000142–62–1	17	123
Heptanoic acid	000111–14–8	3.5	52
Octanoic acid	000124–07–2	3	51
Pentanoic acid	000109–52–4	5.5	34
Propanoic acid	000079–09–4	15	30
Nonanoic acid	000112–05–0		27
Nonanal	000124–19–6	4	24
Octanal	000124–13–0	3	21
Butanoic acid	000107–92–6	3	14
1-Heptanol	000111–70–6		12
Fatty acid			12
Acetic acid	000064–19–7	7.5	9
1-Hexanol, 2–ethyl-	000104–76–7		8
Cyclopropane	000075–19–4		8
3-Methylpentane	000096–14–0		8
Hexanal	000066–25–1	12.5	8
Heptanal	000111–71–7	1	7
Xylenes	001330–20–7		7

[1)] Data from Jensen et al. (1995)
[2)] Data from VTT (1999)

Table 3.2-6. (Continued)

Compound	CAS	New material FID-area [%]	Used material SER_A [µg/m² h]
Ethanol, 2-(2-butoxyethoxy)-	000112–34–5	8	7
Propanoic acid, 2-methyl-	000079–31–2		6
2-Dodecenal	004826–62–4		6
Fatty acid			6
Fatty acid			6
Toluene	000108–88–3	2.5	6
Decanal	000112–31–2		6
Pentanal	000110–62–3		5
2-Nonanone	000821–55–6		5
Isolongifolene	001135–66–6		5
2-Decanone	000693–54–9		5
Hexadecanoic acid	000057–10–3		4
2(3H)-Furanone, 5-ethyldihydro-	000695–06–7		4
Benzene, methyl-	000108–88–3		3
1-Pentene	000109–67–1		3
Benzaldehyde	000100–52–7		3
2(3H)-Furanone, 5-butyldihydro-	000104–50–7		3
2-Butanol	000078–92–2		3
1-Butanol	000071–36–3		2
(-)-(R)-5-Ethyl-2(5H)-furanone	076291–90–2		2
Hexadecane	000544–76–3		2
Butanoic acid, butyl ester	000109–21–7		2
Ethanol, 2-butoxy-	000111–76–2		2
Pentadecane	000629–62–9		1
Octane	000111–65–9		1
Hydrocarbon			1
α-Pinene	000080–56–8		1
Cyclohexane	000110–82–7		1
Benzene, trimethyl-	000526–73–8		1
Xylenes	001330–20–7		1
Δ^3-Carene	013466–78–9		1

Emission from a PVC-coated cork material is shown in Table 3.2-7. The principal emissions derive from the cork material and the binder. Horn et al. (1997) have additionally reported phenol, furfural, cyclohexanone, methyl benzoate, benzophenone and BHT emissions. Some of these, e.g. BHT, cyclohexanone and possibly phenol derive from the PVC coating.

Emissions from PVC materials (Table 3.2-8) are largest in the soft, cushion vinyl products because the process uses a lot of additives and because of the physical properties of the product. The most frequent emission even in new materials is 2-ethyl-

Table 3.2-7. VOC emission from a PVC-coated cork tile.

Compound	CAS	SER_A [µg/m² h]
1,2-Propanediol	000057–55–6	713
Crotonaldehyde	000123–73–9	59
4-Methyldioxalan		37
2,2,4,4,6-Pentamethylheptane	013475–82–6	35
Formaldehyde	000050–00–0	15
Pentanal	000110–62–3	12
2-Furancarboxaldehyde	000098–01–1	6
Docosane c20	000629–97–0	4
Acetaldehyde	000075–07–0	3
Hydrocarbon		2
Diethylenglycol *n*-monobutylether	000112–34–5	1
Acetone	000067–64–1	1

1-hexanol, which is probably not only a degradation product but is also contained in the additives. Other emissions are of the solvent type and derive from the additives and finishing agents. Typical emissions for soft cushion vinyl products are also the co-solvents or viscosity modifiers. Typical of these are high-boiling mineral oil fractions and TXIB, which was frequently used in the 1980s. In the material selection shown in Table 3.2-8 TXIB does not appear, but it is present in many other PVC floorings contained in the data base.

Table 3.2-8. VOC emission, SER_A [µg/m² h] from ten PVC cushion vinyls three days and four weeks old.

Compound	CAS	Max 3 days	Average 3 days	Occur./10 3 days	Max 28 days	Average 28 days	Occur./10 28 days
Ethanol, 2-(2-butoxyethoxy)-	000112–34–5	193	122	3	91	63	3
1-Hexanol, 2-ethyl-	000104–76–7	108	30	10	42	14	10
Cyclohexanone	000108–94–1	91	56	4	47	31	4
2-Pyrrolidinone, 1-methyl-	000872–50–4	89	66	3	47	33	3
Toluene	000108–88–3	46	18	5	18	14	3
Xylenes	001330–20–7	38	20	3	10	7	3
1-Butanol	000071–36–3	35	17	4	6	5	4
2-Hexanone, 5-methyl-5-phenyl-	014128–61–1	32	13	4	16	9	4
Phenol	000108–95–2	29	14	5	16	8	5
Nonanal	000124–19–6	21	12	4	17	9	4

Table 3.2-8. (Continued)

Compound	CAS	Max 3 days	Average 3days	Occur./10 3 days	Max 28 days	Average 28 days	Occur./10 28 days
Heptane	000142–82–5	18	12	2	7	5	3
Cyclohexanol	000108–93–0	16	16	1	9	9	1
2-Ethoxyethanol acetate	000111–15–9	15	7	3	7	4	3
Acetic acid, 2-ethylhexyl ester	000103–09–3	14	14	1	7	7	1
1-Dodecanol	000112–53–8	13	13	1	7	7	1
Benzene, trimethyl-	000526–73–8	13	9	2	24	14	3
Benzene, ethyl-	000100–41–4	11	10	2			
4-Methyl-2-pentanol	000108–11–2	11	11	1	6	6	1
Benzaldehyde	000100–52–7	10	9	2	6	4	3
Cyclododecane	000294–62–2	10	10	1	9	9	1
Benzenemethanol	000100–51–6	9	9	1	5	5	1
Decanal	000112–31–2	9	7	4	16	7	3
1,2-Propanediol	000057–55–6	7	6	2	1	1	2
2-Pentanone, 4-methyl-4-phenyl-	007403–42–1	7	5	2	7	5	2
1-Tetradecanol	000112–72–1	6	6	1	7	7	1
2-Propenoic acid, 6-methylheptyl ester	054774–91–3	6	6	1	3	3	1
Hexanal	000066–25–1	5	5	1	3	3	1
2-Hexanol	000626–93–7	5	5	1	2	2	1
Cyclohexane, methyl-	000108–87–2	5	5	1	1	1	1
Ethanol, 2-ethoxy-	000110–80–5	5	5	1	3	3	1
Hexanoic acid	000142–62–1	5	5	1	9	9	1
Ethanol, 2-(2-ethoxyethoxy)-	000111–90–0	5	5	1	2	2	1
Hexane	000110–54–3	5	2	3			
Tridecane	000629–50–5	3	3	1	2	2	1
N,n-Diethylethanamine	000110–68–9	3	3	1	1	1	1
Benzene, 1,3,5-trimethyl-	000108–67–8	3	3	1	2	2	1
Benzene, (1,1-dimethylbutyl)-	001985–57–5	3	3	1	3	3	1
1-Undecene	000821–95–4	2	2	1			
Undecane	001120–21–4	2	2	1	5	5	1
1,2-Dimethyl-cyclopentane	002452–99–5	1	1	1			
1,1-Dimethyl-cyclopentane	001638–26–2	1	1	1	1	1	1
Hexane, 3-methyl-	000589–34–4	1	1	1	0	0	1
Hexanoic acid, 2–ethyl-	000149–57–5	1	1	1			

Table 3.2-8. (Continued)

Compound	CAS	Max 3 days	Average 3days	Occur./10 3 days	Max 28 days	Average 28 days	Occur./10 28 days
Benzene, 1,2,4-trimethyl-	000095–63–6	1	1	1			
Benzene, 1-ethyl-4-methyl-	000622–96–8	1	1	1			
1,3-Dimethyl-cyclopentane	002532–58–3	1	1	1			
Pentane, 2,3-dimethyl-	000565–59–3	1	1	1			
Cyclohexane	000110–82–7	1	1	1			
Hexane, 2–methyl-	000591–76–4	1	1	1	0	0	1
1-Propanol, 2-(2-hydroxypropoxy)-	000106–62–7	1	1	1			
Octanal	000124–13–0				4	4	1
Benzene, ethylmethyl-	000611–14–3				7	7	1

Emissions from semihard (Table 3.2-9) calendered materials and especially from PVC tiles are much fewer and lower. The primary emissions from a calendered PVC flooring are 1-methyl-2-pyrrolidinone, 2-(2-butoxyethoxy)-ethanol and 1,2-propanediol. Semi-hard PVC flooring also frequently emits phenol, which may derive from the stabilizers (Saarela et al., 1989).

Table 3.2-9. VOC, SER_A [$\mu g/m^2$ h] emission from calendered PVC-flooring material.

Compound	CAS	3 days	28 days
2-Pyrrolidinone, 1-methyl-	000872–50–4	173	77
Ethanol, 2-(2-butoxyethoxy)-	000112–34–5	88	58
1,2-Propanediol	000057–55–6	33	13
Phenol	000108–95–2	7	5
Ethanol, 2-butoxy-	000111–76–2	2	

3.2.10 Time Dependence of Emissions

Figure 3.2-1 shows the time dependence of the TVOC emission for some of the materials whose emissions have been reported in Tables 3.2-2 to 3.2-8. The TVOC in the figure is the FID response of the GC in toluene equivalents integrated mainly over the C6 to C16 boiling range, but in some cases extended, especially in the higher end, when emissions of e.g. a high-molecular-weight fraction or other high-boiling substances show up on the chromatogram. From the Figure, the wide variation of TVOC even within the same material types can be seen. A textile carpet, a waxed parquet (heat-waxed), a factory-varnished parquet, a parquet waxed on site, and a PVC flooring (cushion vinyl)

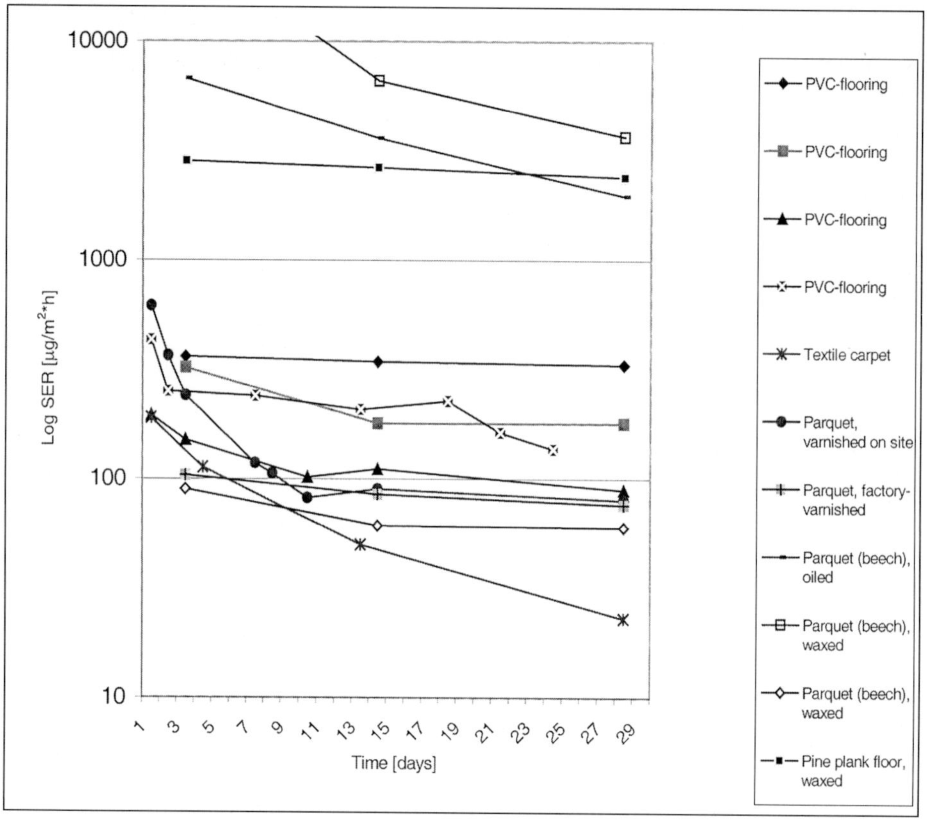

Figure 3.2-1. Variation with time of TVOC emission from flooring materials.

are the lowest emitters. The highest emitters are also found from the same material groups. This shows that if a low-emitting material is wanted it can only be chosen from tested and labelled products.

References

Bluyssen P. M. and van der Wal J. F. (1994): Sorption effects of cigarette smoke in the indoor environment; chemical/physical versus sensory evaluations. Proceedings of Healthy Buildings '94, Budapest, Hungary, Vol. 3, 92–107.

Bluyssen P. M., Cornelissen H. J. M., Hoogeveen A. G., Wouda P. and van der Wal J. F. (1996): The effect of temperature on the chemical and sensory emission from indoor materials. Proceedings of Indoor Air '96, Nagoya, Japan, Vol. 3, 619–624.

Colombo A., De Bortoli M., Knöppel H., Pecchio E. and Vissers H. (1993): Vapor deposition of selected VOCs on indoor surface materials in test chambers. Proceedings of Indoor Air Quality '93, Helsinki, Finland, Vol. 2, 407–412.

EC – European Commission (1991): Guideline for the characterization of volatile organic compounds emitted from indoor materials and products using small chambers, COST Project 613, Report No. 8, EUR 13593 EN, 1991.

EC – European Commission (1997): Evaluation of VOC emissions from building products, indoor air quality and its impact on man, Report No. 18, Luxembourg.

FiSIAQ (1995), Finnish Society of Indoor Air Quality and Climate, Finnish Association of Construction Clients, RAKLI, Finnish Association of Architects, SAFA, Finnish Association of Consulting Firms, SKOL (1995): Classification of indoor climate, construction, and finishing materials. FiSIAQ Publication 5E, Espoo, Finland.

GBR, Trade Standard (1992): Measurement of Chemical Emission from Flooring Materials. Swedish National Flooring Trade Association (GBR) and the Swedish National Testing and Research Institute, Stockholm.

Gunnarsen L., (1991): Air velocities and indoor emissions. New Haven, Conn., CIB-W77.

Gustafsson H., (1990): Kemisk emission från byggnadsmaterial. Swedish National Testing and Research Institute (SP), Rapport 1990 : 25, Borås.

Horn W., Ullrich D. and Seifert B. (1997): VOC emission from cork products for indoor use – chamber and test house measurements. Proceedings of Healthy Buildings '97, Washington, USA, Vol. 3, 533–538.

Jensen B., Wolkoff P., Wilkins C. K. and Clausen P. A. (1995): Characterization of linoleum. Part 1: Measurement of volatile organic compounds by use of the field and laboratory emission cell (FLEC). Indoor Air, *5*, 38–43.

Jørgensen R. B., Knudsen H. N. and Fanger P. O. (1993): The influence on indoor air quality of adsorption and desorption of organic compounds on materials. Proceedings of Indoor Air Quality '93, Helsinki, Finland, Vol. 2, 383–388.

Kirchner S., Manpetit F., Quenard D., Rouxel P. and Girand D. (1995): Characterization of adsorption/desorption of volatile organic compounds on indoor surface materials. Proceedings of Healthy Buildings '95, Milano, Italy, Vol. 2, 953–958.

Kjaer U. and Nielsen P. (1991): Adsorption and desorption of organic compounds on fleecy materials. Proceedings of IAQ '91, Healthy Buildings, 285–288.

Levsen K. and Sollinger St. (1993): Textile floor coverings as sinks for indoor pollutants. Proceedings of Indoor Air '93, Helsinki, Finland, Vol.2, 395–400.

Matthews T. G., Wilson D. L., Thomson A. J., Mason M. A., Bailey S. N. and Nelms L. H. (1987): Interlaboratory comparison of formaldehyde emissions from particle board underlayment in small scale environmental chambers. Journal of the Air Pollution Control Association, *37*, 1320–1326.

Neretnieks I., Christiansson J., Romero L., Dagerholt L. and Yu J.-W. (1993): Modelling of emission an re-emission of volatile organic compounds from building materials with indoor air applications. Indoor Air, 1993, *3*, 2–11.

Nordic Wood (1996): Trä och Miljö, Delprojekt "Emission, träprodukter", NIF, TEKES, Finnish industry, Dansk Teknologisk Institut, Teknologiske Institutt, Trätek, VTT, DTI Draft Report in Danish, October 1996.

Nordtest (1990): NT-Build 358, Building materials, emission of volatile compounds, chamber method. Nordtest, PO Box 116 FIN-02151, Espoo, Finland.

Nordtest, NT build 438 (1995): Building materials: emission of volatile compounds – field and laboratory emission cell (FLEC). P.O. Box 116, FIN-02151 Espoo.

Produktstandard (1998a): Halvhårde gulvbelaegninger, laminatgulve og traegulve. Foreningen Dansk Indeklima Maerkning, 1. Udgave 6. Januar 1998, DTI, Taastrup, Denmark.

Produktstandard (1998b): Traegulvolier. Foreningen Dansk Indeklima Maerkning, 1. Udgave 6. Januar 1998, DTI, Taastrup, Denmark.

Saarela, K. (1982): Plastics additives (in Finnish). Ingeniörsorganisationernas Skolninscenter, INSKO 121–82 VI. 1–51.

Saarela K., (1992): Organic emissions from building materials, Report 3, Ministry of the Environment (in Finnish), English translation 1994.

Saarela K., Kaustia K. and Kiviranta A. (1989): Emissions from materials; the role of additives in PVC. Elsevier Science Publishers B.V., Present and Future of Indoor Air Quality, Amsterdam.

Saarela K., Tirkkonen T. and Tähtinen M. (1994): Preliminary Data Base for Material Emissions. Nordic Committee on Building Regulations (NKB), Report No. 1994 : 4E, Helsinki.

Saarela K., Clausen G., Pejtersen J., Tirkkonen T., Tähtinen M. and Dickson D. (1996): European data base on indoor air pollution sources in buildings; principles of the protocol for testing of building materials. Proceedings of Indoor Air '96, Nagoya, Japan, Vol. 3, 83–88.

Seifert B., Ullrich D. and Nagel R. (1989): Volatile organic compounds from carpeting. Proceedings of the 8th World Clean Air Congress, The Hague, The Netherlands, Vol. 1, 253–258.

Sjöberg A. (1997): Ongoing research. Floor systems, moisture and alkali, Proceedings of Healthy Buildings '97, Washington, USA, Vol. 3, 567–572.
Sparks L.E., Tichenor B. A., Chang J. and Guo Z. (1996): Gas-phase mass transfer model for predicting volatile organic compound (VOC) emission rates from indoor pollutant sources. Indoor Air, *6*, 31–40.
Tirkkonen T. and Saarela, K. (1997): Adsorption of VOCs on interior surfaces in a full scale building. Proceedings of Healthy Buildings '97, Washington, USA, Vol. 3, 551–556.
Tirkkonen T., Mroueh U.-M. and Orko J. (1995): Tenax as a collection medium for volatile organic compounds. Nordic Committee on Building Regulation (NKB), Report No. 1995 : 6 E, Helsinki.
Van der Wal J., Hoogeveen A.W. and Wouda P. (1997): The influence of temperature on the emission of volatile organic compounds from PVC flooring, carpet, and paint. Indoor Air, *7*, 215–221.
VTT (1999): Database for Material Emissions – DAME (ongoing research project).
Weschler C. J. and Shields H.C. (1996): Chemical transformations of indoor air pollutants. Proceedings of Indoor Air '96, Nagoya, Japan, Vol. 1, 909–924.

3.3 Indoor Air Pollution by Release of VOCs from Wood-based Furniture

Tunga Salthammer

3.3.1 Introduction

Lacquers, varnishes and paints have always been important materials used by man to protect and embellish different kinds of surfaces. Even in ancient times, shellac was used for the improvement of wooden surfaces. Nowadays, mainly synthetic lacquers have to meet the increased requirements for mechanical and chemical resistance. Wood lacquers are very important as these are preferred as coatings for furniture, windows, doors, parquetry and panelling in the indoor environment. They have not only to be decorative and resistant but also of low emissions (Salthammer and Marutzky, 1995a; Salthammer, 1997a). Lacquers contain several different chemical components such as binders, solvents, light stabilizers, antioxidants, radical starters and other additives (Pecina and Paprzycki, 1994; Garrat, 1996; Stoye and Freitag, 1996; Baumann and Muth, 1997). In this connection, the question regarding emissions of residual organic substances during use, their accumulation in indoor air and the eventually resulting unpleasant odor and health risks for dwellers repeatedly occur.

3.3.2 Coating Systems for Wood-based Furniture

In the manufacture of furniture, the physical and chemical properties of wood make a surface treatment necessary in order to protect the surface or to change the color. As well as additives, liquid systems such as lacquers and varnishes contain solvents which generate certain flow properties to allow easy and high-quality application properties using the desired technique (painting, pouring, dipping, spraying or rolling). The binders of the solvent-containing lacquers are, for example, polyurethane (PUR, chemically/physically drying), unsaturated polyesters (UPE, chemically drying) and cellulose nitrate (CN, physically drying). Polyacrylates are the binders mainly used in water-based systems. In Germany, the use of acid-curing lacquers has dramatically decreased because of the subsequent emission of formaldehyde. The so-called high-solid lacquers are conventional systems with a particularly high solids content (60–70 %). Up to now, lacquers based on natural ingredients and having oils or isoaliphatic hydrocarbons as solvents and unsaturated fatty acids as binders have only a small share of the market. Various coating systems for furniture and interior decoration are summarized in Table 3.3-1 (see also Hansemann, 1996).

Table 3.3-1. Coating systems for wood-based furniture (see also Hansemann, 1996).

No.	Type of coating	Solvent (%)	Water (%)
A	Solvent stains	70	25–30
B	NC (colorless)	75	
C	NC (pigmented)	60	
D	Acid-curing (colorless, 2C)	56–60	
E	Acid-curing (colorless, 1C)	70–75	
F	PUR (colorless)	70–75	
G	PUR (medium solid, colorless)	40–50	
H	PUR (pigmented)	35–60	
I	UPE (roller coating, UV-curing)	0–5	
K	UPE (spray coating, colorless)	65–70	
L	UPE (spray coating, UV-curing)	40–70	
M	AC (spray coating, UV-curing)	40–70	
N	AC (roller coating, UV-curing)	0–10	
O	Water-based (conventional)	5–10	60–65
P	Water-based (UV-curing)	5–10	58–60
Q	Water-based (PUR, 2C)	10	60
R	ECO (colorless)	40–70	

NC = nitrocellulose
PUR = polyurethane
UPE = unsaturated polyester
AC = acrylate
ECO = ecological (here: synonym for a coating system, which is based on natural resins)
1C/2C = one/two component

3.3.3 Test Methods for Lacquered Surfaces

3.3.3.1 Material Analysis

The residual emissions from lacquers after application can be investigated by different methods of varying costliness. As a qualitative screening method, dynamic headspace gas chromatography (purge and trap) (Bruner, 1993) is quick and informative, especially in conjunction with mass-selective detection. Approximately 10–20 mg of solid material is placed between two quartz wool plugs in a Tenax glass tube. Sampling of finished products is by removal of the lacquer coating using a sharp blade. Liquid samples are first applied as a thin film to a glass plate, and the sampling is carried out after curing. The volatile ingredients are thermally desorbed at 70 °C in a scavenging gas stream and cryofocused in a cooling trap. Injection into the gas chromatographic column is performed by rapid heating to 280 °C. Figure 3.3-1 gives a comparison of total ion chromatograms obtained by dynamic headspace GC/MS and Tenax sampling, with GC/MS analysis of chamber air for a water-based and UV-cured acrylate system.

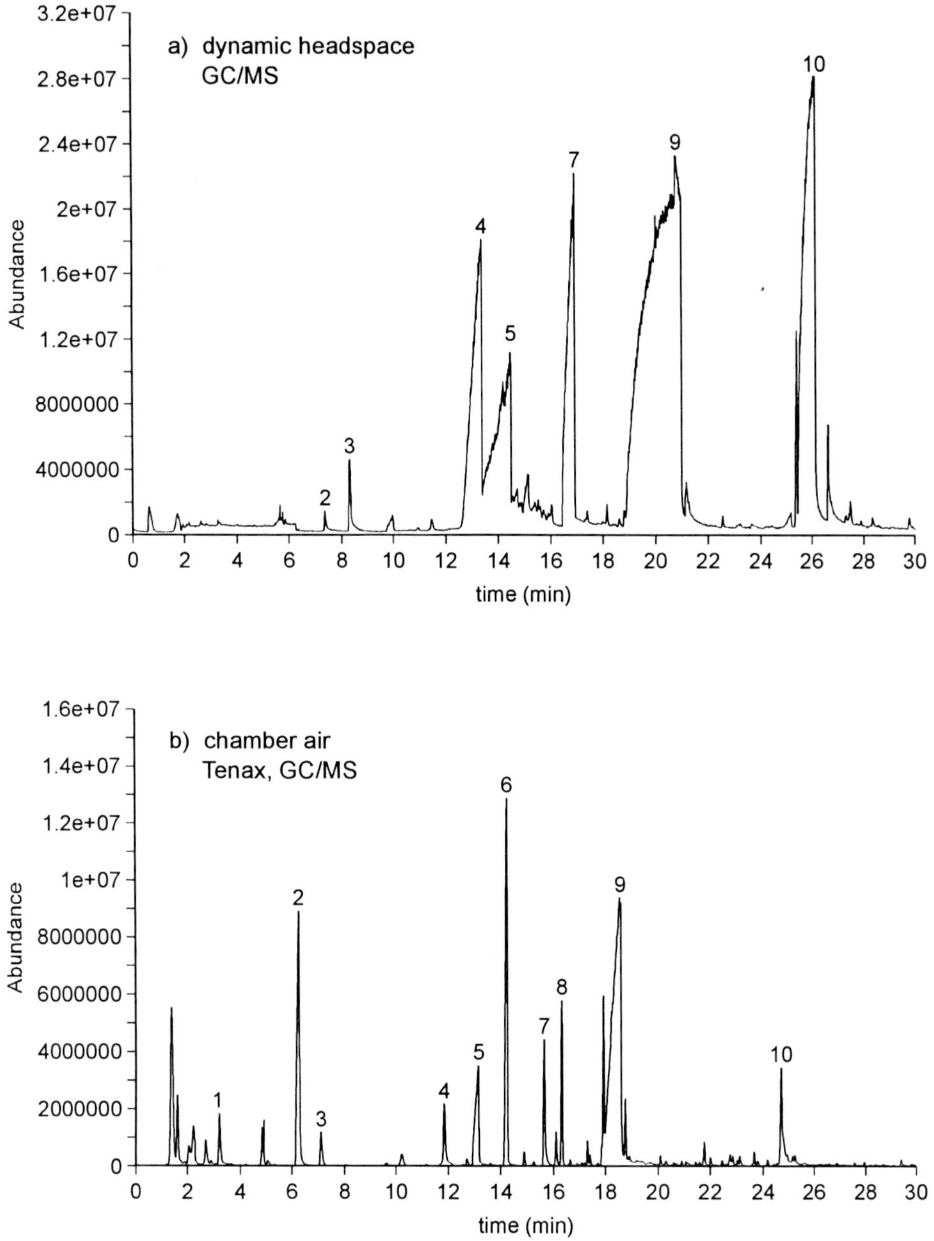

Figure 3.3-1. Comparison of *a* dynamic headspace GC/MS analysis and *b* chamber air (Tenax sampling, GC/MS) for a water-based and UV-cured coating system: *1* pentanal; *2* hexanal; *3* *n*-butyl acetate; *4* cyclohexanone; *5* 2-butoxyethanol; *6* α-pinene; *7* benzaldehyde; *8* β-pinene; *9* ethyldiglycol; *10* 2-(2-butoxyethoxy)-ethanol.

3.3.3.2 Emission Test Chambers

As a result of the increased accumulation of VOCs in indoor air and the related complaints of users regarding nuisance caused by odor and health risks, the investigation of indoor air quality is increasing (Maroni et al., 1995). The determination of VOC and SVOC pollution is only the first step towards an effective reduction of emissions. Emission sources and their contribution to indoor air pollution, depending on the room loading and on climatic parameters such as temperature, humidity and air exchange, must be considered as well. For the evaluation of the emission potential of individual products under practice-related conditions and over defined periods of time, conditioned test chambers are being used. The size varies between some cm^3 and several m^3 according to different applications (CEC, 1993; Salthammer and Marutzky, 1995b). The selection of the chamber, the sample preparation and the performance of the test depend on the individual problem. Application of the FLEC (Wolkoff, 1996) yields only limited information about the emission behavior of finished furniture. Nevertheless, the FLEC is very suitable for non-destructive testing of coated surfaces with sufficient homogeniety, as demonstrated in Fig. 3.3-2. Test chambers allow time-dependent monitoring of emission profiles according to the following balance equation:

$$\mathrm{d}C/\mathrm{d}t = N\ \mathrm{SER_A}(t) - L\ C(t) \qquad (1)$$

where C = chamber concentration in $\mu g/m^3$, L = air exchange rate in h^{-1}, N = loading in m^2/m^3 and $\mathrm{SER_A}$ = area-specific emission rate in $\mu g/(m^2h)$.

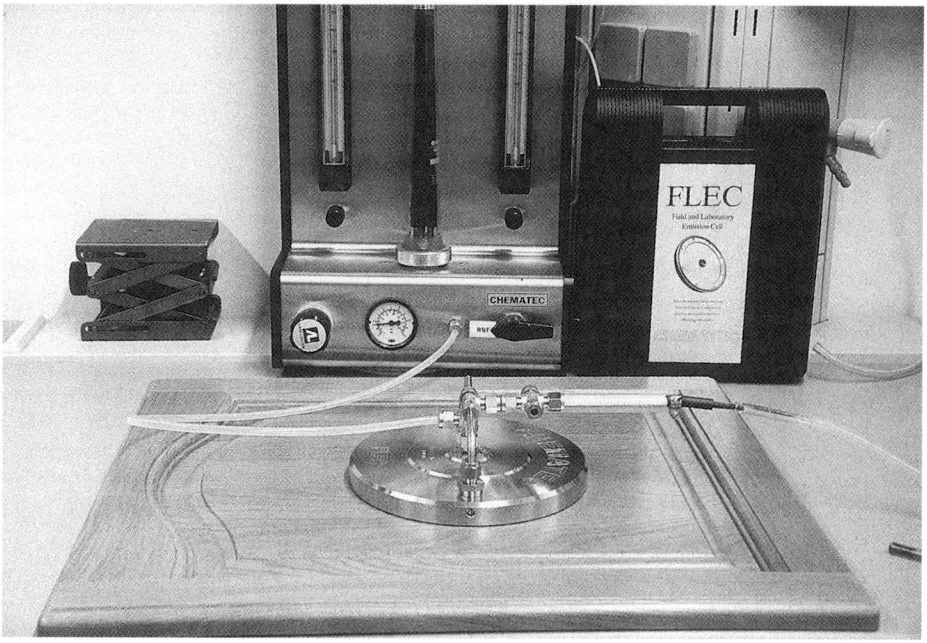

Figure 3.3-2. Application of the FLEC for VOC emission testing of furniture coatings.

For a decaying concentration vs time function, $SER_A(t)$ results from Eq. (1) by transfer to differential quotients according to Eq. (2):

$$SER_A(t) = [\Delta C/\Delta t + L\,C(t)]/N \quad (2)$$

In the steady state ($\Delta C/\Delta t \rightarrow 0$), Eq. (2) is reduced to Eq. (3):

$$SER_A = (LC)/N \quad (3)$$

which is generally applied for conversion from chamber concentrations to emission rates (ECA, 1997). In case of significant sink effects, the mathematical procedure is more sophisticated (Guo, 1993; Guo et al., 1996).

3.3.3.3 Sampling and Analysis

State of the art is discontinuous active sampling with suitable adsorbents (CEC, 1994), but passive samplers are occasionally used for chamber tests on furniture. Tenax TA is a commonly used adsorbent as it provides high recoveries for a wide range of VOCs (Tirkkonen et al., 1995). Monoterpenes are preferably trapped on Tenax/Carbotrap combinations or on activated charcoal (see Chapter 1.1). Sampling on Tenax is fast and requires only 10–20 min. For continuous recording of sum parameters, real-time monitoring using photoacoustics, photoionization or flame ionization can be used (see Chapter 1.6). However, such techniques do not yield information about single compounds. The method of choice for aldehydes and ketones is collection on filter cartridges coated with dinitrophenylhydrazine (DNPH) and analysis of the resulting hydrazones by HPLC and UV detection (see Chapter 1.2). Sensitive determination of airborne diisocyanate monomers requires derivatization to stable chromophores and analysis by HPLC with fluorescence detection. It was previously shown that 2-pyridylpiperazine (2PP) is a very suitable derivatization agent (Schulz and Salthammer, 1998). In particular cases, special techniques are necessary for sampling and analysis of VVOCs, SVOCs or reactive compounds.

3.3.4 Emission of VOCs from Furniture

Furniture coatings are a potential source of a number of VOCs as is evident from Fig. 3.3-1. The ten main components represented solvents, monomers and photoinitiator fragments. Furthermore, loading rates cover a range of ca. 0.5–7.0 m^2/m^3 in private and occupational living spaces (Fischer and Böhm, 1994). Taking into account the fact that average air exchange rates in non-ventilated rooms are only 0.1–0.4 h^{-1} (Salthammer et al., 1995) it might be expected that VOC emissions from furniture contribute significantly to indoor air pollution. The emission behavior will mainly depend on the manufacturing process, because most furnitures are of complex construction. For example, wardrobes and other cabinets are made of engineered wood such as particle board and medium density fiber board (MDF) for front and back, respectively. Particle board is often wood veneered for decorative reasons and stained and coated with up to three layers of different lacquers. Other common materials for overlay are PVC, decorfoil and melamine. Furniture made from solid wood is often treated with wax or oil.

Several authors have studied VOC emissions from furniture using test chambers. In Table 3.3-2 the most frequently identified compounds are grouped and summarized. Fischer and Böhm (1994) monitored 41 components via static headspace-GC and chamber experiments after application of different lacquers on glass plates and MDF. Area-specific emission rates for ΣVOC after 20 days testing time ranged from 250 $\mu g/(m^2h)$ to 3550 $\mu g/(m^2h)$. EPA (1998) has found SER_A for ΣVOC of up to 7800 $\mu g/(m^2h)$ (28 days) when studying six coatings applied to particle board. Jann et al. (1997) present chamber concentrations for ΣVOC and single compounds after 1 day and 28 days testing time

Table 3.3-2. Volatile organic ingredients of wood-based furniture.

Compounds	Possible source
Aliphatic hydrocarbons (solvents): Hexane, heptane, octane, nonane, decane, undecane, dodecane, tridecane, tetradecane, pentadecane, cyclohexane, methylcyclohexane	All coating systems
Aromatic hydrocarbons (solvents): Toluene, ethylbenzene, *o-,m,-p*-xylene, 1,3,5-trimethylbenzene, 1-methylethylbenzene	Solvent-based systems
Esters (solvents): *n*-Butyl acetate, *iso*-butyl acetate, ethyl acetate, 1-butanol-3-methoxy acetate, 1-ethoxy-2-propyl acetate, 1-methoxy-2-propyl acetate, Texanol[1], TXIB[2], 2-(2-butoxyethoxy)-ethyl acetate	Solvent-based systems
Ketones (solvents): 2-Butanone, 4-methyl-2-pentanone, acetone, 1-methyl-2-pyrrolidone, 4-hydroxy-4-methyl-2-pentanone	Solvent-based, water-based systems
Glycols (solvents): 2-Butoxyethanol, 2-(2-butoxyethoxy)-ethanol, 1-methoxy-2-propanol, 2-(2-ethoxyethoxy)-ethanol, 2,4,7,9-tetramethyl-5-decyne-4,7-diol	Water-based systems
Alcohols (solvents): Ethanol, 1-propanol, 2-propanol, 1-butanol, 2-butanol	Solvent-based systems
Isoprenes: α-Pinene, β-pinene, Δ^3-carene, longifolene, β-phellandrene, camphene, myrcene, carvone, limonene	ECO, NC, softwood
Aldehydes (oxidation products): Pentanal, hexanal, heptanal, octanal, nonanal, decanal, undecanal, 2-pentenal, 2-hexenal, 2-heptenal, 2-octenal, 2-nonenal, 2-decenal, 2,4-heptadienal, 2,4-nonadienal	ECO, NC, softwood
Photoproducts: Benzaldehyde, benzophenone, cyclohexanone, benzil, methylbenzoate, 1-phenyl-2-ethoxy-ethane-1-one, 4-(1-methyl-ethyl)-benzaldehyde, acetophenone, 2,4,6-trimethyl-benzaldehyde	UV-cured systems
Monomers: Styrene, vinyltoluene, 2-phenyl-1-propane,	UPE
1,6-hexanediol-diacrylate, 2-ethylhexyl-acrylate,	AC
butyl acrylate, methyl methacrylate, butyl methacrylate,	AC
phenoxyethyl acrylate,	AC
1,6-hexane diisocyanate	PUR
Miscellaneous: Acetic acid, phenol, BHT, DMP, DBP, naphthalene	

[1] 2,2,4-trimethyl-1,3-pentanediol mono-*iso*-butyrate.
[2] 2,2,4-trimethyl-1,3-pentanediol di-*iso*-butyrate.

from their investigation of nine furniture materials. Salthammer (1997a, b) has studied emissions from coated surfaces as a function of time. The complete study included 44 furniture samples of different types, and 150 VOCs could be detected. ΣVOC-values (20 days) covered a range from 4 μg/m³ to 1280 μg/m³. The arithmetic mean value was 174 μg/m³ and the median was 60 μg/m³. Detailed data and chamber conditions are given in Tables 3.3-3 and 3.3-4.

Table 3.3-3. Area-specific emission rates for different types of furniture coating. The substrates are given in parentheses.

Type of coating	Main compounds/ΣVOC	SER_A μg/(m^2h)	Ref.
D (PB)	2-(2-Butoxyethoxy)-ethanol	1700	
	1-Butanol	800	1
	ΣVOC	7800	
Q (PB)	Acetone	490	
	Hexanal	120	1
	ΣVOC	1100	
I (PB)	Acetone	380	
	2-(2-Butoxyethoxy)-ethanol	610	1
	ΣVOC	2300	
N (PB)	Acetone	390	
	m,p-Xylene	110	1
	ΣVOC	1000	
P (PB)	Acetone	430	
	Hexanal	93	1
	ΣVOC	900	
Q (PB)	Acetone	510	
	1-Methyl-2-pyrrolidone	2400	1
	ΣVOC	4100	
	Acetone	420	
PB (uncoated)	Hexanal	410	1
	ΣVOC	1600	
B (MDF)	ΣVOC	1200	2
B (MDF)	ΣVOC	350	2
F (MDF)	ΣVOC	1900	2
F (MDF)	ΣVOC	670	2
L (MDF)	ΣVOC	3550	2
L (MDF)	ΣVOC	230	2
N (MDF)	ΣVOC	510	2
M (MDF)	ΣVOC	250	2
P (MDF)	ΣVOC	370	2
P (MDF)	ΣVOC	640	2

PB = particle board (wood-veneered).
MDF = medium density fiberboard.
1) 23 °C, 45 % r.h., N=1.0 h^{-1}; L=1.0 m^2/m^3, 28 days (EPA, 1998).
2) 23 °C; 45 % r.h.; N=1.0 h^{-1}; L=2.4 m^2/m^3; 20 days (Fischer and Böhm, 1994).

Table 3.3-4. Test chamber concentrations for different types of furniture coating. The substrates are given in parentheses.

Type of coating	Main compounds/ΣVOC	μg/m³ (1d)	μg/m³ (28d)	Ref.
	n-Butyl acetate	4969	221	
F (alder)	4-Methyl-2-pentanone	1885	86	1
	ΣVOC	9472	447	
	Benzophenone	89	54	
N (PB)	n-Butyl acetate	139	5	1
	ΣVOC	296	68	
	Benzophenone	106	32	
N (alder)	Cyclohexanone	312	3	1
	n-Butyl acetate	65	3	
	ΣVOC	565	52	
	Limonene	7738	288	
R (pine)	Carvone	338	20	2
	ΣVOC	9104	383	
	2-Butoxyethanol	14456	40	
Q (PB)	1-Methyl-2-pyrrolidone	8909	142	3
	ΣVOC	23817	187	
	Hexanal	33	12	
I (PB)	2-(2-butoxyethoxy)-ethylacetate	112	9	3
	ΣVOC	213	34	
	n-Butyl acetate	963	48	
C (PB)	*m,p*-Xylene	87	4	3
	ΣVOC	1265	78	
	Benzaldehyde	72	5	
M (PB)	*m,p*-Xylene	60	3	3
	ΣVOC	231	29	
	1-Butanol-3-methoxy-acetate	4493	434	
F (MDF)	*n*-Butyl acetate	13705	736	3
	ΣVOC	18980	1273	
	1-Butanol-3-methoxy-acetate	850	180	
H (MDF)	*n*-Butyl acetate	6221	127	3
	ΣVOC	8083	527	
	2-Butoxyethanol	1176	60	
M (PB)	Ethyldiglycol	4484	156	4
	ΣVOC	6866	251	
	1-Butanol	1080	197	
N (PB)	*n*-Butyl acetate	11899	647	4
	ΣVOC	13615	936	

PB = particle board (wood-veneered).
MDF = medium density fiberboard.
1) 23 °C, 45 % r.h., $N=1.0\ h^{-1}$, $L=1.0\ m^2/m^3$ (Jann et al., 1997).
2) 23 °C, 45 % r.h., $N=6.25\ h^{-1}$, $L=6.25\ m^2/m^3$ (Jann et al., 1997).
3) 23 °C, 45 % r.h., $N=1.0\ h^{-1}$, $L=1.0\ m^2/m^3$ (Salthammer, 1997b).
4) 23 °C, 45 % r.h., $N=1.0\ h^{-1}$, $L=1.0\ m^2/m^3$ (Salthammer, unpublished data).

3.3.4.1 Solvents

Organic solvents are indispensable ingredients of liquid coating systems, and most reported emission data originate from solvent residues in finished products. It is obvious from Table 3.3-1 that the percentage is strongly dependent on the type of lacquer and on the application technique. Spray coating requires solvent contents of up to 75 %. Typical components are aliphatic hydrocarbons, aromatic hydrocarbons, esters, ketones and alcohols. Water-based coatings, which are applied by pouring or rolling, may contain alcohols, glycols or 1-methyl-2-pyrrolidone in amounts of up to 10 %. The main identified compounds are also summarized in Tables 3.3-2 to 3.3-4. Emission profiles of solvent components are characterized by high initial values. A steep descent over a period of between 1 and 10 days is generally observed, after which only a slight decay is evident (see Figs. 3.3-3A and 3.3-3B). Most solvents do not undergo chemical reactions under environmental conditions. An important exception is 2,4,7,9-tetramethyl-5-dicyne-4,7-diol (T4MDD), which is used as a wetting and defoaming agent in several water-based lacquer systems. Under environmental conditions T4MDD decomposes to 4-methyl-2-pentanone (MIBK) and 3,5-dimethyl-1-hexyne-3-ol (Salthammer et al., 1999).

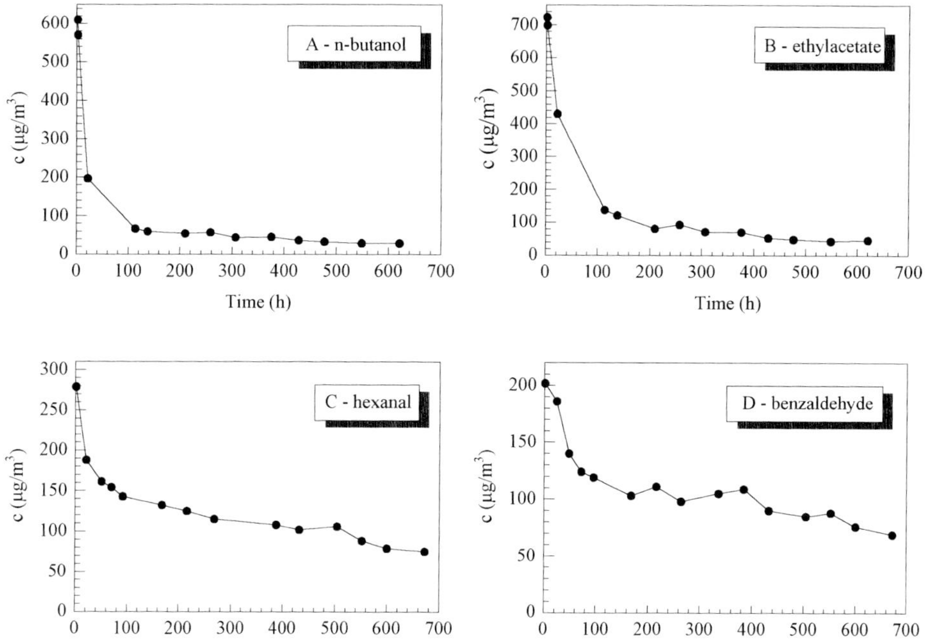

Figure 3.3-3. Concentration vs time profiles for emission of selected VOCs from coated surfaces: *A* *n*-butanol (UV-cured, spray coating); *B* ethyl acetate (UV-cured, roller coating); *C* hexanal (NC); *D* benzaldehyde (UV-cured, roller coating). The test chamber conditions were T=23 °C, r.h.=45 %, N=1.0 h^{-1}; L=1.0 m^2/m^3.

3.3.4.2 Aliphatic Aldehydes

Saturated and unsaturated aldehydes from C5 to C11 belong to one of the problematic and undesired classes of compounds in indoor air. Aliphatic aldehydes are highly odorous, the odor being generally regarded as unpleasant. Reported odor thresholds are in the range of some $\mu g/m^3$, e.g. hexanal (57 $\mu g/m^3$) and nonanal (13 $\mu g/m^3$) (Devos et al., 1990). Odor thresholds of unsaturated aldehydes such as 2-heptenal and 2-nonenal are about a decade lower (Devos et al., 1990). Common sources in indoor air are unsaturated fatty acids such as linoleic acid, linolenic acid and oleic acid, which are ingredients of many building products such as linoleum (Jensen et al., 1995) and furniture coatings containing alkyd or natural resins (Salthammer, 1997b). On oxidation, many types of volatile aldehydes are formed. Typical degradation products of oleic acid are saturated aldehydes from heptanal to decanal, while linoleic acid gives mainly hexanal. Oxidation of linolenic acid leads to unsaturated compounds such as 2,4-heptadienal. Figure 3.3-3C shows the emission of hexanal from a nitrocellulose system containing alkyds. The emission rate of aldehydes from furniture coatings may remain on high levels over months to years. This stems from a continuous degradation of unsaturated fatty acids in the material under living conditions. It has been reported that some types of furniture start to emit odorous aldehydes after more than 5 years, when the coating film splits and penetration of oxygen is enabled.

3.3.4.3 Isoprenes

Organic substances with an isoprenic structure such as α-pinene, β-pinene, limonene, Δ^3-carene, phellandrene, camphene, myrcene, ocimene, longifolene, caryophyllene and others have been well described in indoor air. Terpenes belong to the class of biogenic hydrocarbons which are ingredients of essential oils and are produced and emitted by living sources such as plants. Common sources in indoor air are solvents of eco-lacquers (turpentine balsam), softwood and various household products. Emission studies on freshly produced wooden shelfboards (untreated pine) yielded very high initial area specific emission rates of 3800 $\mu g/(m^2h)$ and 1100 $\mu g/(m^2h)$ for α-pinene and Δ^3-carene, respectively (Salthammer and Fuhrmann, 1996). Other main components were β-pinene and limonene. Jann et al. (1997) have investigated eco-lacquers on pine boards in test chambers (specific air flow rate $Q = 1$ m/h) and have found high limonene concentrations of 7738 $\mu g/m^3$ and 288 $\mu g/m^3$ after 1 and 28 days, respectively. In addition, α-pinene, Δ^3-carene and carvone were monitored in significant concentrations. Limonene is the main component in most solvents for eco-lacquers. A typical emission profile is shown in Fig. 3.3-4A.

3.3.4.4 Photoinitiator Fragments

The reason for the increasing attractiveness of UV curing in coating technology is based on a number of benefits (Decker and Moussa, 1993). The photoinitiator, which starts the polymerization process by formation of radicals, is an essential ingredient

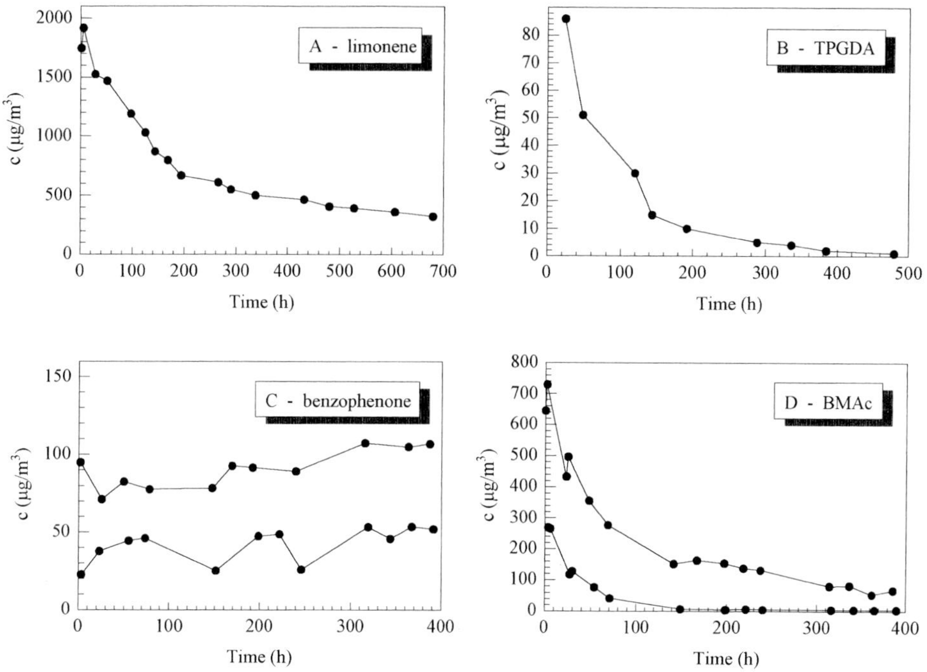

Figure 3.3-4. Concentration vs time profiles for emission of selected VOCs from coated surfaces and cabinet furniture. *A* limonene (ECO, painting); *B* tripropyleneglycol diacrylate (UV-cured, roller coating); *C* benzophenone (UV-cured, roller coating); *D* 1-butanol-3-methoxy-acetate (NC, spray coating). The upper and lower curves in *C* and *D* represent cabinet air and chamber air, respectively. The test chamber conditions were $T = 23\,^{\circ}C$, r.h. = 45 %, $N = 1.0\ h^{-1}$; $L = 1.0\ m^2/m^3$.

of UV-curable systems and has to fulfill a number of requirements (Fouassier, 1995). The photochemistry of most common photoinitiators is a chemistry of the carbonyl group. There are three important fragmentation processes forming radical species. Benzil ketals (Sandner and Osborn, 1974), 2-hydroxy-acetophenones (Phan, 1986) and acylphosphine oxides (Baxter et al., 1988) generate benzoyl radicals via the Norrish-I reaction (α-cleavage), which initiate the polymerization process. Dialkoxy-acetophenones undergo both Norrish-I and Norrish-II cleavage. On excitation of benzophenone in the presence of tertiary amines, an electron transfer complex is formed, and this is followed by proton transfer to a ketyl radical and an aminoalkyl radical. More detailed information about fragmentation processes of photoinitiators can be obtained from Hagemann (1989), Chang et al. (1992) and Dietliker (1991).

Fragmentation processes of photoinitiators form a number of volatile products, which may contribute to indoor air pollution. Benzaldehyde and alkyl-substituted benzaldehydes are usual components, because Norrish-I is the most important reaction for cleavage. A well known example is 1-phenyl-2-hydroxy-2-methyl-propane-1-one (PHMP). α-Cleavage generates two radicals in the first step. The benzoyl radical may recombine to form benzil, reduction of PHMP leads to 1-phenyl-2-methyl-1,2-propane and acetone, and recombination of the 2-hydroxypropyl radical gives 2,3-dimethyl-2,3-butanediol (pinacol). A concentration vs time curve is shown for benzaldehyde in Fig. 3.3-3D. Cyclohexanone is formed from the hydroxycyclohexyl radical upon α-cleavage of 1-hydroxycyclohexylphenone (HCPK), followed by hydrogen abstraction and keto-enol tautomerization. More typical fragmentation products are listed in Table 3.3-2. The results demonstrate that photoinitiator fragments may reach considerable concentrations in indoor air (Salthammer, 1996a). Fischer and Böhm (1994) have performed chamber experiments with an increased loading rate of 2.4 m^2/m^3. Chamber concentrations of 126 $\mu g/m^3$, 156 $\mu g/m^3$ and 522 $\mu g/m^3$ were measured after 20 days testing time for benzaldehyde, methyl benzoate and cyclohexanone, respectively. The release of benzaldehyde can be avoided by introducing substituents in position 4 of the phenyl ring, as was shown for 2-hydroxyacetophenones (Köhler, 1997). Another common photoinitiator is benzophenone, the content in coating formulations being ca. 2–4 %. This compound has a high boiling point of 305 °C, and the emission from materials is slow (see Fig. 3.3-4C). Increasing chamber concentrations can be observed even over a period of 700 h (Salthammer, 1996b; Jann et al., 1997). It must be mentioned that some photoinitiator fragments such as benzaldehyde and acetophenone are also typical degradation products of the polymeric Tenax TA on reaction with radicals, and careful analysis is required to avoid artifacts (Clausen and Wolkoff, 1997).

3.3.4.5 Monomers (Reactive Solvents)

In most UV-cured lacquer systems, properties such as viscosity and drying processes are controlled via liquid monomers or so-called reactive solvents. Such monomers are reaction products for cross-linking of oligomers and prepolymers. They are part of the cured layer and should not evaporate. Acrylic esters (acrylates) have a number of benefits and are now widely used in UV technology. Acrylic monomers are classified as mono-, bi- and polyfunctional, according to the number of terminal $–CH{=}CH_2–$ groups. Typical monofunctional monomers are methyl methacrylate (MMA), butyl acrylate (BA) and 2-ethylhexyl acrylate (EHA). These compounds are very volatile and are now to an increasing extent replaced by substances of higher molecular weight such as propylene glycol monoacrylate (PGMA). Well-known bifunctional monomers are 1,6-hexanediol diacrylate (HDDA) and tripropyleneglycol diacrylate (TPGDA). The most common polyfunctional agent is trimethylolpropane triacrylate (TMPTA). Airborne acrylates may cause various problems as they irritate skin and mucous membrane, and are corrosive to the eyes. Acrylate monomers remaining in the cured layer may lead to dermal burns on contact (Ohara et al., 1985). Furthermore, some acrylates are of very strong odor. Hence, even low quantities may significantly contribute to indoor air pollution. A concentration vs time curve is shown for TPGDA in Fig. 3.3-4B. The emission resulted from a UV-cured acrylate system. The initial value was 86 $\mu g/m^3$, which decreased to 2 $\mu g/m^3$ within 500 h testing time. The non-acrylic monomer styrene is used for copolymerization of UPE systems. Styrene has a low odor threshold and a low threshold limit value (TLV) for workplace air concentrations. Nevertheless, styrene is commonly used in UPE systems. Only a few substitution products such as 2-phenyl-1-propene, *n*-vinylpyrrolidone and vinyltoluene are known.

3.3.4.6 Diisocyanates

Diisocyanates are an important class of chemicals of commercial interest, which are frequently used in the manufacture of indoor materials such as adhesives, coatings, foams and rubbers (Ulrich, 1989). In some types of particle board, the diisocyanates have replaced formaldehyde. Isocyanates are characterized by the electrophilic –N=C=O group, which can easily react with molecules containing hydroxy groups, such as water or alcohols. On hydrolysis with water, primary amines are formed, while a reaction with alcohols leads to carbamates (urethanes). Polyurethane (PUR) products are then obtained from a polyaddition of diisocyanate and diol components. Compounds commonly used in industrial surface technology are 4,4'-diphenylmethane diisocyanate (MDI) and hexamethylene diisocyanate (HDI). The diisocyanate monomers are known as respiratory sensitizers and cause irritation of eyes, skin and mucous membrane. Therefore, polyisocyanates such as HDI-biuret and HDI-isocyanurate with a monomer content <0.5 % are used for industrial applications, and isocyanate monomers will not achieve high concentrations in ambient air. Nevertheless, it is desirable to measure even trace emissions from materials in private dwellings.

Salthammer et al. (1999) have performed chamber experiments with furniture coatings manufactured with MDI and HDI, respectively. The chambers were loaded within 5 h after application of the coating material. Air sampling was performed over a time period of 8 h. With a detection limit of 15 ng/m^3, MDI could not be monitored in the chamber air (Schulz and Salthammer 1998). In contrast, HDI, which is much more volatile than MDI, was detected in a chamber concentration of 1024 ng/m^3 when sampling started immediately after loading. The chamber value decreased to 36 ng/m^3 within 48 h, and after 72 h the HDI concentration was below the detection limit of 10 ng/m^3. Schmidtke and Seifert (1990) have investigated HDI emissions from freshly manufactured PUR coatings. A sampling volume >1m^3 and a sophisticated analytical procedure enabled detection limits <1ng/m^3. Up to 2 days after the coating process, 1800 ng/m^3 HDI could be detected. After 13–14 days only 5 ng/m^3 was found. The concentrations of HDI-biuret and HDI-isocyanurate were below the detection limit throughout.

3.3.4.7 Miscellaneous

It has already been mentioned that a large number of components are used for the construction of wood-based furniture. In addition to the compounds described earlier, a large spectrum of volatile organics can be found in emission studies (Salthammer, 1997b). The main sources for formaldehyde (which is not considered here), phenol and acetic acid are substrates such as particle board and MDF. BHT is a common antioxidant. Volatile plasticizers include dimethyl phthalate (DMP), dibutyl phthalate (DBP), as well as esters of adipic acid and sebacic acid. Further important compound groups are amines, siloxanes, carboxylic acids and naphthalenes. The identification of special substances does not only require suitable analytical equipment. Both experience and detailed knowledge of the chemical composition of furniture are also necessary.

3.3.5 Accumulation of VOCs in Cabinet Furniture

In cabinet furniture, the problem of interior accumulation of VOCs arises. In contrast to the living area, almost static conditions prevail in the interior, and accumulated compounds may diffuse into textiles or escape abruptly when the door is opened. This effect has been investigated in test chambers under standard conditions (Salthammer, 1996b). Five cubic boxes (40-cm edge) with front doors were manufactured using different lacquers and put into the chamber within 48 h after the coating process. For sampling of cabinet air, a Teflon hose was led through a bored hole to the chamber exit. As expected, the concentrations of all detected organic components were significantly higher in the cabinet air than in the chamber air. Concentration vs time curves were monitored for 20–30 days; two profiles are shown in Figs. 3.3-4C and 3.3-4D for benzophenone and 1-butanol 3-methoxy-acetate, respectively. ΣVOC-values up to 50 000 $\mu g/m^3$ were measured in the interior. The maximum concentrations of single compounds in the cabinet air were found for *n*-butyl acetate (14 590 $\mu g/m^3$) and 2-butoxyethanol (30 604 $\mu g/m^3$) after 1 day (plots not shown). In the case of solvents the concentrations decreased rapidly. Only for the photoinitiator benzophenone did the values increase with time, and even after 400 h a decay was not obvious. The cabinet/chamber concentration ratios were spread over the range ca. 2–10. Typical factors were about 3–6.

3.3.6 Conclusion

Wood and wood-based furniture are potential sources of a variety of organic compounds in the indoor environment. Most emitted compounds are typical solvents, but monomers and reaction products are also of importance. The source strength of a product will strongly depend on the substrate, the type of coating and the quality of the manufacturing process. It seems that one main reason for frequent complaints about VOC emissions from furniture is the common practice of tight wrapping in non-permeable foil immediately after manufacture and fast delivery to the customer. VOC residuals are reabsorbed in the interior of cabinets and problems such as a sudden release and contamination of textiles may arise.

References

Baumann W. and Muth A. (1997): Lacke und Farben. Vol.1/2. Springer, Berlin.

Baxter J.E., Davidson R.S., Hagemann H.J. and Overeem T. (1988): Photoinitiators and photoinitiation, 8. The photoinduced α-cleavage of acylphosphine oxides. Identification of the initiating radicals using a model substrate. Makromol. Chem., *189*, 2769–2780.

Bruner F. (1993): Gas chromatographic environmental analysis. VCH, New York.

CEC–Commission of the European Communities (1993): Determination of VOCs emitted from indoor materials and products, COST Project 613, Report No. 13, Brussels.

CEC–Commission of the European Communities (1994): Sampling strategies for volatile organic compounds (VOCs) in indoor air, COST Project 613, Report No. 14, Brussels.

Chang C.-H., Mar A., Tiefenthaler A. and Wostratzky D. (1992): Photoinitiators: mechanisms and applications. In: Calbo L.J. (ed) Handbook of coatings additives, Vol. 2. Dekker, New York.

Clausen P.A. and Wolkoff P. (1997): Degradation products of Tenax TA formed during sampling and thermal desorption analysis: indicators of reactive species indoors. Atmos. Environ., *31*, 715–725.

Decker Ch. and Moussa K. (1993): Recent advances in UV-curing chemistry. J. Coatings Technol., *65*, 49–57.

Devos M., Patte F., Rouault J., Laffort P. and van Gemert L.J. (1990): Standardized human olfactory thresholds. Oxford University Press, New York.

Dietliker K.K (1991): Photoinitiators for free radical and cationic polymerization. In: Oldring P.K.T. (ed) Chemistry and technology of UV & EB formulation for coatings, inks & paints. SITA Technology, Vol. 3, London.

ECA–European Concerted Action (1997): Evaluation of VOC emissions from building products. Report No. 18 (EU 17334 EN), Luxembourg.

EPA–United States Environmental Protection Agency (1998): Reducing emissions from engineered wood products. Inside IAQ, EPA/600/N-98/002, 8–11.

Fischer M. and Böhm E. (1994): Erkennung und Bewertung von Schadstoffemissionen aus Möbellacken, Reihe Schadstoffe und Umwelt, Band 12. Erich Schmidt, Berlin.

Fouassier J.-P. (1995): Photoinitiation, photopolymerization and photocuring. Hanser, Munich.

Garrat P.G. (1996): Strahlenhärtung. Vincentz Verlag, Hannover.

Guo Z. (1993): On validation of source and sink models: problems and possible solutions. In: Nagda N.L. (ed) Modeling of indoor quality exposure. ASTM STP 1205, 131–144, Philadelphia.

Guo Z., Tichenor B.A., Krebs K.A. and Roache N.F. (1996): Considerations on revisions of emissions testing protocols. In: Tichenor B.A. (ed) Characterizing sources of indoor air pollution and related sink effects. ASTM STP 1287, 225–236, West Conshohocken.

Hagemann H.J. (1989): Photoinitiators and photoionization mechanisms of free-radical polymerization processes. In: Allen, N.S. (ed) Photopolymerization and photoimaging science and technology 1–53. Elsevier, Crown House.

Hansemann W. (1996): Wood, surface treatment. Ullmanns Encyclopedia of Industrial Chemistry, Vol. 28A, 385–393. VCH, Weinheim.

Jann O., Wilke O. and Brödner D. (1997): Procedure for the determination and limitation of VOC emissions from furnitures and coated wood based products. In: Woods J.E., Grimsrud D.T. and Boschi N. (eds) Proceedings of Healthy Buildings/IAQ '97, Vol. 3, 593–598, Washington D.C.

Jensen B., Wolkoff P., Wilkins C.K. and Clausen P.A. (1995): Characterization of linoleum. Part 1: measurement of volatile organic compounds by use of the field and laboratory emission cell, FLEC. Indoor Air, *5*, 38–43.

Köhler M. (1997): A versatile α-hydroxyketone photoinitiator. Eur. Coat. J., *12/97*, 1118–1120.

Maroni M., Seifert B. and Lindvall T. (1995): Indoor air quality–a comprehensive reference book. Elsevier, Amsterdam.

Ohara T., Sato T., Shimizu N., Prescher G., Schwind H., Weiberg O. and Marten K. (1985) Acrylic acid and derivatives. Ullmann's Encyclopedia of Industrial Chemistry, Vol. A1, 161–176. VCH, Weinheim.

Pecina H. and Paprzycki O. (1994): Lack auf Holz. Vincentz Verlag, Hannover.

Phan X.T. (1986): Photochemical α-cleavage of 2,2-dimethoxy-2-phenyl-acetophenone and 1-hydroxy-cyclohexyl-phenyl ketone photoinitiators. Effect of molecular oxygen on free radical reaction in solution. J. Radiation Curing, *13*, 18–25.

Salthammer T. (1996a): Emission of photoinitiator fragments from UV-cured funiture coatings. J. Coatings Technol., *68*, 41–47.

Salthammer T. (1996b): VOC emissions from cabinet furnitures. Comparison of concentrations in the test chamber and in the cabinet. In Yoshizawa S., Kimura K., Ikeda K., Tanabe S. and Iwata T. (eds) Proceedings of the 7th International Conference on Indoor Air and Climate, Vol. 3, 567–572, Nagoya.

Salthammer T. (1997a): Holzlacke für Innenräume. Farbe und Lack, *103*, 142–150.

Salthammer T. (1997b): Emission of volatile organic compounds (VOC) from furniture coatings. Indoor Air, *7*, 189–197.

Salthammer T. and Fuhrmann F. (1996): Emission of monoterpenes from wooden furniture. In: Yoshizawa S., Kimura K., Ikeda K., Tanabe S. and Iwata T. (eds) Proceedings of the 7th International Conference on Indoor Air and Climate, Vol. 3, 607–612, Nagoya.

Salthammer T. and Marutzky R. (1995a): Emissionen organischer Verbindungen aus Möbeloberflächen. Holz-Zentralblatt, No. 144/95, 2405 and No. 6/96, 57–58.

Salthammer T. and Marutzky R. (1995b): Kammerverfahren zur Bestimmung der Emissionen organischer Substanzen aus Materialien. In: Bagda E. (ed) Emissionen aus Beschichtungsstoffen und deren Einfluß auf die Innenraumluft. Expert, 75–94, Renningen.

Salthammer, T., Fuhrmann, F., Kaufhold, S., Meyer, B. and Schwarz, A. (1995): Effects of climatic parameters on formaldehyde concentrations in indoor air. Indoor Air, *5*, 120–128.

Salthammer T., Schwarz A. and Fuhrmann F. (1999): Emission of reactive compounds and secondary emission products from wood-based furniture coatings. Atmos. Environ., *33*, 75–84.

Sandner M.R. and Osborn C.L. (1974): Photochemistry of 2,2,-dimethoxy- 2-phenyl-acetophenone–triplet detection via spin-memory. Tetrahedron Lett., 415–418.

Schmidtke F. and Seifert B. (1990): A highly sensitive high-performance liquid chromatographic procedure for the determination of isocyanates in air. Fresenius J. Anal. Chem., *336*, 647–654.

Schulz M. and Salthammer T. (1998): Sensitive determination of airborne diisocyanates by HPLC: 4,4'-dimethylene-diisocyanate (MDI). Fresenius J. Anal. Chem., 362, 289–293.

Stoye D. and Freitag W. (1996) Lackharze. Hanser, München.

Tirkkonen T., Mroueh U.M. and Orko, I. (1995): Tenax as a collection medium for volatile organic compounds–literature survey. Nordic Committee on Building Regulations (NKB), Indoor Climate Committee, Report No. 1995 : 6 E, Helsinki.

Ulrich H. (1989): Isocyanates, organic. Ullmann's Encyclopedia of Industrial Chemistry, Vol. A14, 611–625 VCH, Weinheim.

Wolkoff P. (1996): An emission cell for measurement of volatile organic compounds emitted from building materials for indoor use–the Field and Laboratory Emission Cell FLEC. Gefahrst. Reinhalt. Luft, *56*, 151–157.

3.4 Volatile Organic Ingredients of Household and Consumer Products

Tunga Salthammer

3.4.1 Introduction

People living in a modern industrial society inevitably come into contact with a variety of chemical products. The spectrum of different utilities for hobbies, cleaning, cosmetics, profession etc. is enormous (see Fig. 3.4-1), and the necessity for their use is suggested to the customer by tireless advertising (Selinger 1988; Emsley 1997; Wolkoff et al., 1998). Many of these consumer products contain pesticides, solvents, reactive monomers and other hazardous materials, which are released during use and which may affect human welfare (Ott and Roberts 1998). Nowadays, numerous organic pollutants can be monitored in indoor air (Krause et al. 1987; Brown et al. 1994; Salthammer 1994; Pluschke 1996) and on house dust (Wolkoff and Wilkins 1994; Wilkins et al., 1993) and other adsorptive media.

The pollution of indoor air by volatile organic compounds (VOCs) has for a long time been considered as a hygienic problem, so that systematic investigations have been carried out since the early 1980s. As far as the evaluation of analytical results is concerned, it has to be differentiated strictly between *immission* and *emission*. Investigation of indoor

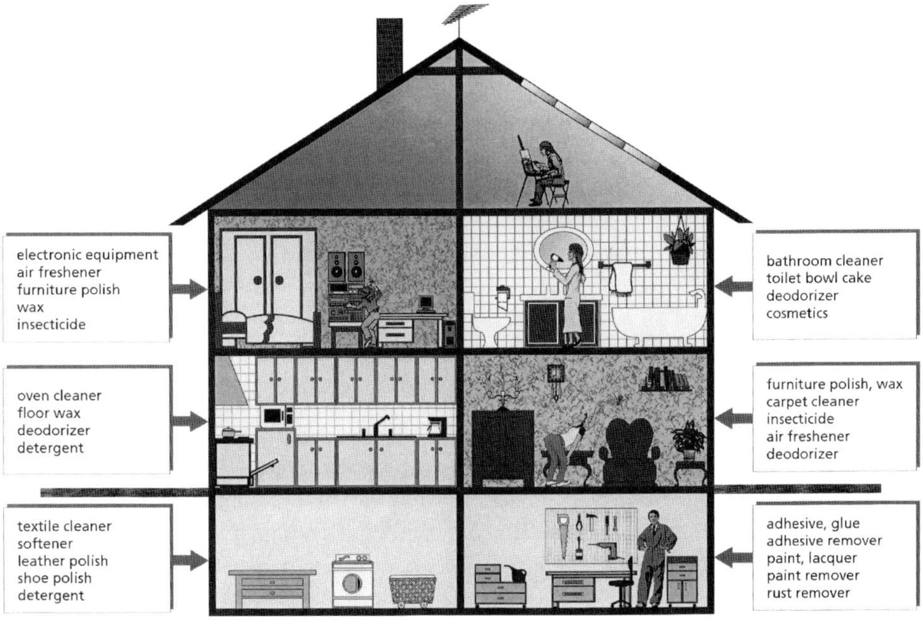

Figure 3.4-1. Typical activities in dwellings and related applications of household products.

air is a typical measure of immission. The result, usually expressed in μg/m^3, represents the average concentration of a certain substance in the indoor air over the sampling period. In contrast, an emission measurement, for example in test chambers or emission cells (see also Chapters 2.1 and 2.2), defines the release from a defined emission source as a function of time. The result is generally expressed as area-specific emission rate[1] (SER_A) in μg/(m^{2}h) or unit specific emission rate (SER_u) in μg/h. In chamber experiments, the result can also be expressed as test chamber concentration in μg/m^3. In this case the air exchange and the loading factor have also to be mentioned.

As far as building materials and furniture are concerned, the determination of the emission vs time behavior under living conditions can be modeled by use of test chambers or emission cells. Decay curves are generally observed, whose shape can be described by multiexponential functions under constant climatic conditions. In contrast, according to their utilization, substances from household products are related to human activities and will show strongly varying emission properties. Moth crystals and toilet deodorizers for example are designed for continuous use and emit volatile components at a constant emission rate. In comparison, the spontaneous release of VOC from sprays, waxes, liquid cleaners and other detergents soon leads to high concentrations which decay rapidly.

This Chapter summarizes the results from emission studies with different household products carried out in a 23.5l test chamber and a model room. Additional consideration of literature data allows a qualitative and quantitative view on the emission behavior and on the exposure during use.

3.4.2 Experimental

3.4.2.1 Test Samples

Household products were selected from a commercial store. Sprays, waxes, adhesives and liquid agents were preferred, because their usage would be expected to cause high VOC emissions. The ingredients were applied on a clean glass plate; the release of VOCs was monitored qualitatively in a test chamber at 23 °C, 45 % r.h. and air exchange $N = 1.0\ h^{-1}$. Two sprays were applied in a model room.

3.4.2.2 Test Chamber

The 23.5l glass chamber has a diameter of 30 cm and a height of 29.3 cm, and is sealed with an arched cover. The surface-to-volume ratio is 16 m^2/m^3. The chamber is purged with compressed air, which is passed through an oil separator and activated charcoal for purification. The air exchange rate is steadily controlled by a mass-flow controller (MKS). The required humidity is regulated by mixing dry and wet air. For heating and cooling, a thermostat is used (Haake). Air velocities are adjusted by use of a rotating

1 The term *"area-specific emission rate"* is used in parallel with *the term "emission factor"*.

paddle. The paddle has a diameter of 8 cm and is driven by a magnetic stirrer (Salthammer et al., 1995).

3.4.2.3 Model Room

Field measurements were performed in a model room (office) with a surface of 17.3 m^2 and a height of 3.7 m ($V = 64$ m^3). The air exchange rate is $N = 0.31h^{-1}$ (door and windows closed). The floor is covered with textile carpet and the walls are of gypsum board coated with water-based paint. The room is equipped with desk, chair, bookcase and wooden shelf board.

3.4.2.4 Sampling and Analysis

Air samples were collected on Tenax TA (60–80 mesh) (0.5–1.0 l total volume at a flow rate of 50 ml/min). The analysis of Tenax tubes was carried out using a GC/MS-system (Hewlett-Packard 6890) or a GC/FID-system (Hewlett-Packard 5890) with a 25-m HP-5 column, each equipped with a thermal desorber – cold trap injector (Perkin Elmer ATD 400). Identification of the compounds was based on a PBM library search. Moreover, mass spectra and retention data were compared with those of reference compounds. The monoterpenes and sesquiterpenes were sampled on active charcoal (NIOSH standard, SKC 226–01, 60–80 l total volume at a flow rate of 0.5 l/min). The terpenes were extracted by use of carbon disulfide under constant shaking for 1 h and analyzed via GC/FID. Aldehydes were trapped on filter cartridges coated with dinitrophenylhydrazine (DNPH) (Macherey & Nagel). The dinitrophenylhydrazone derivatives were extracted with acetonitrile and analyzed via HPLC/UV.

3.4.3 Literature Survey

The spectrum of human activity-related emissions of VOCs is wide. Wolkoff (1995) pointed out that these activities show an even greater variety of different VOC classes than emissions from building materials. His publication summarizes 17 selected examples of human activities and related VOC emissions from 37 references. Furthermore, Wolkoff (1995) refers to Rodes et al. (1991) who stated that human activity-related sources may be regarded as point sources and are located close to occupants, forming spatial and temporal concentration patterns. A total of 1159 different household products were analyzed by Sack et al. (1992) for 31 VOCs. The results have been compiled into a data base. Brown et al. (1994) present area-specific emission rates for 13 different types of wet household products. Braungart et al. (1997) have determined VOC emission rates from 19 building materials, electronic devices and consumer products in a small test chamber. Knöppel and Schauenburg (1987; 1989) could identify more than 80 VOCs when investigating eight waxes and two detergents. More data and references can be obtained from Maroni et al. (1995), Pluschke (1996), Witthauer et al. (1993), Raaf (1992) and Selinger (1988).

3.4.4 Product Classes

In the following, different product classes are discussed in separate Sections. Table 3.4-1 shows VOCs which have been identified by emission testing or material analysis and are representative of the assigned products. In Table 3.4-2, area-specific (SER_A) and unit-specific emission rates (SER_u) are summarized for selected products and compounds to give an overview. As mentioned earlier, the emission behavior of household products will strongly depend on the type of application and on the test conditions. For a detailed understanding it is therefore necessary to consult the cited references.

Table 3.4-1. Volatile ingredients of household and consumer products. See original reference for analytical details.

Product	Ingredients	Ref.
Newspaper, Newspaper journal	Toluene, *o-,m-,p*-xylene, α-/β-pinene, limonene, aliphatic hydrocarbons (C10-C20)	2
Scientific journal	Formaldehyde, toluene, pentanal, hexanal, nonanal, aliphatic hydrocarbons (C8-C18)	7
Schoolbook	Acetone, cyclohexanone, C3-benzenes, 1-butanol-3-methoxyacetate, vinylacetate	8
Electric shaver	1-Ethoxy-2-propanol, cyclohexanone, toluene, naphthalene, methylnaphthalenes (isomers), cyclohexanone, C3-C4 benzenes, aliphatic hydrocarbons, BHT	2
Portable CD player	Toluene, C2-C4 benzenes, cyclohexanone, 2-ethyl-1-hexanol	2
Insect spray	Aliphatic hydrocarbons (C4-C12), branched alkanes, cycloalkanes, tetramethrin, d-phenothrin, piperonyl butoxide	8
Furniture beetle agent	Aliphatic hydrocarbons (C10-C14), branched alkanes, cycloalkanes, acetone, dipropylene glycol monomethyl ether, cyfluthrin	8
Pest control agent	*cis-/trans*-Permethrin, deltamethrin, pyrethrum, piperonyl butoxide	1
Air freshener	α-Pinene, limonene, myrcene, linalool, octanal, nonanal, α-terpineol, decanal, ocimene, linalyl acetate	8
Room freshener	Aliphatic hydrocarbons (C9–C11), branched alkanes, limonene, substituted aromatic hydrocarbons	9
Toilet deodorizer	Myrcene, limonene, terpinene, terpinolene	8
Adhesive (spray)	Branched alkanes, cycloalkanes, toluene, C2-/C3-benzenes	8
Adhesive (liquid, high stability)	Branched alkanes, cycloalkanes, Ethyl acetate, 1,1-diethoxyethane, *n*-butyl acetate, C2-benzenes	8
Adhesive (liquid, all purpose)	Acetone, methyl acetate, ethyl acetate, 2-butanone, 1-ethoxy-1-methoxyethane, 1,1-diethoxyethane, 2,2-diethoxypropane, *n*-butyl acetate	8
Adhesive remover	1-Propanol, 1-propoxy-2-propanol, 2-propanol, tripropylene glycol	8

Table 3.4-1. (Continued)

Product	Ingredients	Ref.
Oven cleaner (spray)	Aliphatic hydrocarbons (C4), 1-methyl-2-pyrrolidone, propylene glycol	8
Specialized cleaner	Acetone, ethanol, 2-butanone, 3-methyl-2-butanone, 2,2-diethoxypropane	8
Cleanser/detergent	Ethanol, 2-propanol, 2-methyl-1-propanol, 3-butenyl propyl ether, 3-methylbutyl acetate, 2-propanol, limonene, tridecane	4, 5, 6
Floor cleanser	Fatty acids, fatty acid salts	10
Paint remover	Acetone, 1-methoxy-2-propanol, 2-(2-methoxyethoxy)-ethanol, 1-methyl-2-pyrrolidone, C2-C6-benzenes	8
Carpet cleanser (spray)	1-Methoxy-2-propanol, 2-methoxy-1-propanol	3
Leather polish	2-Propanol, aliphatic hydrocarbons, branched alkanes	8
Furniture polish (spray)	Branched alkanes, cycloalkanes, 1-butanol-3-methyl acetate, tetrahydronaphthalene	8
Furniture polish (liquid)	1-Methoxy-2-propanol, *n*-butyl acetate, 2-propanol, aliphatic hydrocarbons, branched alkanes, cycloalkanes, 1-butanol-3-methyl acetate	8
Shoe polish	Aliphatic hydrocarbons, branched alkanes, cycloalkanes, C2-C4 benzenes	8
Furniture polish	Aliphatic hydrocarbons, ethylbenzene, limonene	9
Furniture polish (spray)	Aliphatic hydrocarbons (C7/C8), branched alkanes	3
Floor wax paste	Limonene, α-/β-pinene, Δ^3-carene	3,6
Liquid wax	Acetone, 2-butanone, toluene, 1-hexanol, 3,7-dimethyl-1,7-octanediol, α-pinene, linalool, camphor, linalyl acetate	4,5,6
Hair lacquer (spray)	Ethanol, ethyl acetate, branched alkanes, limonene, *n*-octylether	8

References: 1) Berger-Preiss et al., 1997; 2) Braungart et al., 1997; 3) Colombo et al., 1990; 4) Knöppel and Schauenburg H., 1987; 5) Knöppel and Schauenburg, 1989; 6) Person et al., 1990; 7) Salthammer et al., 1997; 8) This work; 9) Tichenor and Mason, 1988; 10) Clausen et al., 1998.

3.4.4.1 Newspaper and Journals

For the manufacture of paper products such as newspapers, journals and magazines, a variety of chemicals such as resins, optical brighteners, fillers, dye pigments, solvents and others are required (Baumann et al., 1993). As far as the printing inks are concerned, solvent-based systems, which may contain ketones, esters, aliphatic hydrocarbons and aromatic hydrocarbons, are still widely used. However, water-based dispersions, whose application has grown continuously over the past years, also contain small quantities of solvents, mostly alcohols and glycols.

Table 3.4-2. Area-specific (SER_A) and unit-specific (SER_u) emission rates for different household and consumer products. See original references for analytical details.

Agent	Compound	SER_A µg/(m²h)	SER_u µg/h	Lit.
Toilet deodorizers	TVOC	$1.3\ 10^6 - 3.7\ 10^6$		5
Room fresheners	TVOC	$1.6\ 10^5 - 2.0\ 10^6$		5
Waxes	TVOC	$1.0\ 10^6 - 9.4\ 10^7$		5
Adhesives (solvent)	TVOC	$5.1\ 10^6 - 1.7\ 10^7$		5
Adhesives (water)	TVOC	$1.0\ 10^4 - 2.1\ 10^6$		5
Floor cleaners	TVOC	$1.0\ 10^4 - 1.5\ 10^5$		5
Moth crystals	TVOC	$1.4\ 10^7$		7
Wax (floor)	TVOC	$2.0\ 10^7$		7
Cleaning agents/	1,1,1-Trichloroethane	$2.2\ 10^3$		6,8
Insecticides	Carbon tetrachloride	$4.3\ 10^3$		6,8
	p-Dichlorobenzene	26.4		6,8
	m-Dichlorobenzene	33.6		6,8
Wax paste (furniture)	TVOC	$2.6\ 10^8$		3,4
Liquid wax (floor)	TVOC	$9.6\ 10^7$		3,4
Wax paste (leather)	TVOC	$3.3\ 10^6$		3,4
Cleanser/detergent	TVOC	$1.1\ 10^6$		3,4
Liquid wax (marble)	TVOC	$4.8\ 10^5$		3,4
Liquid wax (wood)	TVOC	$3.0\ 10^5$		3,4
Detergent	TVOC	$2.4\ 10^5$		3,4
Liquid wax (marble)	TVOC	$1.8\ 10^5$		3,4
Liquid wax (floor)	TVOC	$1.8\ 10^5$		3,4
Liquid wax (ceramic)	TVOC	$1.2\ 10^5$		3,4
Desinfectant	TVOC		$3.5\ 10^4$	2
	Bornyl acetate		$2.9\ 10^4$	2
Detergent (floor)	TVOC		$2.2\ 10^3$	2
	Menthol		$1.3\ 10^3$	2
Spray cleanser (carpet)	1-Methoxy-2-propanol		$5.0\ 10^4$	2
Spray polish (furniture)	TVOC		$2.7\ 10^4$	2
	n-Octane		$3.7\ 10^3$	2
Wax paste (floor)	TVOC		$1.9\ 10^3$	2
	α-Pinene		$6.4\ 10^2$	2
Answering machine	TVOC		51	1
Computer mouse	TVOC		26	1
Telephone	TVOC		34	1
Electric shaver	TVOC		207	1
CD-player	TVOC		61	1

References: 1) Braungart et al., 1997; 2) Colombo et al., 1990; 3) Knöppel and Schauenburg, 1987; 4) Knöppel and Schauenburg, 1989; 5) Person et al., 1990; 6) Sheldon et al., 1988a; 7) Sparks et al., 1990; 8) Wallace et al., 1987.

Braungart et al. (1997) have studied VOC emissions from newspapers and news magazines in a desiccator purged with nitrogen. Main components detected in the chamber air were toluene, hexanal, aliphatic hydrocarbons and α-pinene. The unit-specific emission rates SER_u for the sum of detected VOCs 1 day after starting the tests were 128 µg/h and 143 µg/h for newspaper and news magazine, respectively. The highest emission rate of a single component was 88 µg/h (toluene, news magazine). However, this study lacks details about test conditions (temperature, humidity, air exchange).

Table 3.4-3. VOC emissions from a school-book with PVC cover. Chamber concentrations in a 23.5-l chamber 24 h after loading (T = 23 °C, r.h. = 45 %, N = 1.0 h^{-1}).

Compound	Concentration (μg/m^3)
Acetone	67
Cyclohexanone	412
ΣC3-benzenes[1]	1643
1-Butanol-3-methoxyacetate	1272
2-Butoxyethyl acetate	644
Vinyl acetate	72

[1] Calculated in equivalents of 1,3,5-trimethylbenzene.

It is a known fact that many freshly printed books and journals are of strong odor. In a severe case, school children complained about adverse health effects on touching a school book with a PVC cover. To find out the reason, the book was conditioned for 24 h and put into the 23.5l chamber with open pages at T = 23 °C, r.h. = 45 % and N = 1.0 h^{-1}. Sampling was performed after 24 h and 48 h. As shown in Table 3.4-3, very high emissions of aromatic hydrocarbons, glycols and other compounds were apparent. Even the quarterly delivered Indoor Air journal is of very unpleasant smell. Therefore, an additional chamber test was carried out with issue 4/96. The main components in the chamber air were formaldehyde (335 μg/m^3), hexanal (15 μg/m^3), toluene (123 μg/m^3) and aliphatic hydrocarbons (>C10) (100–150 μg/m^3). Higher aliphatic aldehydes (C7–C11) appeared in concentrations <10 μg/m^3. The results also demonstrate that books and journals may release VOCs in significant amounts and must be regarded as typical point sources located close to occupants (Salthammer et al., 1997). As a consequence, chamber concentrations are presented instead of unit-specific emission rates.

3.4.4.2 Electronic Devices

Braungart et al. (1997) have also investigated emissions from 11 electronic devices in their desiccator. All products were operated during or immediately before the test. Unit-specific emission rates ranged from 5.9 μg/h (cellular phone) to 206 μg/h (electric shaver). Emission rates of 7 products were below 27 μg/h. Typical VOCs detected in the chamber air were aliphatic hydrocarbons (C10–C18), aromatic hydrocarbons (toluene, *o*-, *m*- and *p*-xylene, C3-/C4-benzenes), 2-ethyl-1-hexanol, BHT (2,6-di-*tert*- butyl-4-methylphenol) and cyclohexanone. One electric shaver emitted large amounts of methylnaphthalenes. Emission rates of single compounds were well below 10 μg/h apart from one case in which the emission of 23 μg/h naphthalene from an electric hair drier was measured.

3.4.4.3 Insecticides

In the indoor environment, many types of products such as crystals, sprays and liquids are used for active and preventative protection from insects. Insect sprays are very popular because they are easy to handle and can be combined with air fresheners. Commonly, the amount of active agents is well below 2 %. For example, a commercially available product for indoor use contained 0.25 % tetramethrin, 0.05 % *d*-phenothrin and 1 % of the synergist piperonyl butoxide. Volatile ingredients were mainly aliphatic hydrocarbons (C4–C12), cycloalkanes and branched alkanes. Pyrethroids are also used as active agents in liquid products against furniture beetle. A commercially available insecticide contained 0.1 % cyfluthrin and a variety of volatile components such as acetone, aliphatic hydrocarbons (C10–C14), branched alkanes, cycloalkanes, C3 benzenes and dipropylene glycol monomethyl ether. It should also be mentioned that materials containing natural fibers are often equipped with synthetic pyrethroids as a precaution.

Berger-Preiss et al. (1997) have applied different pest control agents for professional use in a model house. The concentrations of active substances were monitored in the gas phase, on suspended particles, on house dust and on furniture surfaces over a period of 24 months. Permethrin and deltamethrin were detected on suspended particles immediately after application in concentrations of 2 $\mu g/m^3$ and 40 $\mu g/m^3$, respectively. In house dust, the initial concentrations of permethrin and deltamethrin were 50 mg/kg and 150–800 mg/kg, respectively.

Common ingredients of moth crystals are *m*-dichlorobenzene and *p*-dichlorobenzene (Tichenor and Mason, 1988). Area-specific emission rates up to 1.4×10^7 $\mu g/(m^2h)$ were measured (Sparks et al., 1990). Sheldon et al. (1988a) and Wallace et al. (1987) have determined 33.6 $\mu g/(m^2h)$ and 26.4 $\mu g/(m^2h)$for *m*-dichlorobenzene and *p*-dichlorobenzene, respectively.

3.4.4.4 Air Fresheners/Deodorizers

Deodorizers certainly belong to the most important group of consumer products in the modern world. Care products for personal hygiene, which are known as antiperspirants, help to avoid the bacterial decomposition of sweat by use of preservatives with bacteriocidal activity. In contrast, odorous VOCs are the ingredients of room air fresheners and are released in order to mask unpleasant smells. Solid room deodorizers and toilet bowl cakes may contain *m*- and *p*-dichlorobenzene and are mainly used in bathrooms. Shields et al. (1996) have measured concentrations up to 20 $\mu g/m^3$ *p*-dichlorobenzene in commercial buildings. Modern products are free of halogenated hydrocarbons and are based on isoprenes such as myrcene, limonene, terpinene and terpinolene (Ohloff, 1990). However, most air fresheners for living spaces are sprayed as aerosols. Customer demand is immense, as is demonstrated by the variety of available fragrances, from primrose to violet, in commercial stores. Person et al. (1990) identified monoterpenes, ketones, alcohols and aldehydes as the main components of deodorizers. Measured emission rates ranged from 1.3×10^6–3.7×10^6 $\mu g/(m^2h)$ (toilet deodorizers) to 1.6×10^5–2.0×10^6 $\mu g/(m^2h)$ (room fresheners). Shields et al. (1996) regard room fresheners and personal care

Table 3.4-4. VOC concentrations in a model room after application of 8.1 g room freshener.

Compound		Time (h)					
		0.5	1.0	1.5	2.5	3.3	4.0
Limonene	μg/m³	29	24	19	18	16	15
Linalool	μg/m³	38	13	9	6	4	<1
Nonanal	μg/m³	13	6	<1	<1	<1	<1
α-terpineol	μg/m³	23	7	4	<1	<1	<1
Decanal	μg/m³	18	5	<1	<1	<1	<1
Linalyl acetate	μg/m³	96	29	27	14	12	<1
Ocimene	μg/m³	13	<1	<1	<1	<1	<1
Diethyl phthalate	μg/m³	64	15	7	4	<1	<1
TVOC	μg/m³	502	149	100	72	63	45

products as the main sources of *α*-pinene and limonene found indoors. Tichenor and Mason (1988) have monitored nonane, decane, undecane, ethylheptane, limonene and substituted aromatics as representative organic compounds in room fresheners.

In order to determine VOC concentrations after application of an air freshener (flowers and fruits), 8.1 g was sprayed into the model room described earlier. The door and windows were kept closed during the test. Sampling was performed in the time range from 0.5 h to 4 h (see Table 3.4-4). Detected propellents were aliphatic hydrocarbons (C4 and C5). Artificial fragrance was composed of a mixture of terpenes (myrcene, limonene, linalool, *α*-terpineol, linalyl acetate, ocimene) and aliphatic aldehydes (octanal, nonanal, decanal). Other terpenes were detected in trace amounts. The TVOC value after 0.5 h was 502 $\mu g/m^3$ and decreased to 45 $\mu g/m^3$ after 4 h. The background concentration of 33 $\mu g/m^3$ was measured immediately before the test. The highest concentration of a single compound was 96 $\mu g/m^3$ (linalyl acetate).

3.4.4.5 Adhesives

Adhesives are present in every household and are used for manifold purposes. Depending on application, adhesives may be regarded as point sources with short-term exposure (repair, hobbies) or as area sources with long-term exposure (fixing of floor coverings). Many solvent-based adhesives are narcotic and of agreeable smell. This explains their misuse for sniffing. Volatile main components are alkanes, ethyl acetate, toluene, ethyl benzene, *o-,m-,p*-xylene and *n*-butyl acetate. Person et al. (1990) have determined area-specific emission rates of $5.1 \times 10^6 - 1.7 \times 10^7$ $\mu g/(m^2h)$ and $1.0 \times 10^4 - 2.1 \times 10^6$ $\mu g/(m^2h)$ for solvent-based and water-based adhesives, respectively. VOC emission rates from adhesives were reviewed by Maroni et al. (1995). Tichenor and Mason (1988), Sheldon et al. (1988a;b) and Wallace et al. (1987) have studied VOC emissions from adhesives, glued carpet and glued wallpaper using different analytical techniques.

3.4.4.6 Cleaners

The number of available cleaners for floor coverings, furniture, textiles, kitchen, bathroom etc. is enormous. Most products are based on inorganic compounds such as phosphoric esters, peroxides and surfactants (Raaf, 1992). Nevertheless, organic dirt such as oil, grease or tar requires special organic agents. Sack et al. (1992) have investigated more than 100 household cleaners. *m*-Xylene was found in 33 % of the samples. Other main analytes were acetone, 1,1,1-trichloroethane, *n*-octane and methylene chloride. The study of Person et al. (1990) included nine floor cleaning products. Area-specific emission rates covered a range from 1.0×10^4 to 1.5×10^5 µg/(m^{2}h). Detected main components were aliphatic hydrocarbons, aromatic hydrocarbons, oxygenated compounds and terpenes. Colombo et al. (1990) have determined an initial TVOC emission rate of 3.5×10^4 µg/h for a liquid cleaner. A spray cleaner for carpet emitted 5.0×10^4 µg/h 1-methoxy-2-propanol. Knöppel and Schauenburg (1987; 1989) identified numerous VOCs when investigating a cleanser/detergent. Main components were 2-methylpropanol and 2-propanol; the area-specific emission rate was 1.1×10^6 µg/(m^{2}h). Sack et al. (1992) found that cleaners for electronic equipment contain mostly chlorinated solvents, 1,1,2-trichlorotrifluoroethane, 1,1,1-trichloroethane and methylene chloride being the commonest. Ingredients of household products are frequently not declared on the label. This was especially the case for an invesigated paint remover. The only information given to the customer was a content of 1 – 2 % 1-methyl-2-pyrrolidone. However, this product also contained about 10 % of aromatic hydrocarbons (see Fig. 3.4-2A).

Emissions from cleaners may result in high VOC values in the indoor environment, as demonstrated by Seifert et al. (1989). In a private dwelling, limonene concentrations up to 450 µg/m^3 were measured when a rack used to dry washed linen was kept inside during the winter season. Limonene is often added to textile softeners used in the washing process. Clausen et al. (1998) have investigated house dust for detergents. Fatty acids and fatty acid salts as common ingredients of floor cleaning agents were mainly found.

3.4.4.7 Polishes

Polishes are used for cleaning, conservation or esthetic reasons. In most cases, products are applied on large surfaces such as wood, ceramic, marble, linoleum or furniture and may therefore lead to high emissions. Waxes for treatment of leather clothes can cause irritation on inhalation, oral or dermal contact. Person et al. (1990) have investigated several products and claim that waxes emit large quantities of VOCs (1.0×10^6 to 9.4×10^7 µg/(m^{2}h)) when spread on surfaces. Knöppel and Schauenburg (1987, 1989) have identified 84 VOCs in ten different waxes and detergents. Area-specific emission rates ranged from 1.2×10^5 µg/(m^{2}h) to 2.6×10^8 µg/(m^{2}h). Colombo et al. (1990) have determined unit-specific emission rates of 2.7×10^4 µg/h and 1.9×10^3 µg/h for furniture spray polish and floor wax paste, respectively. Sparks et al. (1990) have published an emission rate of 1.4×10^7 µg/(m^{2}h) for a wood floor wax.

Organic components of polishes are widespread. Most products emit complex VOC mixtures, as shown in Fig. 3.4-2B for a furniture polish and Fig. 3.4-2C for a shoe polish.

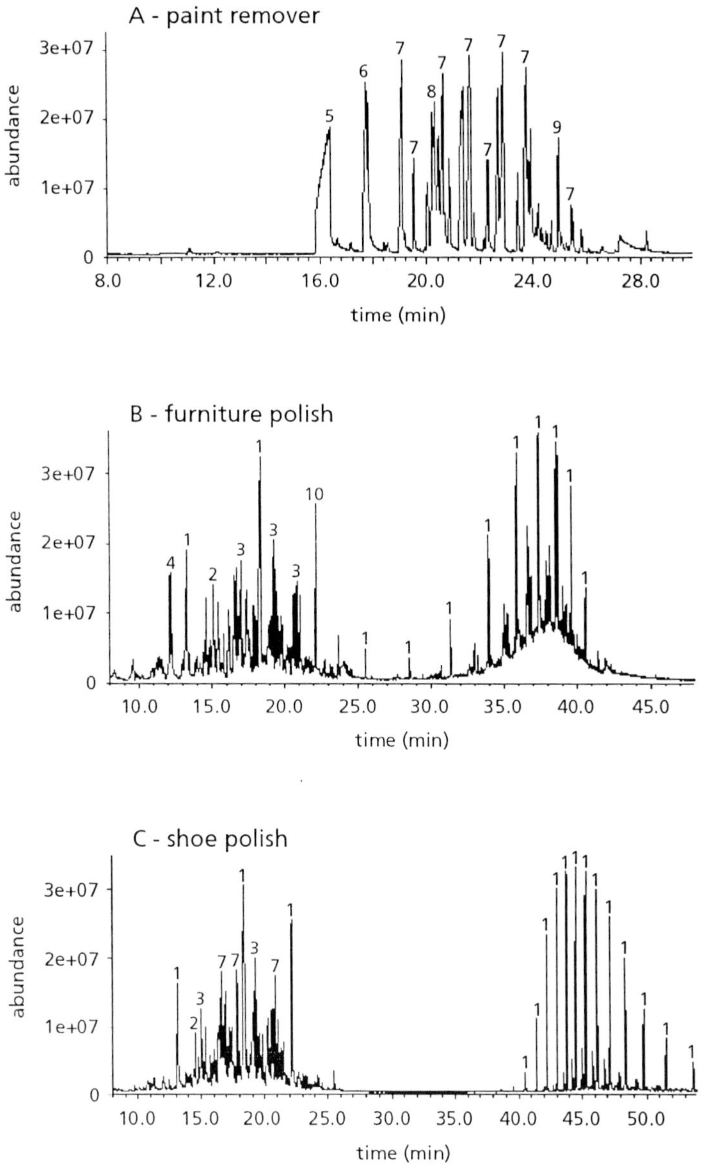

Figure 3.4-2. Volatile ingredients of different household products. *A* paint remover, *B* furniture polish, *C* shoe polish. *1* aliphatic hydrocarbons (C9-C32); *2* cycloalkanes;*3* branched alkanes; *4* 1-butanol-3-methylacetate; *5* diethylene glycol monomethyl ether; *6* 1-butanol-3-methoxy acetate; *7* C2-C5 benzenes; *8* 1-methyl-2-pyrrolidone; *9* naphthalene; *10* hydronaphthalene.

Alkanes, various alcohols, acetates, C2-C4 benzenes, terpenes and derivatives of naphthalene are frequently detected. Many modern floor waxes are based on natural ingredients such as alkyd resins. On oxidative degradation of unsaturated fatty acids, volatile aliphatic aldehydes (C5-C11) of disagreeable smell (Ruth, 1986) are formed,

Table 3.4-5. VOC concentrations in a hair-dressing shop.

Compound	Concentration (μg/m^3)
Toluene	288
Limonene	218
Hexamethyldisiloxane	253
Octamethyltrisiloxane	257
Octamethylcyclotetrasiloxane	19

and the emission rates may remain on high levels over months and even years (Salthammer et al., 1999).

3.4.4.8 Products for Personal Hygiene and Cosmetics

The exposure of humans in the private bathroom or public beauty shops is often underestimated. The major sources of exposure to chloroform ($CHCl_3$) in the United States are showers, boiling water and clothes washers. $CHCl_3$ is formed from the chlorine used to treat water supplies (Ott and Roberts, 1998). Many cosmetic products contain a number of VOCs such as 2-phenoxyethanol, 2-butanone, acetone, terpenes, 2-hydroxy-4-methoxybenzophenone or phenylmethanol. *However, in most product categories we are not selling a product, we are selling a proposition*, as Selinger (1988) pointed out.

A potential source of indoor pollutants is hair sprays. In order to estimate VOC concentrations, a female test person was placed in the model room described earlier and sprayed with 16.1 g hair lacquer. Propellant gases (butane, pentane), ethanol, limonene and tripropylene glycol (isomers) were detected. Thirty minutes after application, highest concentrations were measured for ethanol (> 100 μg/m^3) and limonene (22 μg/m^3). Van der Wal et al. (1997) have studied the exposure of hairdressers to chemical agents. The main component was always ethanol, stationary indoor concentrations in the winter months ranging from 4100 μg/m^3 to 10 400 μg/m^3 directly after application of sprays. The range of $TVOC_{(C6\text{-}C16)}$ values was 360–660 μg/m^3. Measurements in a German hair-dressing shop were carried out in April 1998 (door and windows closed). Increased concentrations of toluene, limonene and siloxanes were measured (see Table 3.4-5). Further components found in indoor air were ethanol, 2-propanol, *n*-butanol and various terpenes. Siloxanes are common ingredients of deodorants and antiperspirants. Shields et al. (1996) pointed out that personal care products are the dominant source of siloxanes, especially decamethylcyclopentasiloxane (D5).

3.4.5 Conclusion

Volatile organic compounds are emitted from a wide variety of household and consumer products. Emission rates are strongly dependent on the type of application and are distributed over several orders of magnitude. Most products emit VOC mixtures, but in many cases typical organic components can be related to product classes (Tichenor and Mason 1988). It is difficult to identify sources of indoor air pollution and to estimate human exposure, because volatile ingredients are only rarely declared on the label. Chamber experiments and simple headspace analysis are helpful to characterize chemical composition (Colombo et al., 1990). Nowadays, the customer has the choice between numerous products. Selection, necessity and application should be considered very critically. Ott and Roberts (1998) stated: *Yet, people cannot take the simple steps required without adequate knowledge. So increased education is needed. Law requiring more detailed information would also help: If a product contains a dangerous pollutant, should not the manufacturer be required a least to list the chemical by name on the package? Armed with a better understanding of toxic substances found in common products and in other sources at home, people could then make their own informed choice.*

References

Baumann W. and Herberg-Liedtke B. (1993): Papierchemikalien. Springer, Berlin.

Berger-Preiss E., Preiss A., Sielaff K., Raabe M., Ilgen B. and Levsen K. (1997): The behavior of pyrethroids indoors: a model study. Indoor Air, *7*, 248–261.

Braungart M., Bujanowski A., Schäding J. and Sinn C. (1997): Poor design practices: gaseous emissions from complex products. Hamburger Umwelt Institut, Project Report, Hamburg.

Brown S.K., Sim M.R., Abramson M.J. and Gray C.N. (1994): Concentrations of volatile organic compounds in indoor air–a review. Indoor Air, *4*, 123–134.

Clausen P.B., Wilkins K. and Wolkoff P. (1998): Gas chromatographic analysis of free fatty acids and fatty acid salts extracted with neutral and acidified dichloromethane from office floor dust. J. Chromatography A, *814*, 161–170.

Colombo A., De Bortoli M., Knöppel H., Schauenburg H. and Vissers H. (1990): Determination of volatile organic compounds emitted from household products in small test chambers and comparison with headspace analysis. In: Walkinshaw D.S. (ed) Proceedings of the 5th International Conference on Indoor Air and Climate, Vol. 3, 599–604, Toronto.

Emsley J. (1997): Parfum, Portwein, PVC. Wiley-VCH, Weinheim.

Knöppel H. and Schauenburg H. (1987): Screening of household products for the emission of volatile organic compounds. In: Seifert B., Esdorn H., Fischer M., Ruden H. and Wegner J. (eds) Proceedings of the 4th International Conference on Indoor Air and Climate, Vol. 1, 27–31, Berlin.

Knöppel H. and Schauenburg H. (1989): Screening of household products for the emission of volatile organic compounds. Environ. Int., *15*, 413–418.

Krause C., Mailahn W., Nagel R., Schulz C., Seifert B. and Ullrich D. (1987): Occurrence of volatile organic compounds in the air of 500 homes in the Federal Republic of Germany. In: Seifert B., Esdorn H., Fischer M., Ruden H. and Wegner J. (eds) Proceedings of the 4th Conference on Indoor Air and Climate, Vol. 1, 102–106, Berlin.

Maroni M., Seifert B. and Lindvall T. (1995): Indoor air quality–a comprehensive reference book. Elsevier, Amsterdam.

Ohloff G. (1990): Riechstoffe und Geruchsinn. Springer, Berlin.

Ott W.R. and Roberts J.W. (1998): Everyday exposure to toxic pollutants. Scientific American, *278*, 72–77.

Person A., Laurent A.M., Louis-Gavet M.C., Aigueperse J. and Anguenot F. (1990): Characterization of volatile organic compounds emitted by liquid and pasty household products via small test chamber. In: Walkinshaw D.S. (ed) Proceedings of the 5th International Conference on Indoor Air and Climate, Vol. 3, 605–610, Toronto.

Pluschke P. (1996): Luftschadstoffe in Innenräumen. Springer, Berlin.

Raaf H. (1992): Chemie des Alltags. Herder, Freiburg.

Rodes C., Kamens R. and Wiener R.W. (1991): The significance and characteristics of the personal activity cloud on exposure assessment measurements for indoor contaminants. Indoor Air, *1*, 123–145.

Ruth J.H. (1986): Odor thresholds and irritation levels of several chemical substances: a review. J. Am. Ind. Hyg. Soc. Assoc., *47*, A/142-A/151.
Sack T.M., Steele D.H., Hammerstrom K. and Remmers J. (1992): A survey of household products for volatile organic compounds. Atmos. Environ., *26A*, 1063–1070.
Salthammer T. (1994): Luftverunreinigende organische Substanzen in Innenräumen. Chemie in unserer Zeit, *28*, 280–290.
Salthammer T., Meininghaus R., Jacoby A. and Bahadir M. (1995): Distribution of air velocities in small test chambers. Fresenius Envir. Bull., *4*, 695–700.
Salthammer T., Fuhrmann F. and Schulz M. (1997): Indoor Air–Letter to the Editor, unpublished.
Salthammer T., Schwarz A. and Fuhrmann F. (1999): Emission of reactive compounds and secondary emission products from wood-based furniture coatings. Atmos. Environ., *33*, 75–84.
Seifert B., Ullrich D. and Nagel R. (1989): Seasonal variation of concentrations of volatile organic compounds in selected German homes. Environ. Int., *15*, 397–408.
Selinger B. (1988): Chemistry in the marketplace. Harcourt Brace Jovanovich, Sydney.
Sheldon L.S., Handy R.W., Hartwell T.D., Whitmore R.W., Zelon H.S. and Pellizzari E.D. (1988a): Indoor Air Quality in Public Buildings: Vol. 1. EPA Project Summary No. 600/S6–88/009a, Research Triangle Institute, Washington, D.C.
Sheldon L.S., Zelon H.S., Sickles J., Eaton C., Hartwell T. and Wallace L. (1988b): Indoor Air Quality in Public Buildings: Vol. 2. EPA Project Summary No. 600/S6–88/009b, Research Triangle Institute, Washington, D.C.
Shields H.C., Fleischer D.M. and Weschler C.J. (1996): Comparisons among VOCs measured in three types of U.S. commercial buildings with different occupant densities. Indoor Air, *6*, 2–17.
Sparks L.E., Jackson M., Tichenor B., White J., Dorsey J. and Stieber R. (1990): An integrated approach to research on the impact of sources on indoor air quality. In: Walkinshaw D.S. (ed) Proceedings of the 5th International Conference on Indoor Air and Climate, Vol. 4, 219–224, Toronto.
Tichenor B.A. and Mason M.A. (1988): Organic emissions from consumer products and building materials to the indoor environment. J. Air Poll. Contr. Assoc., *38*, 264–268.
Van der Wal J.F., Hoogeveen A.W., Moons A.M.M. and Wouda P. (1997): Investigation on the exposure of hairdressers to chemical agents. Environ. Int., *23*, 433–439.
Wallace L.A., Pellizzari E.D., Leaderer B.P., Zelon H. and Sheldon L. (1987): Emissions of volatile organic compounds from building materials and consumer products. Atmos. Environ., *21*, 385–393.
Witthauer J., Horn H. and Bischof W. (1993): Raumluftqualität. C.F. Müller, Karlsruhe.
Wilkins C.K., Wolkoff P., Gyntelberg F., Skov P. and Valbjørn O. (1993): Characterization of office dust by VOCs and TVOC-release–identification of potential irritant VOCs by partial least squares analysis. Indoor Air, *3*, 283–290.
Wolkoff P. (1995): Volatile organic compounds–sources, measurements, emissions, and the impact on indoor air quality. Indoor Air, Suppl. No. 3.
Wolkoff P. and Wilkins C.K. (1994): Indoor VOCs from household floor dust: comparison of headspace with desorbed VOCs; method for VOC release determination. Indoor Air, *4*, 248–254.
Wolkoff P., Schneider T., Kildesø J., Dergerth R., Jaroszewski M., Schunk H. (1998): Risk in cleaning: chemical and physical exposure. Sci. Total Environ., *215*, 135–156.

3.5 Occurrence of Biocides in the Indoor Environment

Werner Butte

3.5.1 Introduction

Biocides are often used in homes and buildings against various household pests and for protection against pests that attack wood and textiles. Some of the compounds used are identical to those used in formulations for protection against forest or agricultural pests (Tomlin, 1994). Some authorities do not require preparations for indoor use to be officially registered, e.g. in Germany, and the indoor use of biocides is often carried out without adequate expert knowledge and in excessive quantities.

Some biocides, especially substances having a low volatility, tend to be persistent; indoor air and dust are significant sources of nonoccupational biocide (pesticide) exposure for the general population, especially children. Indoor pollution has been ranked by the United States Environmental Protection Agency Advisory Board and the Center for Disease Control as a high environmental risk (Roberts and Dickey, 1995). An association between agricultural pesticide exposure and risk of leukemia was reported for some insecticides, including organophosphates, natural product pyrethrins and methoxychlor (Brown et al., 1990; Clavel et al., 1996). There are also indications that the indoor use of biocides may result in a risk to the occupants of adverse health effects (Hoffmann and Hostrup, 1997).

Biocides are either semivolatile or "non-volatile" organic compounds. Since vapor pressures of "non-volatile" biocides, e.g. pyrethroids, are small, these biocides are mainly particle-bound. Their analysis in air might not be appropriate for the detection of indoor contamination, and the investigation of household dust may be used alternatively. Other biocides are semivolatile, e.g. chlorpyrifos, lindane (γ-HCH) and pentachlorophenol; indoor contamination may be detected by analyzing either air or dust.

Methods of determining the most commonly used biocides in the indoor environment, i.e. in air and dust, are discussed here. Furthermore, concentrations in air and household dust obtained under defined conditions as well as results from representative studies are compiled. Data from the literature and our own results show widespread contamination even in homes where no biocides were used by the occupants. In view of these data, it is emphasized that the use of toxic substances should be minimized to the greatest possible extent.

3.5.2 Analytical Methods for Biocides in the Air and Dust of the Indoor Environment

3.5.2.1 Air

General aspects of strategies to measure biocides in indoor air are given in guidelines, e.g. by the VDI [VDI 4300 part 1 (1995), part 4 (1997)].

Sampling of biocides in air is generally performed by passing air through an adsorbent which traps the chemicals. Different materials have been utilized to collect biocides, e.g. Chromosorb 102, Orbo 42, Tenax, silica gel and polyurethane foam. Particle-bound biocides may be collected by glass, quartz or activated carbon fiber filters. Experiments by Roper and Wright (1984), who generated vapors containig 100 μg/m^3 of chlorpyrifos, chlordane, diazinon, propoxur and resmethrin in an air stream, showed no significant difference for the retention efficiency of all five sorbents tested. Regarding the sampling efficiency, Chromosorb 102 and polyurethane foam appeared to be superior to sorbents such as Tenax or Carbowax 20 M on Gas Chrom Q or Porpak C18, especially for volatile pesticides such as diazinon, chlorpyrifos and chlordane.

After air sampling, the sorbents are extracted and the biocides are analyzed using gas chromatrography (GC) or high pressure liquid chromatography (HPLC). For HPLC, quantification is done by UV absorption at an appropriate wavelength, whereas for GC either the electron capture detector (ECD) or a mass spectrometer (MS) is used for quantification. Depending on the sampled volume, air detection limits for HPLC methods are about 0.1 – 1 μg/m^3, for GC-ECD methods about 0.01 – 0.02 μg/m^3, and for MS methods 0.001 μg/m^3 (= 1 ng/m^3). A review of analytical methods for biocides in indoor air, including adsorbents, sampling conditions and detection limits is given in Table 3.5-1.

Especially for higher concentrations of biocides in air, impinging methods are an alternative. Pentachlorophenol concentrations can be analyzed after passing the air through an aqueous alkaline solution. A similar method has been reported for carbaryl and baygon. These biocides were absorbed in a sodium hydroxide solution and the hydrolysis products coupled to *p*-aminoacetophenol. The resulting dye could be spectrophotometrically quantified at $\lambda = 580$ or 555 nm with a detection limit of about 0.5 μg/m^3 (Das et al., 1994)

A method of estimating concentrations of pentachlorophenol, tetrachlorophenol and lindane in the indoor atmosphere of rooms after application of wood protection agents was given by Zimmerli and Zimmermann (1979). Filter papers were soaked with paraffin oil and exposed in the room. The amount of chemical absorbed per unit of time was proportional to the concentration in air.

Passive sampling methods can only be used when sampling volatile or semivolatile organics. On the other hand it is important in active sampling to choose appropriate sorbents and methods. Lindane (γ-HCH), for example, when measured with other pesticides, is only found to be present in the vapor phase and not in the particle phase when concentrations in these two phases are analyzed (Lane et al., 1992).

Table 3.5-1. Methods of determining biocides in indoor air.

Compound(s)	Adsorbent/filter (flow, volume)	Analytical apparatus	Detection Limit	Reference
Allethrin	Chromosorb 102 (100 l/h)	GC-ECD[1)]	2 ng/m^3	Eitzer (1991)
Cypermethrin	Orbo 42 (2 l/min, 240 l)	HPLC[2)] ($\lambda = 214$ nm)	0.1 $\mu g/m^3$	Wright et al. (1993)
Organophosphates (fenitrothion, diazinon)	Quartz-fiber filter/activated carbon-fiber filter (10 l/min, 7.2 m^3)	GC-MS[3)]	0.5 ng/m^3 (diazinon) 1 ng/m^3 (fenitrothinon)	Kawata and Yasuhara (1994)
Pentachlorophenol	0.1 M K_2CO_3–solution (70 l/h, 8 h)	GC-ECD[1)] derivatization	0.5 $\mu g/m^3$	Woiwode et al (1980)
Pentachlorophenol	Glass capillary column coated with SE54 (20 ml/min, 10 l)	GC-ECD[1)] derivatization	?	Grob and Neukom (1984)
Permethrin	Glass fiber filter (1 l/min, ≈ 15 l)	HPLC[2)] ($\lambda = 234$ nm)	8 $\mu g/m^3$	Rando and Hammad (1985)
Pesticides (23 commonly used indoor pesticides)	Tenax (≈ 4 l/min, 1 m^3)	GC-MS (CI)[4)]	1 ng/m^3 (bendiocarb) - 30 ng/m^3 (o-phenylphenol)	Roinestadt et al. (1993)
Pyrethroids (allethrin, cyfluthrin, tetramethrin)	Silica gel (0.5 m^3/h)	GC-ECD[1)]	0.1 $\mu g/m^3$	Claas and Kintrup (1991)
Pyrethroids (cyfluthrin, cypermethrin, deltamethrin, permethrin)	Glass fiber filter followed by plugs of polyurethane foam (1 m^3/h, 2 m^3)	GC-MSD[5)]	2 ng/m^3 (permethrin) - 10 ng/m^3 (deltamethrin)	Ball et al. (1993)
Wood preservatives (pentachlorophenol, tetrachlorophenol)	0.2 M K_2CO_3–solution (100 l/h, 300–600 l)	GC-ECD[1)] derivatization	0.05 $\mu g/m^3$ (pentachlorophenol)	Dahms and Metzner (1979)
Wood preservatives (pentachlorophenol, lindane)	Chromosorb 102 (1 l/min, 100 l)	GC-ECD[1)] PCP: derivatization	0.05 $\mu g/m^3$	Butte (1987)
Wood preservatives (pentachlorophenol, lindane)	Silica gel (2–2.5 m^3/h, 4–5 m^3)	GC-MSD[5)] PCP: derivatization	0.01 $\mu g/m^3$	Blessing and Derra (1992)
Wood preservatives (dichlofluanide pentachlorophenol, lindane)	Chromosorb 102 (20 l/min, 450 l)	GC-ECD[1)]/GC-MSD[5)] PCP: derivatization	?	Kasel et al. (1995)

[1)] GC-ECD: Gas chromatography-electron capture detector [2)] HPLC: High pressure liquid chromatography [3)] GC-MS: Gas chromatography-mass spectrometry [4)] GC-MS (CI): Gas chromatography-mass spectrometry (chemical ionization) [5)] GC-MSD: Gas chromatography-mass selective detector (single ion modus)

As mentioned above, "non-volatile" biocides are mainly particle bound. Analysis of these compounds in dust seems to be superior to air analysis if indoor contamination is to be detected.

3.5.2.2 Dust

A short review of methods for the analysis of biocides in household dust is given in Table 3.5-2. Most methods use either toluene or ethyl acetate for the extraction of organic compounds. In some cases, especially for methods having low detection limits, an additional clean-up with Florisil or silica gel is added. Methods of quantifying the substances of interest in the (cleaned) extract are either GC-ECD or GC-MS. All procedures are rather similar for carbamates, organochlorine compounds, organophosphates and pyrethroids. For the analysis of some phenols, e.g. pentachlorophenol, a derivatization step (alkylation, acetylation) is necessary.

Whereas the analytical part of the determination of biocides in dust displays only minor differences, samples referred to as "household dust" vary significantly. So far, no standard protocol for dust sampling has been reported. In the United States a "High Volume Small Surface Sampler" (HVS3) collecting dust samples of 2–100 g from carpets and bare floors in about 15 min has been developed (Roberts and Dickey, 1995). In Germany, predominantly dust samples from vacuum cleaner bags collected by commercial vacuum cleaners are used for evaluation of biocides; in some cases samples were analyzed in passively deposited suspended particulate (PDSP) (Krause et al.,

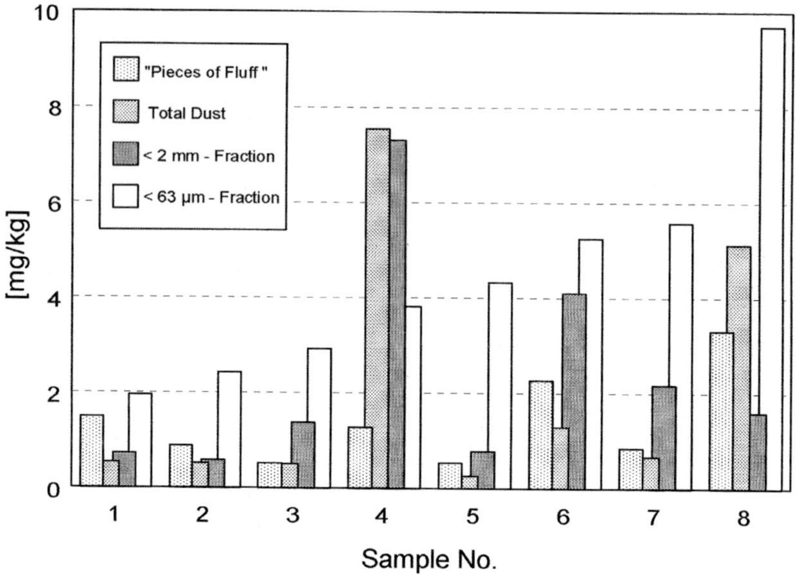

Figure 3.5-1. Pentachlorophenol in different fractions of household dust: Comparison of concentrations in "total dust", "pieces of fluff", the "< 2-mm fraction", and the "< 63-µm fraction" for 8 representative dust samples.

Table 3.5-2. Methods of determining biocides in dust.

Compound(s)	Dust Used for Analysis	Analytical apparatus	Detection limit[a]	Reference
Carbamates (bendiocarb, carbaryl etc.)	Homogenized dust from vacuum cleaner bags	GC-MS (CI)[b]	0.05–0.1	Roinestad et al. (1993)
Carbamates (bendiocarb, propoxur etc.)	63-µm-fraction of dust from vacuum cleaner bags	GC-MSD (SIM)[c]	0.1–0.5	Walker et al. (1994)
Organochlorine compounds (aldrine, endosulfane etc.)	Homogenized dust from vacuum cleaner bags	GC-MS (CI)[b]	0.05–0.1	Roinestad et al. (1993)
Organochlorine compounds (lindane)	63-µm-fraction of dust from vacuum cleaner bags	GC-ECD[d]	0.1	Butte and Walker (1994)
Organophosphates (dichlorvos, dimethoat etc.)	63–µm fraction of dust from vacuum cleaner bags	GC-MSD (SIM)[c]	1	Walker et al. (1994)
Phenols (o-phenylphenol)	Homogenized dust from vacuum cleaner bags	GC-MS (CI)[b]	0.075	Roinestad et al. (1993)
Phenols (pentachlorophenol)	63-µm fraction of dust from bags of vacuum cleaners	GC-ECD[d] (derivatization)	0.1	Butte and Walker (1994)
Phenols (pentachlorophenol)	Passively deposited suspended particulate	GC-ECD[d] (derivatization)	1	Meißner and Schweinsberg (1996)
Pyrethroids (cyfluthrin, cypermethrin etc.)	Dust from vacuum cleaner bags, discarding coarse particles	GC-MSD (SIM)[c]	1–2	Ball et al. (1993)
Pyrethroids (permethrin, resmethrin)	Homogenized dust from vacuum cleaner bags	GC-MS (CI)[b]	0.05–0.1	Roinestad et al. (1993)
Pyrethroids (cyfluthrin, cypermethrin etc.)	Dust from vacuum cleaner bags, discarding coarse particles	GC-ECD[d]	1–2	Stolz et al. (1994)
Pyrethroids (bioallethrin, cypermethrin etc.)	63–µm fraction of dust from vacuum cleaner bags	GC-MSD (SIM)[b]	0.1–0.5	Walker et al. (1994)

[a)] (mg/kg) [b] GC-MS (CI); Gas chromatography-mass cpectrometry (chemical ionization)
[c] GC-MSD: Gas chromatography–mass selective detection (single ion mode) [d] GC-ECD: Gas chromatography-electron capture detector

1991; Meißner and Schweinsberg, 1996). However, even dust obtained from vacuum cleaner bags is not a standard material, as different groups use different fractions of this dust for analysis. Some analysts take the complete contents of the dust bag, either after discarding coarse particles (sand, hair etc.) (Ball et al., 1993; Stolz et al., 1994) or after homogenizing the sample (Roinestadt et al., 1993), some take the "< 2-mm fraction" after sieving the sample (Krause et al., 1995; Friedrich et al., 1998) and some take the "< 63-µm fraction" after sieving (Butte and Walker, 1994; Walker et al., 1994). Results from different dust fractions differ significantly, as reported for permethrin (Butte and Walker, 1994; Stolz et al., 1996) and as shown in Fig. 3.5-1 for pentachlorophenol. In most cases, results for the "< 63-µm fraction" of household dust are higher than those for the other fractions evaluated. However, a systematic comparison, e.g of results for PDSP and dust from vacuum cleaner bags, is still lacking. Regarding precision, the "< 63-µm fraction" gives the most reproducible results on consecutive analyses of the same sample.

As there is no constant factor for the comparison from one fraction with another it is difficult to compare results from different studies. Thus, reference values for the assessment of indoor contamination are only useful if there is a description of the dust fraction or dust sample for which these are obtained.

3.5.3 Occurence of Biocides in the Air and Dust of the Indoor Environment

Biocides applied in the indoor environment differ significantly depending on climatic and cultural factors (Davis et al., 1992). In the United States, compounds to fight insects, cockroaches and termites, e.g. chlorpyrifos, diazinon, chlordane, propoxur and heptachlor, are mainly used (Lewis et al., 1988; Roinestad et al., 1993; Wright et al., 1996). For other regions, e.g. developing countries, organochlorine compounds such as DDT and hexachlorocyclohexanes (HCHs) to fight mosquitoes and malaria are still marketed (Singh et al., 1992). In Germany, public discussion is mainly focused on wood preservatives such as pentachlorophenol, lindane and pyrethroids, mainly permethrin. Thus, different biocides are detected in the air and dust of different countries.

3.5.3.1 Air

Very high concentrations of biocides in air were measured while they were being applied, especially with spraying or nebulizing pesticide formulations or during electroevaparation. For pyrethroids, concentrations may reach some hundred µg/m^3 during spray operations (Claas and Kintrup, 1991) and on electroevaporation some µg/m^3, respectively (Claas and Kintrup, 1991). For organophosphates such as dichlorvos even concentrations in the mg/m^3 range may be reached, as shown by Sagner and Schöndube (1982) in experimental chambers resembling living rooms. During spraying, persons may suffer absorption by inhalation and percutaneously. This has been shown for workers spraying permethrin on cabbage by monitoring its metabolite 3-phenoxybenzoic acid (3PBA) in urine (Asakawa et al., 1996).

Some time after spraying insecticides or painting wood, the highest possible concentrations in air are saturation concentrations. The dependence of saturation concentrations on temperature for some biocides was given by Zimmerli (1982). Regarding PCP, Zimmerli (1982) calculated a saturation concentration of 222 $\mu g/m^3$, whereas Warren et al. (1982) obtained concentrations for PCP in air vaporized from wood up to 580 $\mu g/m^3$. These values are of the same magnitude as concentrations (up to 160 $\mu g/m^3$) measured by Gebefügi et al. (1978) for PCP in indoor air some days after painting timber with wood preservatives. The amount of biocide vaporized from wood into the surrounding air, i.e. the quantity released per unit of time, is dependent on the formulation of the wood preservative and the timber species (Petrowitz, 1986). The concentration of PCP in indoor air is further controlled by the quantity of wood present in the room in relation to its volume, by the exchange rate of the indoor air and the temperature in the room. However, no correlation was observed between the temperature and levels of cypermethrin in air (Wright et al. 1993). The half life of allethrin in indoor air was reported to be only 1.9 days (Eitzer 1991). Furthermore, a spatial distribution in air was observed for allethrin for different sampling sites in the same room (Eitzer 1991).

Depending on the volatility of the biocide, some weeks or months after application vaporization will lead to levels not higher than some hundred ng/m^3 for "non-volatile" biocides such as permethrin, whereas semivolatile compounds such as PCP may be present in concentrations of some $\mu g/m^3$. Concentrations of PCP in indoor air are < 0.025 $\mu g/m^3$ (Eckrich, 1989), < 0.05 $\mu g/m^3$ (Butte & Walker, 1994), respectively, if no wood preservatives have been applied. In our experience, concentrations of PCP in indoor air of rooms where PCP has been used as a wood preservative more than 20 years ago are nowadays in many cases of the same magnitude as in rooms where PCP has never been applied.

Typical concentrations for biocides in indoor air obtained under defined conditions either directly during or after application and up to 20 years after an incident have been compiled by Pluschke (1996) and are given in Table 3.5-3. For non-volatile biocides, i.e. pyrethroids, base levels in indoor air are < 20 ng/m^3, and even for semivolatile biocides such as pentachlorophenol and lindane they are < 50 ng/m^3.

If biocides are present in buildings, indoor air is the major pathway of exposure for the occupants. Therefore, objective values for the decision whether or not to sanitize rooms are of interest for both authorities and the public. In Germany the "Innenraumkommission" (indoor commission) of the "Umweltbundesamt" (Federal Office of the Environment) issues "Richtwerte" ("standard values") for biocides in indoor air (IRK, 1996). These values are normally much lower than the MAK values ("Maximale Arbeitsplatz-Konzentration = maximum concentration in air at the workplace") and will replace the rather old MIK values (Maximum Imission Concentration) of the VDI (VDI, 1974). The assessment of the "Richtwerte" is a two step definition, declaring "Richtwert I" and "Richtwert II". For concentrations above the "Richtwert II" a sanitation of rooms bearing biocides is necessary to reduce the exposure, e.g. reconstruction, removing wood treated with PCP; for concentrations below the "Richtwert I" it is not recommended to take further action as no adverse health effects are thought to occur.

Table 3.5-3. Concentrations of biocides in indoor air.

Compound	Maximum Concentration	Application	Reference
Allethrin	5 μg/m³	After 12 h application via electro-evaporator	Claas and Kintrup (1991)
Allethrin	48 ng/m³	0,1 d after application (distribution by vending machines)	Eitzer (1991)
Chlordane	45 ng/m³	Randomly selected homes	Roinestadt et al. (1993)
Chlorpyrifos	150 ng/m³	Randomly selected homes	Roinestadt et al. (1993)
Cyfluthrin	90 μg/m³	15 s after spraying	Claas and Kintrup (1991)
Cyfluthrin	< 5 ng/m³	After fighting cockroaches (up to 4 times per year in up to 5 years)	Ball et al. (1993)
Cypermethrin	5 ng/m³	→ cyfluthrin	Ball et al. (1993)
Cypermethrin	19 μg/m³	Immediately after spraying	Wright et al. (1993)
DDT	5 μg/m³	After using DDT for wood preservation	Lederer and Angerer (1997)
Deltamethrin	< 10 ng/m³	→ cyfluthrin	Ball et al. (1993)
Diazinone	6 ng/m³	Randomly selected homes	Roinestadt et al. (1993)
Dichlorvos	2.5 mg/m³	After spraying in an experimental chamber furnished like a living room	Sagner and Schöndube (1982)
Dichlorvos	250 ng/m³	Randomly selected homes	Roinestadt et al. (1993)
Lindane (γ -HCH)	2 μg/m³	Rooms where wood preservatives were used (at least 7 years after application)	Blessing and Derra (1992)
Pentachlorophenol	82 μg/m³	1.5 years after using wood preservatives	Dahms and Metzner (1979)
Pentachlorophenol	25 μg/m³	< 9 years after using wood preservatives	Krause and Englert (1980)
Pentachlorophenol	2 μg/m³	→ lindane	Blessing and Derra (1992)
Permethrin	4.7 ng/m³	→ cyfluthrin	Ball et al. (1993)
Permethrin	8.8 μg/m³	14 months after fighting cockroaches	Fromme (1991)
o-Phenylphenol	58 ng/m³	Randomly selected homes	Roinestadt et al. (1993)
Propoxur	63 ng/m³	Randomly selected homes	Roinestadt et al. (1993)
Tetramethrin	300 μg/m³	15 s after spraying	Claas and Kintrup (1991)

Roinestadt et al. (1993), analyzing 23 biocides in indoor air and dust, reported that pesticides in air were always found in the dust, with the exception of dichlorvos, *o*-phenylphenol and chlordane. The data suggest that very volatile pesticides such as dichlorvos, *o*-phenylphenol and chloronaphthalene are more appropriately sampled in the air. The majority of household pesticides, however, are preferably detected in the home environment by dust sampling (Roinestad et al., 1993). This holds true particularly for permethrin, which could not be detected in the air (detection limit: 1 ng/m^3), whereas it was present in the dust samples in the mg/kg range (Roinestad et al., 1993). Stolz et al. (1996), reporting results for permethrin in dust samples and air, observed no correlation, although air samples showing higher concentrations of permethrin had a tendency to high concentrations in dust as well. On the other hand, for pentachlorophenol as a wood preservative, there was a relationship between concentrations in air and in dust during the first two years after painting timber (Krause, 1982). After this time, this relationship was no longer observed (Krause, 1982). Data for PCP in air and dust measured at least 20 years after the application confirm that there is no relationship (Liebl et al., 1996).

3.5.3.2 Dust

Most metals, pesticides, and organic pollutants with relatively low vapor pressures or high polarities are expected to accumulate more in dust than in air (Roberts and Dickey, 1995; Roinestad et al., 1993). Thus, house dust is a reservoir for the biocides used in the indoor environment (Moriske, 1997). Its sampling can give one of the best estimates of recent exposure to pollutants in the home and serves as an indicator of chronic exposure to persistent chemicals (Roberts & Dickey, 1995). Ingestion and inhalation of and contact with house dust can be primary routes of exposure to pesticides, lead, and allergens for adults, pets, and children. Infants and toddlers, who crawl and put their hands and other objects into their mouths, ingest a daily rate that is twice that of adults and is estimated to be 0.02–0.2 g (Calabrese and Stanek, 1991). Toddlers, possibly eating non-food items, may even consume as much as 10 g of soil or dust per day (Calabrese and Stanek, 1991). Potential risks to small children compared to adults are further increased by their smaller size, higher ratio of surface area to body weight, and the stage of development of their organs, nervous and immune systems (ICPS, 1986).

Support for the thesis that household dust leads to contamination comes from correlations between biocides in dust and in samples of human origin. This correlation was reported for PCP in the urine of women and children and dust from vacuum cleaner bags (Krause and Englert, 1980), and in PDSP and urine respectively (Meißner and Schweinsberg, 1996). On the other hand, no correlation was observed between PCP in household dust and blood by Liebl et al. (1996).

Quite a number of publications giving concentrations of biocides in household dust are available, but most of them lack any description of the use of biocides (e.g. Butte and Walker, 1994; Lahl and Neisel, 1989; Pöhner et al., 1997; Roinestad et al., 1993; Stolz and Krooss, 1993; Stolz et al., 1996; Walker et al. 1994). Results for dust samples obtained after certain incidents are compiled in Table 3.5-4A. Furthermore, concentrations for biocides in the dust of households from Germany obtained from representative studies are given in Table 3.5-4B. The medians and the 95 percentiles of two investigations performed in co-operation with the author (Hostrup et al., 1997, LANU, 1997) are shown in Table 3.5-5. Together with the "Umweltsurveys" (Krause et al., 1991; Krause et al., 1995; Friedrich et al., 1998), they give complete data about the residues in household dust from Germany. However, to our knowledge there is only one representative study for which chemical analysis were performed, together with information collected by a detailed questionnaire of the occupants concerning the use of biocides such as wood preservatives, insecticides and agents to combat fleas (Hostrup et al., 1997). This resulted in detailed knowledge of the indoor application of 40 biocides and 2 synergists in Germany. Dust samples were obtained from 336 households from selected regions of Lower Saxony and North Rhine-Westphalia; analyses were performed with the "< 63-µm fraction" of dust. Biocides investigated were wood preservatives (1- and 2-chloronaphthalene, chlorothalonil, dichlofluanid, *α*- and *β*-endosulfane, fenobucarb, furmecyclox, lindane, parathion, pentachlorophenol [PCP], propiconazole, tebuconazole and tolylfluanid) and pyrethroids (bioallethrin, cyfluthrin, *λ*-cyhalothrin, cypermethrin, cyphenothrin, deltamethrin, empenthrin, fenvalerate, D-phenothrin, permethrin, resmethrin and tetramethrin) as well as other insecticides (bendiocarb, chlorpyrifos, *p,p*'-dichlorodiphenyltrichloroethane [DDT], diazinon, dichlorvos, dimethoate, fenitrothion, malathion, methoprene, methoxychlor, pirimiphos-ethyl, pirimiphos-methyl, propetamphos, propoxur, fenchlorphos and tetrachlorvinphos). Furthermore, piperonyl butoxide [PBO] and octachlorodipropyl ether [S421] were included in the analyses, as these synergists are present in various biocide formulations.

Wood preservatives and agents against insects and fleas are used by more than two-thirds of the population; thus their residues can be found in nearly every home. Only in two households (< 1 %) could no biocide or synergist be observed. On the other hand chlorpyriphos, propoxur and lindane were present in more than 10 %, PBO in more than 50 %, DDT, permethrin and methoxychlor in more than two-thirds and PCP in nearly every sample. Results regarding concentrations of biocides in household dust were confirmed by those obtained in Schleswig-Holstein (LANU, 1997), but information about the use of biocides indoors was not obtained for the latter study.

In no case were concentrations of biocides in dust normally distributed. For those biocides present in more than 10 % of the samples the question whether there was any significant difference between the concentration used for application as a wood preservative and that used for application as an agent against insects or fleas was investigated (Walker et al., 1998). These findings were used to calculate reference values for the base contamination with certain biocides in household dust (Walker et al., 1998). Reference values are a tool in detecting an elevated contamination as an indicator of indoor pollution.

For chlorpyrifos and lindane, no differences were observed between the dust of homes where biocides had been used ("users") and dust without application of any biocide by the occupants ("non-users"). DDT, methoxychlor and PBO concentrations in the dust of users of insecticides were significantly higher than in that of non-users (see Fig. 3.5-2 for methoxychlor). Users of wood preservatives showed higher concentrations of PCP and users of biocides against insects and fleas showed higher concentrations of propoxur.

Table 3.5-4. Concentrations of biocides in household dust.

A: Samples analysed because of certain incidents

Compound	Maximum concentration	Application/Collective	Reference
Cypermethrin	8 mg/kg	After fighting cockroaches (up to 4 times per year in up to 5 years)	Ball et al. (1993)
Deltamethrin	10 mg/kg	→ cypermethrin	Ball et al. (1993)
DDT	8500 mg/kg	>8 years after using DDT for wood preservation	Lederer and Angerer (1997)
Pentachlorophenol	1400 mg/kg	>20 years after using PCP for wood preservation	Liebl et al., (1996)
Pentachlorophenol	3000 mg/kg	>15 years after using PCP for wood preservation (PDSP)[1]	Meißner and Schweinsberg (1996)
Permethrin	320 mg/kg	→ cypermethrin	Ball et al. (1993)

[1] PDSP: passively deposited suspended particulate

B: Representative Studies [2]

Compound	Maximum concentration	dust fraction	Reference
Chlorpyrifos	870 mg/kg	"<63 – µm fraction"	Hostrup et al., (1997)
Chlorpyrifos	1300 mg/kg	"<63-µm fraction"	LANU (1997)
DDT	40 mg/kg	"<63-µm fraction"	Hostrup et al., (1997)
DDT	14 mg/kg	"<63-µm fraction"	LANU (1997)
Lindane (γ -HCH)	27.4 mg/kg	"<2-mm fraction"	Krause et al., (1995)
Lindane (γ -HCH)	4.8 mg/kg	"<63-µm fraction"	Hostrup et al., (1997)
Lindane (γ -HCH)	2.2 mg/kg	"<63-µm fraction"	LANU (1997)
Methoxychlor	120 mg/kg	"<63 – µm fraction"	Hostrup et al., (1997)
Methoxychlor	110 mg/kg	"<63-µm fraction"	LANU (1997)
Pentachlorophenol (PCP)	30.9 mg/kg	"<2-mm fraction"	Krause et al. (1995)
Pentachlorophenol (PCP)	40 mg/kg	"<63-µm fraction"	Hostrup et al., (1997)
Pentachlorophenol (PCP)	53 mg/kg	"<63-µm fraction"	LANU (1997)
Permethrin	150 mg/kg	"<63-µm fraction"	Hostrup et al., (1997)
Permethrin	990 mg/kg	"<63-µm fraction"	LANU (1997)
Permethrin	267 mg/kg	"<2 – mm fraction"	Friedrich et al., (1998)
Piperonyl butoxide (PBO)	270 mg/kg	"<63-µm – fraction"	Hostrup et al., (1997)
Piperonyl butoxide (PBO)	50 mg/kg	"<2-mm – fraction"	LANU (1997)
Piperonyl butoxide (PBO)	67 mg/kg	"<2-mm – fraction"	Friedrich et al., (1998)
Propoxur	15 mg/kg	"<63-µm – fraction"	Hostrup et al. (1997)
Propoxur	16 mg/kg	"<63-µm – fraction"	LANU (1997)

[2] Only biocides which were present in more than 10 % of the samples are considered

Table 3.5-5. Occurrence of biocides in dust of households in Germany according to two case control studies.

Compound	Lower Saxony / North Rhine-Westphalia (n = 336)[a]			Schleswig-Holstein (n = 220)[b]		
	Number of samples > DL[c)]	Median (mg/kg)	95 percentile (mg/kg)	Number of samples > DL[c]	Median (mg/kg)	95 percentile (mg/kg)
Chlorpyrifos	35 (10.4 %)	≤ 0.1	0.63	18 (8.2 %)	≤ 0.1	0.5
Cyfluthrin	4 (1.2 %)	≤ 0.1	–*	– (0 %)	–	–
Cypermethrin	5 (1.5 %)	≤ 0.1	–*	1 (0.4 %)	≤ 0.1	–*)
DDT	275 (81.8 %)	0.31	4.2	163 (74.1 %)	0.3	4.4
Diazinon	10 (3.0 %)	≤ 0.1	–*	9 (4.1 %)	≤ 0.1	–*)
Dichlofluanid	9 (2.7 %)	≤ 0.1	–*	9 (4.1 %)	≤ 0.1	–*)
Lindane	68 (20.2 %)	≤ 0.1	0.83	39 (17.7 %)	≤ 0.1	1.1
Methoxychlor	275 (81.8 %)	0.92	27	173 (78.8 %)	0.5	12
Piperonyl butoxide (PBO)	178 (53.0 %)	0.11	13	92 (41.8 %)	≤ 0.1	5.7
Pentachlorophenol (PCP)	325 (96.7 %)	0.95	8.0	215 (97.7 %)	1.4	9.1
Permethrin	243 (72.3 %)	0.67	37	179 (81.4 %)	1.1	73
Propoxur	47 (14.0 %)	≤ 0.1	0.90	24 (10.9 %)	≤ 0.1	0.8
Tetrachlorvinphos	7 (2.1 %)	≤ 0.1	–*	3 (1.4 %)	≤ 0.1	–*)
Tetramethrin	13 (3.9 %)	≤ 0.1	–*	4 (1.8 %)	≤ 0.1	–*)

[a] Hostrup et al. (1997) [b] LANU (1997) [c] DL = Determination Limit * Not enough data to calculate a 95 percentile

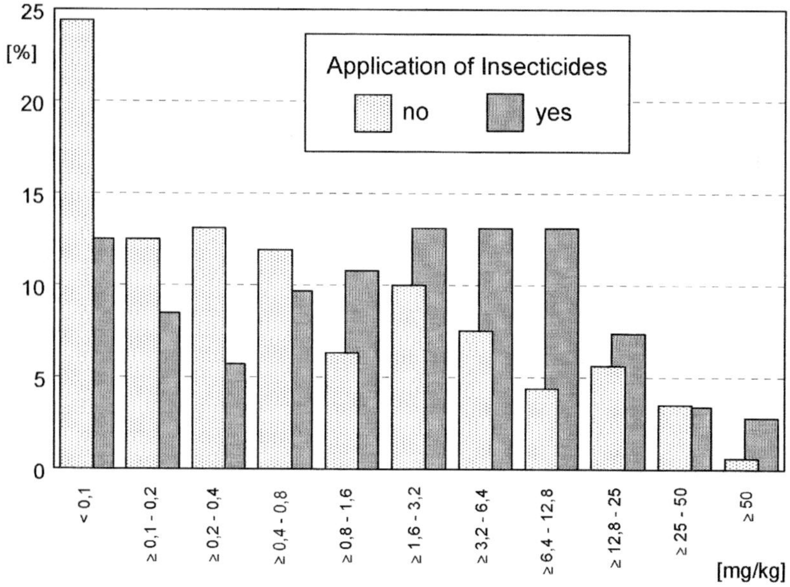

Figure 3.5-2. Frequency distribution for the concentration of methoxychlor in household dust: comparison of users and non-users of insecticides.

The biocides most often occurring in household dust in Germany are methoxychlor, PCP and permethrin. Methoxychlor is rarely found in formulations intended for indoor use; publications reviewing concentrations in dust are so far not available. However, about 80 % of the dust samples showed residues of more than 0.1 mg/kg and up to several mg/kg.

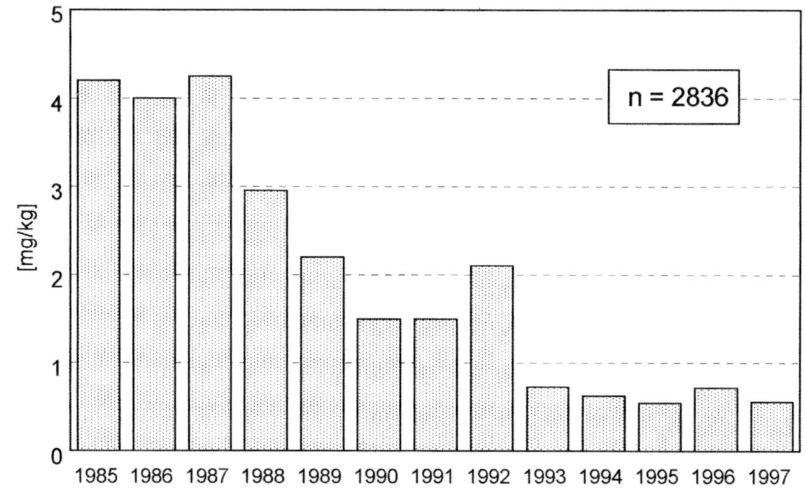

Figure 3.5-3. Pentachlorophenol in household dust: comparison of medians from 1985 to 1997 (incident-related data).

Pentachlorophenol (PCP), used predominantly as a wood preservative, was banned in Germany in 1989 (Pentachlorphenol-Verbotsverordnung, 1989). It displays a high persistence, even though residues in household dust decrease with time (see Fig. 3.5–3). Values in East Germany are lower than those in West Germany (Krause et al., 1995). Declining concentrations in dust are also reported for lindane (Krause et al., 1995). But still there are large quantities of timber treated with PCP or lindane several years ago which are responsible for the occurrence of these compounds in the indoor environment.

For permethrin, high values in dust may either be due to use by the occupant, commercial pest control or the impregnation of carpets or textiles. An impregnation of wool carpets with permethrin may cause concentrations in household dust of the same order of magnitude, up to some hundred mg/kg. As permethrin is replacing other insecticides such as lindane for many applications, an increase of concentrations of this substance in dust is being observed (Friedrich et al., 1998).

3.5.4 Summary

3.5.4.1 Biocides in Indoor Air

The main benefit to be obtained from analytical results of the air in rooms is the possibility of estimating the exposure of the occupants. For semivolatile biocides, e.g. pentachlorophenol, lindane, and chlorpyrifos, analytical methods are available that can be performed with standard equipment (GC-ECD, GC-MS). For non-volatile biocides, e.g. pyrethroids, concentrations obtained for non-contaminated air will in many cases lead to values lower than the detection limit. For some biocides, standard levels deduced from toxicological data have been calculated. Thus, a decision may be reached whether decontamination measures should be taken for the rooms evaluated. However, air analysis has a number of shortcomings: valid analysis is time consuming and expensive; furthermore, concentrations fluctuate considerably due to variations in room ventilation, humidity and temperature. Since sampling conditions may greatly influence the analytical result, it should always be obtained using a proper sampling protocol. Most representative values are obtained from consecutive analyses or under long-term conditions giving a representative average.

3.5.4.2 Biocides in Household Dust

Concentrations in dust are normally much higher than those in air. The advantages and limitations of an analysis of biocides in dust are as follows:

Advantages

- dust is easily accessible; vacuum cleaner bags can be taken by the occupants as samples
- biocides in dust are stable; shipment is simple
- concentrations in dust are high compared to those in air, so that the analysis is more convenient

- for some biocides, i.e. PCP, there is a correlation between concentrations in dust and those in urine (a correlation not yet reported for air).

Limitations

- there may be an uneven spatial distribution of biocides on the floor or in different rooms
- if information about the age of a contaminated dust is lacking, misjudgement of values is possible
- the distribution of biocides in different fractions of dust is variable, so that concentrations obtained for different fractions are not comparable
- so far there is no standard protocol for dust sampling.

Dust may be regarded as the ideal material for detection and identification of indoor biocides and any of their residues still existing. Commercial vacuum cleaners can be used for taking samples. Analysis using the "< 63-µm fraction" of dust lead to results that are more reproducible than those for any other fraction. Reliable results are only obtained under equilibrium conditions in rooms; therefore there should be no cleaning for at least one week before dust samples are taken. Since the "Umweltsurveys" and two case control studies published recently are based on representative samples for household dust of the German population (especially PCP, lindane and pyrethroids), the assessment of results is possible by comparison with reference values.

References

Asakawa F., Jitsunari F., Miki K., Choi J.-O., Takeda N., Kitamado T., Suna S. and Manabe Y. (1996): Agricultural worker exposure to and absorption of permethrin applied to cabbage. Bull. Environ. Contam. Toxicol., *56*, 42–49.

Ball M., Herrmann T., Wildeboer B., Koss G., Sagunski H. and Czaplenski U. (1993): Indoor pollution by pyrethroids: Sampling, analysis, risk evaluation. Indoor Air 93, Proceedings of the 6th International Conference on Indoor Air Quality and Climate, Helsinki, Finland, Vol. 2, 201–206.

Blessing B. and Derra R. (1992): Holzschutzmittelbelastungen durch Pentachlorphenol und Lindan in Wohn- und Aufenthaltsräumen. Staub Reinh. Luft, *52*, 265–271.

Brown L.M., Blair A., Gibson R., Everett G.D., Cantor K.P., Schuman L.M., Burmeister L.F., van Lier S.F. and Dick F. (1990): Pesticide exposures and other agricultural risk factors for leukemia among men in Iowa and Minnesota. Cancer Res., *50*, 6585–6591.

Butte W. (1987): Simultaneous determination of pentachlorophenol and γ-hexachlorocylohexane in indoor air. Fresenius Z. Anal. Chem., *327*, 33–34.

Butte W. and Walker G. (1994): Sinn und Unsinn von Hausstaubuntersuchungen – das Für und Wider. Hausstaub als Meßparameter zum Erkennen einer Innenraumbelastung mit Permethrin, Pentachlorphenol und Lindan. VDI-Berichte, *1122*, 535–546.

Calabrese E.J. and Stanek E. (1991): A guide to interpreting soil ingestion studies: II. Quantitative evidence for soil ingestion. Regul. Toxicol. Pharmacol., *13*, 278–292.

Claas T.J. and Kintrup J. (1991): Pyrethroids as household insecticides: analysis, indoor exposure and persistence. Fresenius J. Anal. Chem., *340*, 446–453.

Clavel J., Hemon D., Mandereau L., Delemotte B., Severin F. and Fandrin, G. (1996): Farming, pesticide use and hairy-cell leukemia. Scan. J. Work. Environ. Health, *22*, 285–293.

Dahms A. and Metzner W. (1979): Zur Analytik von Pentachlorphenol und Tetrachlorphenol in der Luft und im Urin. Holz Roh- und Werkst., *37*, 341–344.

Das J.V., Ramachandran K.N. and Gupta V.K. (1994): Spectrophotometric determination of baygon and carbaryl in air. Fresenius J. Anal. Chem., *348*, 840–841.

Davis J.R., Browson R.C. and Garcia R. (1992): Family pesticide use in the home, garden, orchard, and yard. Arch. Environ. Contam. Toxicol., *22*, 260–266.

Eckrich W. (1989): Innenraumbelastung durch Holzschutzmittel, insbesondere Pentachlorphenol. VDI-Berichte, *745*, 297–308.

Eitzer B.D. (1991): Cycling of indoor air concentrations of D-*trans*-allethrin following repeated pesticide applications. Bull. Environ. Contam. Toxicol. *47*, 406–412.

Friedrich C., Becker K., Hoffmann G., Hoffmann K., Krause C., Nöllke P., Schulz C., Schwabe R. and Seiwert M. (1998): Pyrethroide im Hausstaub der deutschen Wohnbevölkerung–Ergebnisse zweier bundesweiter Querschnittstudien. Gesundh.-Wes., *60*, 96–101.

Fromme H. (1991): Anwendung von Pestiziden in Innenräumen unter besonderer Berücksichtigung der Pyrethroide. Toxikologische Aspekte und Darstellung der Anwendungsproblematik. Teil II. Öff. Gesundh.-Wes., *53*, 662–667.

Gebefügi I., Michna A. and Korte F. (1978): Zur ökologisch-chemischen Bewertung des Pentachlorphenols in geschlossenen Räumen. Chemosphere, *7(4)*, 359–364.

Grob K. Jr. and Neukom H.P. (1984): Concept for the sampling and derivatization of pentachlorophenol from air in a capillary pre-column, followed by gas chromatographic determination. J. Chromatogr., *295*, 49–54.

Hoffmann W. and Hostrup O. (1997): Biozidanwendungen im Haushalt als möglicher Risikofaktor für die Gesundheit der Raumnutzer. Umwelt & Gesundheit, *8*, 140–143.

Hostrup O., Witte I., Hoffmann W., Greiser E., Butte W. and Walker G. (1997): Biozidanwendungen im Haushalt als möglicher Risikofaktor für die Gesundheit der Raumnutzer. Abschlußbericht. Studie im Auftrag des Niedersächsischen Sozialministeriums. Oldenburg/Bremen.

IRK–Ad hoc Arbeitsgruppe aus Mitgliedern der Innenraumlufthygiene-Kommission (IRK) des Umweltbundesamtes und des Ausschusses für Umwelthygiene der AGLMB (1996): Richtwerte für Innenraumluft: Basisschema. Bundesgesundhbl., *39*, 422–426.

Kasel U., Wichmann G. and Juhl B. (1995): Einfache Bestimmung von Dichlofluanid, Pentachlorphenol (PCP) und Lindan in Raumluft. Staub–Reinh. Luft, *55*, 439–440.

Kawata K. and Yasuhara A. (1994): Determination of fenitrothion and diazinon in air. Bull. Environ. Contam. Toxicol., *52*, 419–424.

Krause C. (1982): Wirkstoffe von Holzschutzmitteln im häuslichen Bereich. p. 309–316 In: Aurand K. et al. (eds) Luftqualität in Innenräumen. G. Fischer, Stuttgart-New York.

Krause C. and Englert N. (1980): Zur gesundheitlichen Bewertung pentachlorphenolhaltiger Holzschutzmittel in Wohnräumen. Holz Roh- und Werkst., *38*, 429–432.

Krause C., Chutsch M., Henke M., Kliem C., Leiske M., Schulz C. and Schwarz E. (1991): Umweltsurvey Band IIIa, Wohn-Innenraum: Spurenelementgehalte im Hausstaub. WoBoLu Hefte 2/1991, Bundesgesundheitsamt, Berlin.

Krause C., Becker K., Bernigau W., Hoffmann K., Nöllke P., Schulz C., Schwabe R. and Seiwert M. (1995): Umweltsurvey in den fünf neuen Ländern der Bundesrepublik Deutschland 1991/92 (unter Berücksichtigung der Erhebungen in den alten Ländern 1990/91 und 1985/86). Forschungsbericht 11606088/02 im Auftrag des Bundesministers für Umwelt, Naturschutz und Reaktorsicherheit.

Lahl U. and Neisel F. (1989): Sanierung von holzschutzmittelbelasteten Kindergärten. Gesundheits-Ingenieur-Haustechnik-Bauphysik-Umwelttechnik, *110*, 206–210.

Lane D.A., Johnson N.D., Hanley M.-J.J., Schroeder W.H. and Ord D.T. (1992): Gas- and particle-phase concentrations of α-hexachlorocylohexane, γ-hexachlorocyclohexane and hexachlorobenzene in Ontario air. Environ. Sci. Technol., *26*, 126–133.

LANU–Landesamt für Natur und Umwelt des Landes Schleswig-Holstein (1997): Umwelttoxikologische Studie im Kreis Pinneberg, 1995/96. Flintbek (Kiel).

Lederer P. and Angerer J. (1997): Raumluftbelastung durch DDT als Holzschutzmittel. Umweltmed. Forsch. Prax., *2*, 32.

Lewis R.G., Bond A.E., Johnson D.E. and Hsu J.P. (1988): Measurement of atmospheric concentrations of common household pesticides. Environ. Monitor. Assessm., *10*, 59–73.

Liebl B., Mayer R., Kaschube M. and Wächter H. (1996): Pentachlorphenol–Ergebnisse aus einem bayerischen Human-Monitoring-Programm. Gesundh.-Wes., *58*, 332–338.

Meißner T. and Schweinsberg F. (1996): Pentachlorophenol in the indoor environment: evidence for a correlation between pentachlorophenol in passively deposited suspended particulate and in urine of exposed persons. Toxicol. Lett., *88*, 237–242.
Moriske, H.-J. (1997): Zusammenfassung der Ergebnisse der 4. WaBoLu-Innenraumtage in Berlin vom 26. bis 28.5.1997. Bundesgesundhbl., *40*, 338–340.
Pentachlorphenol-Verbotsverordnung (PCP-V) (1989): Bundesgesetzbl. *I*, 2235.
Petrowitz H.-J. (1986): Zur Abgabe von Holzschutz-Wirkstoffen aus behandeltem Holz. Holz Roh- und Werkstoff *44*, 341–346.
Pluschke P. (1996): Luftschadstoffe in Innenräumen. Springer, Berlin.
Pöhner A., Simrock S., Thumulla J., Weber S. and Wirkner T.(1997): Hintergrundbelastung des Hausstaubs mit mittel- und schwerflüchtigen organischen Schadstoffen. Umwelt & Gesundheit, *8*, 79–80.
Rando R.J. and Hammad Y.Y. (1985): Filter collection of airborne permethrin with determination of HPLC. J. Liq. Chromatogr., *8*, 1869–1880.
Roberts J.W. and Dickey P. (1995): Exposure of children to pollutants in house dust and indoor air. Rev. Environ. Contam. Toxicol., *143*, 59–78.
Roinestad K.S., Louis H.N. and Rosen J.D. (1993): Determination of pesticides in indoor air and dust. J. Assoc. Off. Anal. Chem. Int., *76*, 1121–1126.
Roper E.M. and Wright C.G. (1984): Sampling efficiency of five solid sorbents for trapping air-borne pesticides. Bull. Environ. Contam. Toxicol. *33*, 476–483.
Sagner G. and Schöndube M. (1982): Bestimmung und toxikologische Bewertung von Dichlorvos-Raumluft-Konzentrationen nach Ausbringung von Nebelmitteln. p. 359–368 In: Aurand K. et al. (eds) Luftqualität in Innenräumen. G. Fischer, Stuttgart-New York.
Singh P.P., Udeaan A.S. and Battu S. (1992): DDT and HCH residues in indoor air arising from their use in malaria control programmes. Sci. Total. Environ., *116*, 83–92.
Stolz P. and Krooss J. (1993): Vorkommen pyrethroidhaltiger Insektizide in Innenräumen. Forum Städte-Hygiene, *44*, 205–209.
Stolz P., Meierhenrich U. and Krooss J. (1994): Dekontaminations- und Abbaumöglichkeiten für Pyrethroide in Innenräumen. Staub–Reinh. Luft, *54*, 379–386.
Stolz P., Meierhenrich U., Krooss J. and Weis N. (1996): Messung der Korrelation der Belastung von Hausstaub und Raumluft bei Innenraumbelastungen mit Pyrethroiden. VDI-Berichte, *1257*, 789–796.
Tomlin C. (1994): The pesticide manual, 10th edn, The British Crop Protection Council and The Royal Society of Chemistry. Surrey, Cambridge.
VDI–Verein deutscher Ingenieure (1974): Maximale Immissionswerte. VDI-Verlag, Düsseldorf.
VDI–Verein deutscher Ingenieure (1995): Indoor air pollution measurement. General aspects of measurement strategy. VDI-Richtlinien 4300, part 1. VDI-Verlag, Düsseldorf.
VDI–Verein deutscher Ingenieure (1997): Indoor air pollution measurement. Measurement strategy for pentachlorophenol (PCP) and γ-hexachlorocyclohexane (lindane) in indoor air. VDI-Richtlinien 4300, part 4. VDI-Verlag, Düsseldorf.
Walker G., Keller R., Beckert J. and Butte W. (1994): Anreicherung von Bioziden in Innenräumen am Beispiel der Pyrethroide. Zbl. Hyg. Umweltmed., *195*, 450–456.
Walker G., Hostrup O., Hoffmann W. and Butte W. (1999): Biozide im Haustaub: Ergebnisse eines repräsentativen Monitorings in Wohnräumen. Gefahrst. Reinh. Luft, 59, 33–41.
Warren J.S., Lamparski L.L., Johnson R.L. and Gooch R.M. (1982): Determination of pentachlorophenol volatized from wood via collection on silica gel. Bull. Environ. Contam. Toxicol. *29*, 719–726.
Woiwode W., Wodarz R. Drysch K. and Weichardt H. (1980): Bestimmung von freiem Pentachlorphenol in der Luft und im Blut durch leistungsfähige Routineverfahren. Int. Arch. Occup. Environ. Health *45*, 153–161.
Wright C.G., Leidy R.B. and Dupree H.E.Jr. (1993): Cypermethrin in the ambient air and on surfaces of rooms treated for cockroaches. Bull. Environ. Cotam. Toxicol., *51*, 356–360.
Wright C.G., Leidy R.B. and Dupree H.E.Jr.(1996): Insecticide residues in the ambient air of commercial pest control buidings, 1993. Bull. Environ. Contam. Toxicol., *56*, 21–28.
Zimmerli B. (1982): Modellversuche zum Übergang von Schadstoffen aus Anstrichen in die Luft. p. 235–267 In: Aurand K. et al., (eds) Luftqualität in Innenräumen. G. Fischer, Stuttgart New York.
Zimmerli B. and Zimmermann H. (1979): Einfaches Verfahren zur Schätzung von Schadstoffkonzentrationen in der Luft von Innenräumen. Mit. Gebiete Lebensm.-Hyg., *70*, 429–442.

3.6 Secondary Emission

Lars Gunnarsen and Ulla D. Kjaer

3.6.1 Introduction

Indoor air quality problems often persist over years, although most simple emission tests of construction products in test chambers show a fairly rapid decay of emissions during the first weeks or months after manufacture. Solvents and other compounds initially contained in the products gas off quickly. Deposition of airborne pollutants on surfaces and in structures may play a major role in sustaining poor air quality in buildings. Deposits of air pollution from time-limited events such as paint jobs, cleaning or smoking, are well recognized as an effect of secondary emission processes when they cause poor air quality in a building for an extended period.

Air pollutants may be temporarily or permanently stored in construction products as a result of sorption processes. The structures of a building may adsorb chemicals at times when air concentrations are high and reemit or desorb them later when concentrations are lower. Chemical reactions between compounds added to construction products during their use or between compounds contained in the products or present in the ambient air may lead to lasting poor air quality. Oxidation of chemicals in surface materials by oxidants from indoor sources or outside air pollution may decrease the air quality in certain circumstances. Choice of construction products, periods of ventilation, the use of cleaning and maintenance agents and timing and handling of incidents of high concentrations of air pollution require particular consideration during design and operation of buildings.

This chapter summarizes some findings regarding these aspects of secondary emissions.

3.6.2 Definition

Secondary emission is any process that releases new airborne contaminants from existing sources, changes the total emittable mass of existing contaminants, or results in chemical reactions between compounds on surfaces and in the air. Secondary emission may be based on sorption, oxidation, hydrolysis, decomposition or other chemical reactions in or on a source or the indoor air. A secondary emission process is often highly influenced by past and present environmental conditions. It is not always possible to tell if a compound found in the air is there because of a primary or secondary emission process since a source may emit the same compound by both primary and secondary processes. Purposely added materials such as cleaning products may be a primary emission source, but reactions between constituents of new and existing products may cause a secondary emission process.

3.6.3 Steps in the Emission Process

For an emission process to take place, volatile chemicals must be available or formed. They must be transported to the surface of a source and, finally, they must be able to evaporate from this surface. The emission process may be limited by the availability of emission-prone chemicals. Emission will, as time goes by, reduce the amount of emittable chemicals in materials. Aging of materials will therefore typically reduce primary emission rates, but, for chemicals introduced regularly to the material surface, this may not occur.

If the chemical reactions leading to volatile chemicals are slower than the other steps in a secondary emission process, this formation of the chemicals may limit the emission rate. Temperature and availability of the reactants will influence the rate of reactions.

The diffusion of chemicals in construction products may also be the limiting factor (Dunn and Chen, 1994). It has been common to consider this as the only emission-limiting process. For a given product and relevant air concentration of the diffusing chemical, the diffusion-controlled emission rate is mainly influenced by temperature and availability. Only at very high air concentrations may a reduced emission rate be seen; otherwise, little influence by parameters of the surrounding air is likely for diffusion-controlled emission.

Finally, emissions may be limited by the rate of phase change at the material surface. In an indoor climate context, this evaporation process is limited by the mass transfer through the boundary layer of the air above the surface. Parameters such as temperature, concentration in the room air of the emitted compounds, and velocity and turbulence in the boundary layer of air surrounding the product influence the emission rates. Theoretically, the emission rate will always approach zero when there is no ventilation and the air concentration of chemicals has been allowed to rise to equilibrium with the chemicals in the material. However, it is important to consider whether this equilibrium is very far away from the typical conditions in a building. For the emissions from some large-area materials such as newly painted walls it is quite possible that emissions will increase due to reduced air concentration when the air change rate is increased (Gunnarsen, 1997). For other materials, this effect may not be pronounced at the relevant concentrations (Knudsen et al., 1998).

The properties of the volatile chemicals themselves and the materials in which they are contained are essential for determining which step in the emission process will limit emission rates and how the steps are influenced by environmental parameters. Normally it is hard to tell which step limits emissions of a compound from a given source. It may be a combination of several steps and it may change with environmental conditions. Drying of large surfaces of wet paint is normally limited by evaporation and is highly influenced by several parameters of the environment. Emissions from a thick and relatively small area of material in a large room will normally be limited by diffusion, and the emission rate is influenced mainly by temperature.

3.6.4 Sorption

In an indoor climate context, sorption is the term used for the surface processes adsorption, absorption, and desorption. Adsorption is the accumulation of a chemical substance from the gas phase on the surface of a solid building material. Absorption is the solution of a gas phase substance into the bulk of a solid building material. Finally, desorption is the re-emission of ad- or absorbed substances.

Different mechanisms for adsorption processes exist which result in different degrees of binding strength and reversibility (Sleiko, 1985). Physical adsorption involves the binding of molecules by the relatively weak Van der Waals forces and is therefore expected to be almost fully reversible. Specific adsorption involves the binding of functional groups (e.g. hydrophilic or polar) of the adsorbate (VOC) to similar groups on the adsorbent (construction product), but without chemical transformation. Specific adsorption is also expected to be relatively reversible. Chemical adsorption involves reactions between the adsorbate and the adsorbent which usually lead to a strong bond (e.g. electron sharing) between the two. Chemical adsorption is expected to be the least reversible of the different types of sorption, and in practice it is not easy to distinguish this from absorption.

Sorption equilibria and kinetics are influenced by the nature of the adsorbent and the adsorbate, by the mechanism of adsorption, and by environmental parameters such as temperature, relative humidity, concentration of the adsorbate, and air velocity and turbulence past the adsorbent surface. Air velocity and turbulence only affect sorption kinetics; the other parameters also affect equilibria. In general, low adsorbate saturation vapor pressure, low temperature, and high adsorbate concentration in the air increase adsorption. Relative humidity does not always affect adsorption. Colombo et al. (1993) found a 35 % decrease in adsorbed mass when relative humidity was changed from <10 % to 35 %, but only an 8 % decrease when the humidity was increased from 35 % to 70 %. Building materials, which are exposed to indoor air in the normal humidity range of 35–70 %, will typically already be covered by at least one monolayer of adsorbed water, and the formation of multilayers will only have a limited influence on sorption properties for other airborne substances. Kirchner et al. (1997) found that an increase in air velocity increased the rate of desorption of a VOC mixture from painted gypsum, but not from carpet. The air velocity of air above the tuft may be insignificant for the desorption processes of carpet fibers deeper in the tuft.

The rate of adsorption and desorption for smooth non-porous products is typically governed by mass transfer in the boundary layer of air above the surface material. Sorption processes are therefore typically faster than most other emission processes, which furthermore may be limited by other processes inside the sources. Sorption processes may therefore often be considered to take place near equilibrium. If the concentration of an adsorbate changes in the surrounding air, the direction of the sorption processes at a surface may change. For example, the air-polluting desorption process may be changed to an air-cleaning adsorption process when concentrations rise.

Diffusion of air pollutants into structural materials such as brick, concrete and gypsum may increase the air pollution storage in a building (Meininghaus et al., 1998). The weight of buildings is typically several hundred or thousand times the weight of the air they contain. Diffusion-tight or diffusion-limiting layers or membranes are included in most outer walls in cold climates. This is done to limit diffusion of water from the moist indoor air into the structure, among other reasons, thereby avoiding problems with decay of materials or reduced thermal insulation caused by condensation and build-up of water content. Positioning of these layers may limit the structural mass interacting with the indoor air.

3.6.5 Chemical Reactions in Indoor Air

Oxidizing compounds such as ozone, nitrous oxides and peroxides like PAN may react with chemicals in air (Weschler and Shields, 1997; Jacobi and Fabian, 1997) or surface materials and form reactive species which are very problematical pollutants from a health and comfort point of view. Not only photocopiers and laser printers but also the outside air is a source of oxidants. Here, photochemical air pollution may be formed as a result of reactions involving nitrous oxides, oxygen, methane and many other outside air-polluting hydrocarbons when ultraviolet light from sunshine is available in the troposphere. It may also be hypothesized that ultraviolet radiation from some fluorescent lights and halogenated light bulbs may cause some reactions similar to those causing photochemical air pollution outside.

Comprehensive research into the chemistry of the atmosphere has been conducted in recent years. As summarized by Fenger (1997), photochemical reactions in smog have been shown to exacerbate outdoor air pollution. Ozone and nitrous oxides in the atmosphere may react with organic compounds and other chemicals forming even more reactive compounds with added energy from solar radiation. Ryan and Koutrakis (1990) have suggested that similar reactions among chemicals in the indoor air play a significant role for the composition of indoor air. Major differences between indoor air chemistry and atmospheric chemistry are increased surface-to-volume ratios, absence of sunlight and higher concentrations indoors of, for example, isoprene (a human metabolite) and other compounds with important indoor sources (Salthammer, 1999).

Weschler et al. (1992), Sundell (1994) and Wolkoff (1995) have suggested that chemical reactions between ozone and volatile organic compounds may lead to the formation of irritating compounds such as formaldehyde and organic acids. Sundell (1994) suggests that such reactions may explain that in northern Sweden he found a large number of office buildings having lower room air concentrations of TVOC than in inlet air. He furthermore found that this apparent loss of TVOC was associated with an increased prevalence of annoyance among the office workers.

Ozone reacts primarily with unsaturated chemicals, i.e. those having double or triple bonds. One oxidizing molecule may, in a chain of reactions with indoor air pollutants, form several irritating compounds. Table 3.6-1 presents the time taken for the concentration of selected compounds to be reduced to half the initial concentration. The ozone reaction is in most cases expected to be slower than further reactions between the reaction products.

Because of their double bonds, isoprene and α-pinene are likely to react with ozone (Weschler and Shields, 1999). These reactions may cause a difference between indoor and outdoor atmospheric chemistry, since these compounds have known indoor sources. An important product from these reactions may be the far more reactive OH· radical (Atkinson et al., 1992; Weschler and Shields, 1996).

Indoor sources of oxidants need to be carefully controlled. It is not possible to give further recommendations based on indoor air chemistry before more knowledge on the importance as well as the mechanisms of these reactions has been gathered.

Table 3.6-1. Half-lives (h) for some selected hydrocarbons in reactions with ozone. Numbers are based on an ozone concentration of 30 ppb at 22 °C, which is approximately the mean outdoor concentration in low-polluted temperate regions. Data from Fenger (1997).

Methane	>1 300 000	Formaldehyde	>7 600
n-Octane	>1 300 000	Benzene	8 000 000
Isoprene	13	Toluene	13 000
α-Pinene	2		

3.6.6 Chemical Reactions on or in Sources

Chemical reactions in sources are well known from hydrolysis of urea formaldehyde resins in chipboard resulting in the emission of formaldehyde and from the hardening processes of many sealants, glues, and varnishes (Roffael, 1993). The emission of volatile compounds from these products may in some cases be limited by formation.

Linoleum is hardened at elevated temperature during production, a process that involves the oxidation of the linseed oils with formation of miscellaneous aldehydes and carboxylic acids. A study of emissions from linoleum (Jensen et al., 1995) indicated that further oxidative degradation takes place resulting in a continuous production of these VOCs during the lifetime of the material. If linoleum is water damaged or cleaned with alkaline detergents, the combined effect of oxidation and hydrolysis may result in larger amounts of especially butyric and valeric acids, leading to a very unpleasant odor (Wolkoff et al., 1995). Another study of the autoxidation of linoleic acid (Porter et al., 1980) suggested that peroxy radicals which are formed as oxidation intermediates can provide oxygen for further oxidation, since the rate of oxidation was found to be independent of oxygen pressure.

The probability of reactions between oxidants and organic compounds is typically higher at material surfaces than in the air, since adsorbed compounds are more likely to be oxidized when the intermolecular forces responsible for the adsorption also cause redistribution of electrons. Oxidants may react with non- or semi-volatile organic compounds on the surfaces. The reaction products will typically be more volatile (Reiss et al., 1995). This has been used to speed up the off-gassing process when ozone has been used as an air-cleaning agent for odor removal.

The use of cleaning and maintenance products as well as smoking and other incidents during the use of buildings may introduce many chemicals. Chemical reactions may lead to the emission of new air pollutants. Also, some products may decompose by slow chemical reactions between their constituents. If the chemicals formed are allowed to diffuse and evaporate from the products, these reactions may continue for a long time leading to sustained deterioration of the air quality.

Compatibility between cleaning and maintenance products and surface materials is very important for the air quality in buildings.

3.6.7 Implications for the Use of Buildings

Experiments in small-scale chambers (Gunnarsen et al., 1994; Kephalopoulos et al., 1996; Meininghaus et al., 1998) and the development of mathematical models (Axley, 1991; Dunn and Tichenor, 1988) give credibility to the hypothesis that secondary emissions are important for the air quality in buildings. However, very few scientific results have yet been produced in real buildings to support this hypothesis (Wolkoff et al., 1997). Secondary emission processes add to the importance of doing things right in buildings at all times. A single incident of bad air quality may cause air deterioration for a long period afterwards. Possible reactions involving chemicals added during the operation of buildings should be considered.

Operation of building services and secondary emissions

In a workplace, temperatures will often increase during the day, but at night ventilation is reduced and temperatures are decreased. Sorption processes may compromise this energy-saving operation strategy. At the end of the working day, when both pollution loads and temperatures are at their maximum values, a large mass of pollutants will be adsorbed onto the interior surfaces, as more pollution is adsorbed at higher air concentrations. A reduction in ventilation at this time will maintain the high concentration for a longer time and a decrease in temperature will further increase the adsorption capacities of the surfaces. At the beginning of the next working day, when temperatures are increased again, the direction of the sorption process will shift towards desorption, and desorbed compounds will add to the air pollution generated by people and activities.

For surface materials with a high sorption potential these processes deserve particular attention. With surfaces mostly exposed to low concentrations of pollution it should theoretically be possible to clean the air during incidents of higher concentrations of air pollution. The air quality in rooms with very varying pollution loads may benefit from sorption processes, as sorption tend to reduce peak concentration if desorption is allowed to take place at regular intervals. Assembly rooms and classrooms are examples of spaces where it may be possible to benefit from sorption. However, no scientific demonstration of these benefits yet exists.

Sources of oxidants such as gas stoves and photocopiers may not only place the health of the occupants at risk by increased concentrations of nitrous oxides and ozone, but these oxidants may, in chain reactions with chemicals at surfaces or in the air, form other harmful volatile chemicals. Therefore oxidant source control is essential. It may also be hypothesized that elevated concentrations of oxidants in the outside air, by the same type of reactions, may cause decreased air quality when they enter buildings together with the ventilating air intended for dilution of pollution from interior sources.

Maintenance

The use of low-emitting products is not sufficient to assure good indoor air quality. The choice of compatible products for cleaning and maintenance is equally important. Fast deterioration of surfaces and the risk of increased harmful emissions may be the result if cleaning agents that react with the construction products are used.

Acknowledging that incidents of high concentrations of pollutants will cause extended periods of decreased air quality requires focus on routine maintenance. Careful isolation should be established when painting, and air concentrations in the painted rooms should be kept low. It is important to avoid polluting other surfaces with the solvents, since this may extend the area influenced by the paint job and prolong the time in which there is increased concentration of pollution in the air. Smoke should be "aired out" as fast as possible.

Airing out may not only mean diluting the concentration of pollutants in the air: it may also cause the inner surfaces to be cleaned and their emission rates to be decreased afterwards.

3.6.8 Conclusions

Secondary emissions have the potential to explain continued deterioration of the air quality in many buildings. The small number of studies carried out in real buildings demonstrate that the magnitude of secondary emissions is important for indoor air quality. Future intervention studies involving the diurnal operation of ventilation and heating systems could give more validity to the reported small-chamber experiments and mathematical models giving preliminary credibility to the importance of the processes.

Some studies demonstrate increased emissions at increased ventilation rates, while others fail to do so. Oxidants in the supply air may initiate chains of reactions between the oxidants and otherwise stable chemicals on the inner surfaces or in the indoor air. Partial pressure of VOCs in the air may reduce their emission. When ventilation is increased, the result may be less reduction of air pollution than expected, since the emission rates may increase at the resulting lower VOC concentrations and the increased oxidant concentration may promote secondary emissions.

Sorption processes of significant importance have been demonstrated repeatedly in laboratory investigations. They are essential for the understanding of emission processes in buildings. Most inner surfaces may store chemicals, and equilibrium with the surrounding air will be approached by desorption or adsorption of airborne chemicals. These processes are typically fast and controlled by mass transport in the boundary layer of air. However, adsorbed pollutants may diffuse into the materials, and their reemission is typically slower and also influenced by the speed of diffusion back to the surface.

References

Atkinson R., Aschmann S.M., Arey J. and Shorees B. (1992): Formation of OH radicals in the gas phase reactions of O_3 with a series of terpenes. J. Geophys. Res., *97* , 6065–6073.

Axley J. W. (1991): Adsorption modelling for building contaminant dispersal analysis I. Indoor Air, *2*, 147–171.

Colombo A., De Bortoli M., Knöppel H., Pecchio E. and Vissers H. (1993): Vapour deposition of selected VOCs on indoor surface materials in test chambers. Indoor Air '93: Proceedings of the 6th International Conference on Indoor Air Quality and Climate.

Dunn J. E. and Tichenor B. A. (1988): Compensating for sink effects in emission test chambers by mathematical modeling. Atmos. Environ., *22*, 885–894.

Dunn J. E. and Chen T. (1994): Critical evaluation of the diffusion hypothesis in the theory of porous media volatile organic compound (VOC) sources and sinks. In: Nagda N.L. (ed) Modeling of Indoor Air Quality and Exposure, ASTM STP 1205, Philadelphia, USA, 64–80.

Fenger J. (ed) (1997): Photochemical air pollution, NERI technical report no. 199. National Environmental Research Institute, Roskilde, Denmark.

Gunnarsen L. (1997): The influence of area-specific ventilation rate on the emissions from construction products. Indoor Air, *7*, 116–120.

Gunnarsen L., Nielsen P. A. and Wolkoff P. (1994): Design and characterization of the CLIMPAQ, chamber for laboratory investigations of materials, pollution and air quality. Indoor Air, *4*, 56–62.

Jaakkola J. J. K., Ilmarinen R. and Seppänen O. (eds): Laboratory of Heating, Ventilating and Air Conditioning, vol 2. Helsinki University of Technology, 407–412.

Jakobi G. and Fabian P. (1997): Ozon und Peroxiacetylnitrat (PAN) – Konzentration in Innenräumen. Annal. Meteorologie, 33, 66–71.

Jensen B., Wolkoff P., Wilkins C. K., Clausen P. A. (1995): Characterization of linoleum. Part 1: Measurement of volatile organic compounds by use of the field and laboratory emission cell, "FLEC". Indoor Air, *5*, 38–43.

Kephalopoulos S., Knöppel H. and De Bortoli M. (1996): Testing of sorption and desorption of selected VOCs on/from indoor materials. In: Indoor Air 96. Proceedings of the 7th International Conference on Indoor Air Quality and Climate, Nagoya, Japan, 21–26 July, 2, 61–66.

Kirchner S., Karpe P., Quenard D., Kephalopoulos S., Knöppel H., De Bortoli M., Bluyssen P., Van der Wal J., Cornelissen H. J. M., Hoogeveen A. W., Kjaer U. D. and Tirkkonen, T. (1997): Characterization of adsorption-desorption of organic pollutants on wall and floor coverings surfaces. In: Proceedings of the 4th International Conference on Characterization and Control of Emissions of Odors and VOCs. Montreal, Canada, 270–282.

Knudsen H.N., Kjaer U.D., Nielsen P.A. and Wolkoff P.(1998): Sensory and chemical characterization of VOC emissions from building products: Impact of concentration and air velocity. Atmos. Environ., Vol 33, 1219–1230.

Meininghaus R., Gunnarsen L. and Knudsen H.N. (1999): Measurements of diffusion and sorption properties for construction products. EPIC 98–3rd International Conference on Indoor Air Quality, Ventilation and Energy Conservation in Buildings, Lyon.

Porter N.A., Weber B.A., Weenen H. and Khan J. A. (1980): Autoxidation of polyunsaturated lipids: factors controlling the stereochemistry of product hydroperoxides. J. Am. Chem. Soc., *102*, 5597–5601.

Reiss R., Ryan B., Koutrakis P. and Tibbetts S. J. (1995): Ozone reactive chemistry on interior latex paint. Environ. Sci. Technol., 29, 1906–1912.

Roffael E. (1993): Formaldehyde Release from Particleboard and other Wood-based Panels. Forest Research Institute Malaysia (FRIM), Malayan Forest Records No. 37, Maskha Sdn. Bhd., Kuala Lumpur.

Ryan P. B. and Koutrakis P. (1990): Indoor air chemistry: An emerging field. Proceedings of Indoor Air 90, Toronto, Canada, Vol. 2, 489–494.

Salthammer T. (1999): Contribution of reactive compounds and secondary emission products to indoor air pollution – a review of case studies. VDI-Report No. 1443, VDI-Verlag GmbH, Düsseldorf, 337–346.

Sleiko F.L. (1985): Adsorption technology: a step-by-step approach to process evaluation and application. Dekker, New York.

Sundell J. (1994): On the association between building ventilation characteristics, some indoor environment exposures, some allergic manifestations and subjective symptom reports. Indoor Air, Supplement no. 2.

Weschler C. J., Brauer M. and Koutrakis P. (1992): Indoor ozone and nitrogen dioxide: a potential pathway to the generation of nitrate radicals, dinitrogen pentoxide and nitric acid indoors. Environ. Sci. Tech., *26*, 2371–2377.

Weschler C. J. and Shields H. C. (1996): Production of the hydroxy radical in indoor air, Environ. Sci. Technol., 30, 3250–3258.

Weschler C. J. and Shields H. C. (1997): Potential reactions among indoor air pollutants. Atmos. Environ., 31, 3487–3495, 1997.

Weschler C. J. and Shields H. C. (1999): Indoor ozone/terpene reactions as a source of indoor particles. Atmos. Environ., 33, 2301–2312.

Wolkoff P. (1995): Volatile organic compounds–Sources, measurements, emissions and the impact on indoor air quality. Indoor Air, Suppl. No. 3.

Wolkoff P. (1996): Characterization of emissions from building products: long-term chemical evaluation. The impact of air velocity, temperature, humidity, oxygen, and batch/repeatability in the FLEC. Indoor Air '96. Proceedings of the 7th International Conference on Indoor Air Quality and Climate, Nagoya, Japan, July 21–26, Vol. 1, pp 579–584.

Wolkoff P., Clausen P.A., Jensen B., Nielsen G.D. and Wilkins C. K. (1997): Are we measuring the relevant indoor pollutants. Indoor Air, *7*, 92–106.

Wolkoff P., Clausen P.A. and Nielsen P.A. (1995): Application of the field and laboratory emission cell "FLEC"–performance study, intercomparison study and case study of damaged linoleum in an office. Indoor Air, *5*, 196–203.

3.7 Release of MVOCs from Microorganisms

Jonny Bjurman

3.7.1 Introduction

Molds, bacteria and other micro-organisms (see Fig. 3.7-1) are often found growing in parts of buildings with excess moisture. The micro-organisms may produce several potentially harmful agents such as spores, which might cause different allergic diseases or carry toxins (Samson et al., 1994). Micro-organisms also produce unpleasant odors. Odors produced by common micro-organisms such as *Penicillium* sp, *Aspergillus* sp and *Streptomyces* spp and others, have been characterized as musty, earthy, cat-like, muddy, rotten and fetid (Harris et al., 1986). These compounds often have very low odor thresholds. Volatile organic compounds of this kind as well as other volatile compounds with a less pronounced odor are today suspected to contribute to the "sick building syndrome" (SBS). Several metabolic pathways by which these volatiles may be produced are known from studies on different micro-organisms in different contexts such as detection of off-flavors produced by fungi (Seifert and King, 1982) and production of natural flavors for use as food additives (Lanza et al., 1966; Gatfield, 1986; Welsh et al., 1989). These compounds, now often denoted MVOCs, are usually present in buildings in very low concentrations, at the $\mu g\ m^{-3}$ level or below, even when the attack by micro-organisms is severe (Ström et al., 1994; Dewey et al., 1995).

Generally the emission of VOCs from building materials decreases with time. If, however, SBS symptoms appear in a building formerly without SBS symptoms, agents produced by micro-organisms that have started to grow in the buildings may be suspected to be a cause of the development of the SBS. Even building materials that have been characterized as low emitters of VOCs and therefore considered environmentally safe may, when humidified, support microbial growth and thereby production of MVOCs.

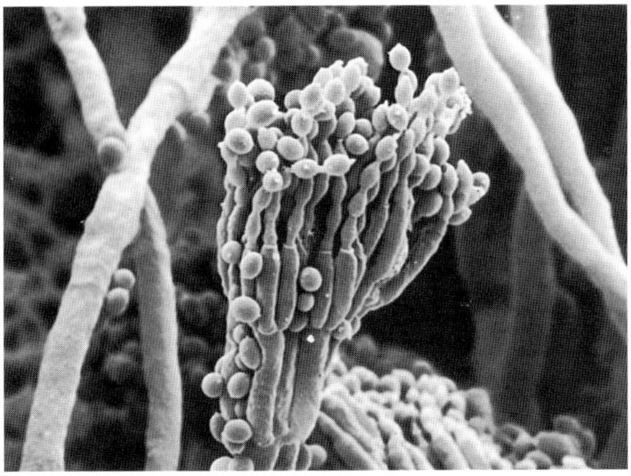

Figure 3.7-1. Scanning electron micrograph of a mold fungus (Photo: Daniels and Jirjis).

A very large number of volatile compounds have been identified or tentatively identified in different micro-organisms growing on different media, including building materials, or under different environmental conditions. It would be out of the scope of this article to mention all these compounds or written reports. Rather, my intention has been to try to outline some general principles which lead to the production, emission and spread of different types of volatile compounds found and methods used for the analysis of such compounds, and thereby develop some continuity in our understanding of the relevance of MVOCs in buildings.

The science of MVOCs is not new. Following the advent of the gas chromatograph and its use particularly in the field of food science, the foundation for studies of flavor compounds produced by micro-organisms became established. Even today, in the field of MVOCs in buildings, we are indebted to the results of these research workers. Since then a vast amount of literature on different aspects of the production of volatile organic compounds by micro-organisms has been published, although the science of MVOCs in buildings is a rather new one, which, however, now deserves its own review. One difficulty with evaluating the current status of this comparatively new field of study is that potentially relevant knowledge has been produced by research efforts in several disciplines such as the detection of undesired fungal growth, where most of the work has been done on cereals (Börjesson et al., 1992), detection of off-flavors caused by fungi (Seifert and King, 1982), effects of volatile fungal metabolites on other micro-organisms and insects, production of natural flavors for use as food additives (Lanza et al., 1966; Gatfield, 1986; Welsh et al., 1989), or the possible relationship between fungal volatiles and the sick building syndrome (Samson, 1985; Ström et al., 1990; Bjurman and Kristensson, 1992b; Bjurman, 1993).

Work aimed at clarifying the role played by MVOCs when growing on different building materials in the development of SBS is today a particularly rapidly evolving field of research.

3.7.2 Organisms Concerned

Several different types of micro-organisms belonging to different taxonomic groups are known to be able to grow on building materials within a building (Grant et al., 1988; McJilton et al., 1990 ; Flannigan et al., 1991; Flannigan, 1992; Andersson et al., 1997). The main types of micro-organisms found in buildings are listed in Table 3.7-1.

Table 3.7-1. Prominent types of micro-organisms in buildings that have been shown to synthesize MVOCs.

Mold fungi
Bacteria
Actinomycetes
Basidiomycetes (Decay fungi)

3.7.3 The Role Played by MVOCs in SBS

In 1990 more than 300 volatile organic compounds had been identified in indoor air (Molhave, 1990). If present in concentrations above a certain limit, these compounds are believed to be an important cause of the development of SBS symptoms. The main sources of these impurities are not micro-organisms.

Different studies have shown that there is often no significant difference in airborne spore levels in buildings affected or unaffected by microbial growth. Growth of micro-organisms in buildings may also occur inside floors, walls etc. The fungal spores are too large to diffuse through such materials. In contrast, many volatile metabolites produced by the fungi can diffuse through substrates such as insulation material, wallpaper and even through the plastic film usually recommended in temperate climates to prevent moisture transport from the rooms into the colder wall (Ström et al., 1994; Dewey et al., 1995). Volatile metabolites from molds or bacteria growing within walls etc. can also reach indoor air by convection.

The possible relationship between microbial volatiles and the sick building syndrome has been proposed by several workers (Ström et al., 1990; Bjurman and Kristensson, 1992a; Bjurman, 1993). However, as was pointed out by Sunesson et al. (1995a), the hypothesis that MVOCs are one cause of the sick building syndrome can be firmly evaluated only if the identities of these compounds are known. In relation to other volatile compounds found in buildings, the concentration of MVOCs is usually comparatively low. In order to be of significance as a cause of SBS, some of the microbially produced volatile compounds must be very potent irritants or accumulate into living cells over longer periods of exposure.

It is known that volatile organic compounds with eight carbon atoms are common metabolites of fungi (Tressl et al., 1982). Such compounds often have inhibitory effects on various micro-organisms, indicating that they have the potential to interfere with cells in an unspecific way (Andersen et al., 1994; Fiddaman and Rossal, 1994; Viegas and Sa Correia, 1995).

3.7.4 Sampling and Analysis of MVOCs

Various procedures have been used by different research workers for the sampling and analysis of MVOCs. These procedures have different selectivities (Parliament, 1986), which have led to different conclusions on the specific compounds produced by the different micro-organisms even when growing under identical conditions. A particular difficulty is that the compounds of relevance in this context also have a very wide variability in boiling point from very volatile compounds to so-called semi-volatile compounds.

A more thorough description of several general analysis procedures for VOC is given in Chapter 1 of this book. It is, however, essential in this context to give a short outline of the procedures used when trying to clarify the production of MVOCs.

A simple way by which analysis for microbial volatile compounds was often done before the general use of adsorbents was the withdrawal, using a gastight syringe, of an air sample, which was then injected into a gas chromatograph (Norrman, 1971) which

Table 3.7-2. Some adsorbents that have been used in the analysis of MVOCs by different workers.

Tenax TA
Tenax GR
Porapak
Charcoal
Chromosorb

limited detection to the major compounds. Today, adsorbents are usually used for sampling of volatile compounds, which enables detection of volatiles present in low concentration. Activated carbon followed by solvent extraction of the adsorbed compounds has been frequently used for MVOCs analysis (Larsen and Frisvad, 1994a). However, there are several potential drawbacks with solvent extraction of adsorbents, such as the possible introduction of volatile impurities, the masking effect of the solvent peak and the loss of very volatile compounds during concentration of the eluate. Today, the use of polymer adsorbents followed by thermal desorption is therefore most frequently used (Table 3.7-2).

Using thermal desorption gas chromatography, eight adsorbents, Tenax TA, Tenax GR, Chromosorb 102, Carbotrap C, Carbopack B, Anasorb 727, Anasorb 747 and Porasil C/n octane (Durapak) were evaluated by Sunesson et al. (1995b) for sampling and quantitative analysis of a selection of known MVOCs at low substance concentrations and under varying humidity conditions. Tenax TA proved to have the best average properties generally, with high recoveries recorded. Tenax TA is also the sorbent that has been chosen most extensively by other workers in this field today. It is, however, important to keep in mind that some polar compounds or low molecular weight compounds are not analyzed optimally with this sorbent, and the amounts of such compounds may therefore be underestimated. The main disadvantage of the carbon type sorbents appears to be the poor kinetics of desorption when employing thermal elution (Sydor and Pietrzyk, 1978).

MVOCs have been sampled with active sampling for short periods or passive (diffusive) sampling (Bjurman and Kristensson, 1992a,b; Larsen and Frisvad, 1994a,b; Larsen and Frisvad, 1995a; Nilsson et al.,1996). The long sampling times necessary when diffusive sampling is used, usually more than a week, give time-weighted average values. This sampling may potentially analyze compounds produced only for a limited time period. However, it may thereby give false indications of time-related production of certain MVOCs.

MVOCs, in the same way as other VOCs, may be firmly adsorbed to different materials during certain conditions. Such adsorbed MVOCs may potentially be released when conditions change, e.g. when the temperature or the moisture level increases. It may therefore be essential to analyze adsorbed volatiles also. A possible method would be application of supercritical extraction employing CO_2 (Nilsen et al., 1991).

Recently "electronic noses" based on an array of comparatively unselective sensors and neural networks or multivariate methods have been introduced for detection of MVOCs in relation to microbial growth (Börjesson et al., 1996; Nilsson, K., 1996). These methods may have a potential also for detection of micro-organisms in buildings.

3.7.5 Biosynthesis of MVOCs

Besides being dependent on species (Zechman and Labows Jr, 1985; Wilkins and Larsen, 1995; Larsen and Frisvad, 1995b,c), environmental factors and substrate compositions are known to have great influence on both qualitative and quantitative production of volatile metabolites and the growth phase related production of these compounds (Norrman, 1971; Yong et al., 1985; Yong and Lim, 1986; Yong, 1992; Fiddaman and Rossal, 1994; Wilkins and Larsen, 1995; Sunesson et al., 1995c; Bjurman et al., 1997). An astonishingly large variety of compounds might be emitted even by a single species under the influence of different conditions. Important environmental factors that may influence the emission of MVOCs are water activity, R.H., pH and atmospheric compositions such as CO_2- and O_2-levels and temperature.

From studies on the production of MVOCs by micro-organisms which contribute to taste and flavor of food, several metabolic pathways are known which might explain the presence of the most prominent types of MVOCs found to be emitted from micro-organisms when they are growing on humidified building materials.

Several studies in particular were made in the early stages of cheese flavor research (Lawrence and Hawke, 1968; Kinsella and Hwang, 1976; Marth, 1982; Law, 1984; Okumura and Kinsella, 1985; Kahradian et al., 1985a,b; Rabie, 1989) and have been a good base for more recent studies related to the production of MVOCs emitted to indoor air by molds growing on humidified building materials.

It is not always quite clear if the production of certain volatile compounds by micro-organisms when growing on a certain substrate depends on a general effect on the metabolism or if compounds in the medium act as precursors.

3.7.6 Effects of Media Composition

It is well known that the medium composition strongly determines the flavor compounds produced by micro-organisms (Gallois et al., 1990). Particularly the composition of volatile compounds produced can vary significantly, depending on the types of carbon and nitrogen sources used and the C/N ratio (Norrman, 1971; Lanza et al., 1966; Lanza and Palmer, 1977; Sprecher and Hanssen, 1982; Yong and Lim, 1986; Yong et al., 1985). Available trace metal ions or phosphate have also been demonstrated to influence the production of volatile compounds by micro-organisms (Bjurman and Kristensson, 1992b). The presence of specific precursors in the media may lead to the production of higher concentrations of certain MVOCs.

Several different types of compounds have been shown to be produced, of which several types may be related to common metabolic pathways. Thus, free fatty acids may be first released as a result of fungal lipase activity. These fatty acids may be oxidized to keto-acids, subsequently being decarboxylated to methyl ketones followed by reduction to secondary alcohols (Hawke, 1966). Different mold fungi possess different types of lipases, which may give rise to clusters of different compounds. Unsaturated fatty acids may be transformed to volatile aldehydes, alcohols and esters by hydroperoxidation by lipoxygenase activity (Eriksson, 1974).

The frequent occurrence of C8 compounds among fungal VOCs might be related to the peroxidation specificity of fungal lipoxygenases (c.f. Gaillard and Phillips, 1976; Eskin et al., 1979; Croft et al., 1993)

3.7.7 Building Materials as a Medium for Production of MVOCs

Studies comparing the production of volatile compounds produced when the micro-organisms are growing on different building materials and when growing on commonly used artificial media have revealed great differences. Unsoiled building materials often contain only low amounts of available nutrients for growth of micro-organisms. When growing on a medium with a very low nutrient concentration, volatile compounds otherwise not so commonly produced on other, richer, media may be produced by certain mold fungi and ascomycetes. Thus, Bjurman and Kristensson (1992b) found that an isolate of *Aspergillus versicolor* produced increasing amounts of 2-ethyl-1-hexanol on a medium containing only water and agar. Sunesson et al. (1995d) found that 2-ethyl-1-hexanol was produced by an isolate of *Streptomyces albidoflavus* when growing on gypsum board which is also a low nutrient medium. The main reason for this is probably that endogenous substrates, such as lipids, are used as substrates for the synthesis of these volatile compounds instead of nutrients assimilated from the substrate.

Studies with the micro-organisms growing on the often much richer artificial media may therefore have a low value for prediction of the potential types of MVOCs produced by the micro-organisms when they grow on building materials. It is therefore essential to study the production of MVOCs by the micro-organisms when they are growing on an array of building materials.

However, even if several building materials are low in potential nutrients there are some exceptions. Available potential nutrients in certain types of building materials are also very variable. Thus, the nutrient content of wood may vary to a great extent (Terziev et al., 1996). Glass wool and mineral wool are in themselves poor substrates, but insulation materials and air filters made of such materials often contain urea-formaldehyde, and dust-binding oils are added which may serve as nutrients for micro-organisms (Morey and Williams, 1990; Bjurman, 1993; Ezeonu et al., 1994 a,b).

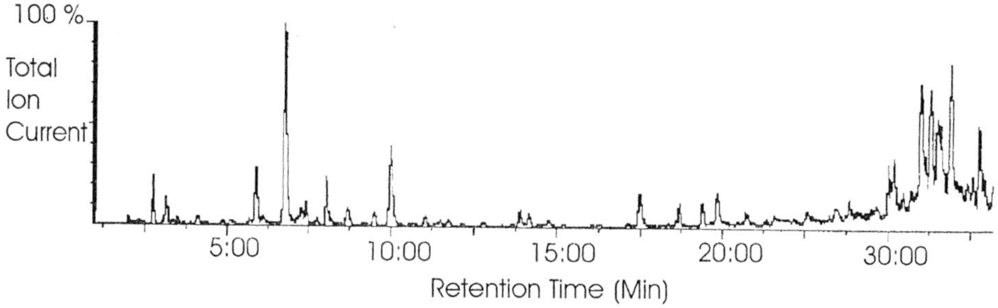

Figure 3.7-2. Gas chromatogram of headspace volatiles released from a culture of *Stachybotrys chartarum* on damp paper. Volatiles analyzed by diffusive sampling on Tenax TA followed by thermal desorption.

Contamination of building materials by soil, dust or other sources of dirt may add potential nutrients for micro-organisms which support production of MVOCs (Chang et al., 1996). Wallpaper may support microbial growth and production of MVOCs (Fig. 3.7-2). Dust may also function as a sink for MVOCs (Wilkins et al., 1997).

3.7.8 Stage of Growth

Volatiles may be produced during most phases of microbial growth. 2-Heptanone and methyl ketones have been shown to be produced by spores of *Penicillium roquefortii* (Dartey and Kinsella, 1973; Creuly et al., 1990).

The types of compounds produced during different growth phases are influenced by substrate changes occurring during the growth of microorganisms, which often leads to nutrient imbalances (Wilkins and Larsen, 1995; Bjurman et al., 1997; Gervais and Sarrette, 1990). Such changes often lead to the expression of secondary metabolism. Thus, several of the most prominent MVOCs are clearly produced as a result of secondary metabolism (Bjurman and Kristensson, 1992a; Bruce et al., 1996). For a general discussion of this concept, see e.g. Herbert (1981), Bennet (1983) or Vining (1992). Adsorbed MVOCs may be further metabolized by other micro-organisms (Kleinheinz and Bagley, 1997), which may change the emission of MVOCs.

3.7.9 Influence of Moisture and Temperature

The different types of micro-organisms have different minimum moisture requirements (Grant et al., 1988). Molds have a lower limit for growth than other micro-organisms growing in buildings; usually the limit is around 80 % R.H. Molds would therefore be expected to be more common sources of MVOCs in buildings than other micro-organisms which need higher moisture levels. It is even possible that MVOCs may be produced by molds when periodically subjected to moisture levels below the limit for growth. In this case, the main substrates would be storage compounds such as fat in the cell.

Moisture also strongly influences the concentration in the air phase of produced volatile compounds by regulating the tendency for sorption and desorption to different building materials as has been shown for other matrices (Buttery et al., 1969a,b; Lanciotti and Guerzoni, 1993; Unger et al., 1996; Börjesson et al., 1994).

Since many of the volatile compounds produced by micro-organisms are secondary metabolites, the production is often stimulated at lower temperatures (Lee et al., 1979; Aoyama et. al., 1993; Dionigi and Ingram, 1994; Sunnesson et al., 1995c). Low-temperature conditions are common in certain parts of buildings in temperate climates, e.g. in crawl spaces, which are very often attacked by mold fungi with concomitant odor production (Bjurman and Kristensson, 1992b). However, low temperature may also lead to increased sorption of produced compounds.

Mold fungal activity is highly dependent on variations in the available moisture, but can tolerate periods of low moisture (Viitanen and Bjurman, 1994). The production of

MVOCs may therefore vary in relation to such moisture variations. It was shown that the production of 2–heptanone by a mold fungus was higher at an aw of 0.98 than at an aw of 1.0 (Gervais and Sarette, 1990).

3.7.10 Marker Compounds

MVOCs have been suggested to be indicators (marker compounds) of microbial growth in buildings if they are present in sufficiently high levels and the only source of the compounds used is micro-organisms growing in the buildings (Miller et al., 1988; Rivers et al., 1992; Sunnesson et al., 1996; Bjurman et al., 1997).

The emission of MVOCs has also been compared with other detection methods (Miller et al., 1988; Börjesson et al., 1992; Börjesson et al., 1994). Compounds whose presence is more independent of environmental conditions or media used may be more suitable markers. Alternatively, an array of compounds that are produced under different conditions might be used as markers.

Detection of microbial growth at an early stage is essential to avoid serious damage. Compounds produced at an early stage may therefore be more valuable as markers. However, such compounds may not reach the detection limit.

Problems concerning volatile metabolites as a method for detection of micro-organisms have been addressed by several workers (Sunesson, 1995a; Wilkins and Larsen, 1995; Bjurman et al., 1997). Such problems include, for example, insufficient knowledge about the metabolites produced by the micro-organisms when growing on building materials under the environmental conditions likely to be encountered in different parts of a building during different growth phases, low concentrations of the metabolites in relation to the concentration of compounds emitted from the uncolonized building materials, and lack of sufficiently specific fragments in the mass spectra of otherwise good candidate compounds (Bjurman et al., 1997), this being a particular problem when selected ion monitoring is used for detection.

3.7.11 Types of Volatile Compounds Produced

3.7.11.1 Alcohols

Several investigations have revealed that 1-octen-3-ol; 3-methyl-1-butanol and 2-methyl-1-propanol are the compounds produced in the largest quantities by molds on substrates mainly containing carbohydrates as the carbon source (Kaminski et al., 1974; Karahdian et al., 1985b)

Commonly produced metabolites on several media are also 1-pentanol, 2-heptanol, 2-nonanol, 1-hexanol, 2-methyl-1-butanol, 1-octen-3-ol, 2-ethyl-1-hexanol and 3-octanol (Eriksson et al., 1992; Börjesson et al, 1993; Larsen and Frisvad 1994a,b; 1995b,c; Sunesson et al., 1995c). Eight carbon alcohols and ketones may be produced via lipoxygenase-mediated conversions of linoleic acid and linolenic acid (Eskin et al., 1979; Harris et al., 1986). Wurzenberger and Gros (1984) demonstrated the formation of

1-octen-3-ol in *Psalliota bispora* from linoleic acid. Many of the alcohols are produced from amino-acids via the Ehrlichs pathway with decarboxylation and reduction of aminoacids (Berry, 1988). Thus, the presence of asparagine leads to increased production of isopropanol. valine, leucine and phenylalanine may lead to the production of 2-methylpropanol, 3-methylbutanol and 2-phenylethanol, respectively (Spinnler and Djian, 1991).

3.7.11.2 Ketones

2-Pentanone, 2-hexanone, 2-heptanone, 3-octanone and methyl ketones are frequently produced (Okumura and Kinsella, 1985; Larsen and Frisvad, 1994a). Lipid degradation leads to the production of methyl ketones (Marth, 1982; Kinderlerer, 1987). C5 and C13 methyl ketones may be produced from saturated fatty acids (Law, 1984). The decrease in methyl ketones is due first to their reduction to the corresponding secondary alcohol (Jollivet and Belin, 1993).

3.7.11.3 Aldehydes

Aldehydes have sometimes been found to be produced by mold fungi. Thus, Law (1984) found aldehydes among volatiles produced by *Penicillium camemberti.*

The possible involvement of a lipoxygenase enzyme system also in the production of aldehydes was shown by Andersen et al. (1994).

3.7.11.4 Acids

Fatty acids are both substrates and products in the production of MVOCs.

Karahdian et al. (1985a,b) demonstrated the production of 1-octen-3-ol, 8-nonen-2-one, 3-octanone, 3-octanol and octanoic acid from linoleic acid and linolenic acid. The use of microbially produced fatty acids for characterization of mold fungi has been suggested (Blomquist et al., 1992). Lipolysis of triglycerides or amino acids may lead to the production of compounds such as 2-methylpropanoic acid, butanoic acid , 2-methylbutanoic acid, pentanoic acid, hexanoic acid and octanoic acid (Jolivet and Belin, 1993). The presence of different lipids and lipases in different fungi (Ha and Lindsay, 1993) and under different environmental conditions may explain a great deal of the variation in the volatile compounds produced.

The production of octanoic acid may increase from mold fungi when growing on insulation materials (Bjurman, 1993) in comparison with growth on more favorable substrates. Secondary metabolism has been shown to be induced by the presence of compounds toxic to micro-organisms (Gueldner et al., 1985; Jones, 1970). This may be one reason for the increased emission of octanoic acid by *Phialophora fastigiata* when growing on wood treated with certain preservatives against decay fungi (Bjurman and Kristensson, 1993).

3.7.11.5 Terpenes and Terpene-related Compounds

Terpenes and terpene-related compounds have been frequently revealed as metabolites of micro-organisms (Collins, 1976; Lanza and Palmer, 1977). Monoterpenes such as β-pinene, limonene, β-phellandrene, α-terpinolene and 1-methoxy-3-methylbenzene are commonly found. Another group of terpenes often produced is the sesquiterpenes, C15H24 compounds (Seifert and King, 1982; Bjurman and Kristensson, 1992b; Gurtler et al., 1994; Zerinque et al., 1993). The terpene-related, earth-smelling compound geosmin and 2-methylisoborneol, also with a distinctive odor, have long been known to be produced by *actinomycetes* (Gerber, 1967; Gerber and Lechevalier, 1965; Gerber, 1986; Wilkins, 1996). Geosmin has also been shown to be produced by the soft rot fungus *Chaetomium globosum* (Kiguchi et al., 1981; Bjurman and Kristensson, 1992b) and several *Aspergillus* and *Penicillium* species (Mattheis and Roberts, 1992; Dionigi and Ingram, 1994). Terpenes and terpene derivatives are produced via the mevalonic acid pathway (Lanza and Palmer, 1977; Sprecher and Hanssen, 1982).

The production of higher levels of terpenes can be ascribed to the onset of secondary metabolism triggered by a lack of nutrients.

3.7.11.6 Other Compounds

Several aromatic compounds, 2,5 – dimethylfuran, 3-methylfuran and styrene have been revealed (Larsen and Frisvad, 1994a). Dimethyl sulfide and other sulfur-containing MVOCs can be produced from methionine (Kadota and Ishida, 1972; Wilkins, 1996). Phenylalanine may give rise to 2-phenylethanol (Spinnler and Djian, 1991). It is clear that some compounds could be formed by transformation from compounds present in the building materials. Thus, trichloroanisoles may be produced by micro-organisms (Curtis et al., 1974; Whitfield et al., 1984; Nyström et al., 1992). Chloroanisoles may be produced from the wood preservative trichlorophenol by methylation, but this preservative has now been abandoned in several countries (Bjurman and Kristensson, 1993). The production of pyrazines may be related to the concentration of certain amino acids, notably cysteine (Riha et al., 1996). Pyrazines have been shown to be produced in a model system reaction of cysteine with glucose (Zhang and Ho, 1991).

3.7.12 Conclusions

This review has drawn attention to the numerous contributions to the understanding of MVOCs in buildings made by chemists and mycologists in different disciplines, often with quite different goals, over a long time period. However, it is essential to realize that knowledge developed within other fields may not be entirely applicable to the problem with MVOCs in buildings, since building materials are often quite different as substrates and micro-organisms may grow for extended periods in a building. To some extent, fungal species found in buildings can differ from those found in other environments, such as storage areas for food and grain.

It is my hope that this review has shed some light on current knowledge in the field of MVOCs in buildings, and particularly that it will encourage new research to fill remaining gaps in our knowledge of the specific effects of building materials as substrates for the production of MVOCs by micro-organisms.

Acknowledgements

This chapter was written while the author received special grants from The Swedish University of Agricultural Sciences, SLU and from "The Healthy Building" administered by The Swedish Building Research Council, BFR.

References

Andersen R.A., Hamilton-Kemp T.R., Hildebrand D.F., McCracken C.T. Jr, Collins R.W. and Fleming P.D. (1994): Structure antifungal activity relationships among aliphatic aldehydes, ketones and alcohols. J. Agric. Food. Chem., *42*, 1563–1568.

Andersson M.A., Nikulin M., Köljalg U., Andersson M.C., Rainey F., Reijula K., Hintikka E.L. and Sakinjoja-Salonen M. (1997): Bacteria, molds and toxins in water-damaged building materials. Appl. Environ. Microbiol., *63*, 387–393.

Aoyama K., Tomita B.I. and Chaya K. (1993): Influence of incubation temperature on production of earthy-musty odor substances by actinomycetes. Japan. Toxicol. Environ. Health., *33*, 207–112.

Bennet J.W. (1983): Differentiation and secondary metabolism in mycelial fungi. In: Bennet J.W. and Ciegler A. (eds) Secondary Metabolism and Differentiation, 1–32. Dekker, New York.

Berry D.R. (1988): Products of primary metabolic pathways. In: Berry D.R. (ed) Physiology of Industrial Fungi. Blackwell, Oxford, 130–160.

Bjurman J. (1993): Thermal insulation materials, microorganisms and the sick building syndrome. In: Kalliokoski P., Jantunen M., Säppenen D. (eds) Proceedings of Indoor Air '93, Helsinki, Vol 4, 339–344.

Bjurman J. and Kristensson J. (1992a): Production of volatile metabolites by the soft-rot fungus *Chaetomium globosum* on building materials and defined media. Microbios, *72*, 47–54.

Bjurman J. and Kristensson J. (1992b): Volatile production by *Aspergillus versicolor* as a possible cause of odor in houses affected by fungi. Mycopathologia, *118*, 173–178.

Bjurman J. and Kristensson J. (1993): Soft rot fungi as possible sources of odor in impregnated wood in buildings. The International Research Group on Wood Preservation, Doc /IRG /WP93–20013.

Bjurman J., Nordstrand E. and Kristensson J. (1997): Growth-phase related production of potential volatile organic compounds by molds on wood. Indoor Air, *7*, 2–7.

Blomquist G., Andersson B., Andersson K. and Brondz I. (1992): Analysis of fatty acids. A new method for characterization of molds. J. Microbiol. Methods., *16*, 59–68.

Börjesson T., Stöllman U. and Schnürer J. (1992): Volatile metabolites produced by six fungal species compared with other indications of fungal growth on cereal grains. Appl. Environ. Microbiol., *58*, 2599–2605.

Börjesson T., Stöllman U. and Schnürer J.L. (1993): Compounds produced by molds on oatmeal agar. Identification and relation to other growth characteristics. J. Agric. Food Chem., *41*, 2104–2111.

Börjesson T., Stöllman U. and Schnürer J. (1994): Adsorption of volatile fungal metabolites to wheat grains and subsequent desorption. Cereal Chem., *71*, 16–20.

Börjesson T., Eklöv T., Jonsson A., Sundgren H. and Schnürer J. (1996): Electronic nose for odor classification of grains. Cereal. Chem., *73*, 457–461.

Bruce A., Kundzewicz A. and Wheatley R. (1996): Influence of culture age on the volatile organic compounds produced by *Trichoderma aureoviride* and associated inhibitory effects of selected wood decay fungi. Mat. Org., *30*, 79–94.

Buttery R.G., Ling L.C. and Guadagni D.G. (1969a): Volatilities of aldehydes, ketones and esters in dilute water solution. J. Agric. Food. Chem., *17*, 385–389.

Buttery R.G., Seifert R.M., Guadagni D.G. and Ling L.C. (1969b): Characterization of some volatile constituents of bell peppers. J. Agric. Food. Chem., *17*, 1322–1327.

Chang,J.C.S., Foarde K.K., Vanosdell D.W. (1996): Assessment of fungal (*Penicillium Chrysogenum*) growth on three HVAC duct materials. Environ. Internat., *22*, 4, 425–431.

Collins R.P. (1976): Terpenes and odoriferous materials from microorganisms. Lloydia, *39*, 20–24.

Creuly C. Larroache C. and Gros J.-B. (1990): A fed-batch technique for 2-heptanone production by spores of *Penicillium roquefortii*. Appl. Microbiol. Biotechnol., *34*, 20–25.

Croft K.P.C., Jüttner F. and Slusarenko A.J. (1993): Volatile products of the lipoxygenase pathway evolved from *Phaseolus vulgaris* (L.) leaves inoculated with *Pseudomonas syringue* pv. *phaseolicola*. Plant Physiol., *101*, 13–24.

Curtis R.F., Dennis C., Gee D.G., Gee M.G., Griffiths N.M., Land D.G., Peel J.L. and Robinson D. (1974): Chloroanisoles as a cause of musty taint in chickens and their microbiological formation from chloroanisoles in broiler house litter. J. Sci. Food. Agric., *25*, 811–828.

Dartey C.K. and Kinsella J.E. (1973): Oxidation of sodium U 14-C palmitate into carbonyl compounds by *Penicillium roqueforti* spores. J. Agric. Food. Chem., *21*, 721–726.

Dewey S., Sagunski H., Palmgren U. and Wildeboer B. (1995): Mikrobielle flüchtige organische Verbindungen in der Raumluft: Ein neuer diagnostischer Ansatz bei feuchten und verschimmelten Wohnräumen. Zbl. Hyg., *197*, 504–515.

Dionigi C.P. and Ingram D.A. (1994): Effects of temperature and oxygen concentration on geosmin production by *Streptomyces tendae* and *Penicillium expansum*. J. Agric. Food Chem., *42*, 143–145.

Eriksson C. (1974): Enzymic and non-enzymic lipid degradation in foods. In: Spencer B. (ed) Industrial Aspects of Biochemistry. North Holland/American Elsevier, Amsterdam, Vol. 30, Part II, 865–890.

Eriksson C.E., Kaminski E., Adamek P. and Börjessson T. (1992): Volatile compounds and off-flavor produced by microorganisms in cereals. In: Charalambous G. (ed) Off-flavors in Food and Beverages. Elsevier, Amsterdam, 37–56.

Eskin M.N.A. (1979): Plant pigments, flavors and texture. The chemistry and biochemistry of selected compounds. Academic, New York, 94–119.

Ezeonu I.M., Price D.I., Simmons R.B., Crow S.A. and Ahearn D.G. (1994a): Fungal production of volatiles during growth on fiberglass. Appl. Environ. Microbiol., *60*, 4172–4173.

Ezeonu I.M., Noble J.A., Simmons R.B., Price D.L., Crow S.A. and Ahearn D.G. (1994b): Effect of relative humidity on fungal colonization of fiberglass insulation. Appl. Environ. Microbiol., *60*, 2149–2151.

Fiddaman P.J. and Rossal S. (1994): Effect of substrate on the production of antifungal volatiles from *Bacillus subtilis*. J. Appl. Bacteriol., *76*, 395–405.

Flannigan B. (1992): Approaches to assessment of the microbial flora of buildings. Proc. IAQ 92: Environments for people. San Francisco, October 18–21, 1992 American Society of Heating, Refrigerating and Air-Conditioning Engineers.

Flannigan B., McCabe E.M. and McGarry F. (1991): Allergenic and toxigenic microorganisms in houses. In: Austin B.(ed) Pathogens in the Environment. J. Bacteriol., Symposium Supplement, 6, 61s-73s.

Gaillard T. and Phillips D.R. (1976): The enzymatic cleavage of linoleic acid to C9 carbon fragment in extracts of cucumber (*Cucumis sativus*) fruit and the possible role of lipoxygenase. Biochem. Biophys. Acta., *431*, 278–287.

Gallois A., Gross B., Langlois D., Spinnler H.E. and Brunerie P. (1990): Influence of culture conditions on production of flavor compounds by 29 lignolytic basidiomycetes. Mycol. Res., *94*, 494–504.

Gatfield I.L. (1986): Generation of flavor and aroma components by microbial fermentation and enzyme engineering technology, Chapter 24. In: Parliament T.H. and Croteau R. (eds) Generation of Flavor and Aroma Compounds by Microbial fermentation and Enzyme Engineering Technology. ALS Symposium series 317, American Chemical Society, Washington, DC.

Gerber N.N. (1967): Geosmin, an earthy smelling substance isolated from *actinomycetes*. Biotechnol. Bioeng., *9*, 321–327.

Gerber N.N. (1986): Volatile substances from *actinomycetes*. Their role in odor pollution of water. Water Sci. Technol., *15*, 115–125.

Gerber N.N. and Lechevalier H.A. (1965): Geosmin, an earthy-smelling substance from *actinomycetes*. Appl. Microbiol., *13*, 935–938.

Gervais P. and Sarrette M. (1990): Influence of age of mycelium and water activity of the medium on aroma production by *Trichoderma viride* grown on solid substrate. J. Ferment. Bioeng., *69*, 46–50.

Grant C., Hunter C.A., Flannigan B. and Bravery A.F. (1988): The moisture requirement of molds isolated from domestic dwellings. Int. Biod., *25*, 259–284.
Gueldner R.C., Wilson D.M. and Heidt A.R. (1985): Volatile compounds inhibiting *A. flavus*. J. Agric. Food Chem., *33*, 411–413.
Gurtler H., Pedersen R., Anthioni U., Christophersen C., Nielsen P.H., Wellington E.M.H., Pederssen C., Bock K. (1994): Albeflavenone, a sesquiterpene with a zizaene skeleton produced by a streptomycete with a rope morphology. J. Antibiotics, *47*, 434–439.
Ha J.K. and Lindsay R.C. (1993): Release of volatile branched-chain and other fatty acids from ruminant fats by various species. J. Dairy Sci., *76*, 677–690.
Harris N.D., Karahdian C. and Lindsay C.R. (1986): Musty aroma compounds produced by selected molds and *actinomycetes* on agar and whole wheat bread. J. Food Protection, *49*, 964–970.
Hawke J.C. (1966): Reviews of the progress of dairy science, Section D, Dairy Chemistry. The formation and metabolism of methyl ketones and related compounds. J. Dairy Res., *33*, 225–243.
Herbert R.B. (1981): The biosynthesis of secondary metabolites. Chapman and Hall, London.
Jollivet N.; Belin J. M. and Vayssier Y. (1993): Comparison of volatile flavor compounds produced by ten strains of *Penicillium camemberti.* Thom. J. Dairy. Sci., *76*, 1837–1844.
Jones G.M. (1970): Preservation of high moisture corn with volatile fatty acids. Can. J. Anim. Sci., *50*, 739–741.
Kadota H. and Ishida Y. (1972): Production of volatile sulfur compounds by microorganisms. In: Annu. Rev. Pharmacol., *12*, 127–138.
Karahdian C., Josephson D.B. and Lindsay R.C. (1985a): Contribution of *Penicillium* sp. to flavors of Brie and Camembert cheese. J. Dairy Sci., *658*, 1865–1877.
Karahdian C., Josephson D.B. and Lindsay R.C. (1985b): Volatile compounds from *Penicillium* sp. contributing musty-earthy notes to Brie and Camembert cheese flavors. J. Agric. Food. Chem., *33*, 339–343.
Kaminski E., Strawick S. and Wasowics E. (1974): Volatile flavor compounds produced by molds of *Aspergillus*, *Penicillium* and fungi imperfecti. Appl. Microbiol., *27*, 1001–1004.
Kiguchi T., Kadota S., Suehara H., Niski A. and Tsubakks K. (1981): Odorous metabolites of a fungus *Chaetomium globosum* Kinze ex Fr. Identification of geosmin, a musty smelling compound. Chem. Pharm. Bull., *29*, 1782–1784.
Kinderlerer J.L. (1987): Conversion of coconut oil to methyl ketones by two *Aspergillus* species. Phytochemistry, *26*, 1417–1420.
Kinsella J.E. and Hwang D.H. (1976): Biosynthesis of flavors of *Penicillium roqueforti*. Biotechnol. Bioeng., *58*, 927–938.
Kleinheinz G.T. and Bagley S.T. (1997): A filter paper method for the recovery and cultivation of microorganisms utilizing volatile organic compounds. J. Microbiol. Methods, *29*, 139–144.
Lanciotti R. and Guerzoni M.E. (1993): Competitive inhibition of *Aspergillus flavus* by volatile metabolites of *Rhizopus arrhizus*. Food Microbiol., *10*, 367–377.
Lanza E. and Palmer J.K. (1977): Biosynthesis of monoterpenes by *Ceratocystis moniliformis*. Phytochemistry, *16*, 1555–1560.
Lanza E., Ko K.H. and Palmer J.K. (1966): Aroma production by cultures of *Ceratocystis moniliformis*. J. Agric. Food. Chem., *24*, 1246–1250.
Larsen T.O. and Frisvad J.C. (1994a): Production of volatiles and presence of mycotoxins in conidia of common indoor *Penicillia* and *Aspergillii*. In: Samson R.A., Flannigan M.E., Verhoeff A.P., Adan O.C.G. and Hoekstra E.S. (eds) Health implications of fungi in indoor environments. Air Quality monographs, Vol 2.
Larsen T.O. and Frisvad J.C. (1994b): A simple method for collection of volatile metabolites from fungi based on diffusive sampling from Petri dishes. J. Microbiol. Methods, *19*, 297–305.
Larsen T.O. and Frisvad J.C. (1995a): Comparison of different methods for collection of volatile chemical markers from fungi. J. Microbiol. Methods, *24*, 135–144.
Larsen, T.O. and Frisvad J.C. (1995b): Characterization of volatile metabolites from 47 taxa in genus *Penicillium*. Mycol. Res., *99*, 1153–1166.
Larsen T.O. and Frisvad J.C. (1995c): Chemosystematics of species in genus *Penicillium* based on profiles of volatile metabolites. Mycol. Res., *99*, 1167–1174.
Law B.A. (1984): Flavor development in cheese. In: Davies F.L. and Law B.A. Advances in the microbiology and biochemistry of cheese and fermented milk. Elsevier Applied Science, London, 187–208.

Lawrence R.C. and Hawke J.C. (1968): The oxidation of fatty acids by mycelium of *Penicillium roqueforti*. J. Gen. Microbiol., *51*, 289–302.

Lee M.L., Smith D.L. and Freeman L.R. (1979): High resolution gas chromatography profiles of volatile organic compounds produced by micro-organisms at refrigerated temperatures. Appl. Environ. Microbiol., *37*, 85–90.

Marth E.H. (1982) Cheese. In: G.Reed (ed) Prescott and Dunn's Industrial Microbiol, 4th edn, Avi., Westpoint, 65–112.

Mattheis J.P. and Roberts R.G. (1992): Identification of Geosmin as a volatile metabolite of *Penicillium expansum*. Appl. Environ. Microbiol., *58*, 3170–3172.

McJilton C.E., Reynolds S.J., Streifel A.J. and Pearson R.L. (1990): Bacteria and indoor odor problems: Three case studies. Am. Ind. Hyg. Assoc. J., *51*, 545–549.

Miller J.D., Laflamme A.M., Sobol Y., Lafontaine P. and Greenhalgh (1988): Fungi and fungal products in some Canadian houses. Int. Biodeter., *24*, 103–120.

Mølhave L. (1990): Volatile organic compounds, indoor air quality and health. Proceedings of Indoor Air '90, Vol. 5, Ottawa, 1990.

Morey P. and Williams C. (1990): Porous insulation in buildings: A potential source of micro-organisms. In: Indoor Air '90 Proceedings, 5th Int. Conf.. Indoor Air Quality and Climate, Toronto, Vol. 4, 529–533.

Nilsen T.S., Jägerstad I.M., Öste R.E. and Siuik B.T.G. (1991): Supercritical fluid extraction coupled with gas chromatography for the analysis of aroma compounds adsorbed by low-density polyethylene. J. Agric. Food. Chem., *39*, 1324–1237.

Nilsson K. (1996): Electronic noses for detection of rot in wood. Int. Res. Group on Wood Pres., Doc. No. IRG/WP/96–20098.

Nilsson T. Larsen T.O., Montarella L. and Madisen J.O. (1996): Application of headspace solid-phase micro-extraction for the analysis of volatile metabolites emitted by *Penicillium* species. J. Microbiol. Methods, *25*, 245–255.

Norrman J. (1971): The influence of different nitrogen sources on the production of volatile compounds by *Dipodascus aggregatus*. Arch. Mikrobiol., *80*, 338–350.

Nyström A., Grimvall A., Krantz-Rülcker C., Sävenhed R. and Åkerstrand K. (1992): Drinking water off-flavor caused by 2,4,6-trichloroanisole. Water Sci. Tech., *25*, 241–249.

Okumura J. and Kinsella J.E. (1985): Methyl ketone formation by *Penicillium camemberti* in model systems. J. Dairy. Sci., *68*, 11–16.

Parliament T.H. (1986): Sample preparation techniques for gas-liquid chromatographic analysis of biologically derived aromas, Chap. 4. In: Parliament T.H. and Croteau R. (eds) Biogeneration of Aromas, ACS Symposium Series 312, 34–52.

Rabie A.M. (1989): Acceleration of blue cheese ripening by cheese slurry and extracellular enzymes of *Penicillium roqueforti*. Lait., *69*, 305–314.

Riha W.E. III., Hwang C.F., Kerme M.V., Hartman T.G. and Ho C.T. (1996): Effect of cysteine addition on the volatiles of extruded wheat flour. J. Agric. Food. Chem., *44*, 1847–1850.

Rivers J.C., Pleil J.D., Russel W. and Winer W. (1992): Detection and characterization of volatile compounds produced by indoor air bacteria. J. Exp. Environ. Epidem., Suppl. 1.

Samson R.A. (1985): Occurrence of molds in modern living and working environments. European J. Epidemiol., *1*, 54–61.

Samson R.A., Flannigan M.E., Verhoeff A.P., Adan O.C.G. and Hoekstra E.S (eds) (1994): Health implications of fungi in indoor environments. Air quality monographs, Vol. 2, Elsevier, Amsterdam.

Seifert R.M. and King A.D. Jr (1982): Identification of some volatile constituents of *Aspergillus clavatus*. J. Agric. Food. Chem., *30*, 786–790.

Spinnler H.E. and Dijan A. (1991): Bioconversion of amino acids into flavoring alcohols and esters by *Erwinia carotova* subsp. *atroseptica*. Appl. Microbiol. Biotechnol., *35*, 264–269.

Sprecher E. and Hanssen H.P. (1982): Influence of strain specificity and culture condidtions on terpene production by fungi. Planta Medica., *44*, 41–43.

Ström G., Palmgren U., Wessén B., Hellström B. and Kumlin A. (1990): The sick building syndrome – an effect of microbial growth in building constructions. In: Walkinshaw D.S. (ed) Proceedings of Indoor Air '90, Ottawa, Vol. 1, 173–178.

Ström G., West J., Wessén B. and Palmgren U. (1994): Quantitative analysis of microbial volatiles in damp Swedish houses. In: Samson R.A., Flannigan B., Flannigan M.E., Verhoel A.P., Adan O.G.L. and Hoekstra E.S. (eds) Health Implications of Fungi in Indoor Environments. Amsterdam.

Sunesson A.L. (1995a): Volatile metabolites from microorganisms in indoor environments–sampling, analysis and identification. Thesis, Umeå University, Sweden.

Sunesson A.L., Nilsson C.A. and Andersson B. (1995b): Evaluation of adsorbents for sampling and quantitative analysis of microbial volatiles using thermal desorption gas chromatography. J. Chromatogr., *A 699*, 203–214.

Sunesson A.L., Vaes W.H.J., Nilsson C.A., Blomquist G. Andersson B. and Carlson R. (1995c): Identification of volatile metabolites from five fungal species cultivated on two media. Appl. Environ. Microbiol., *61*, 2911–2918.

Sunesson A.L., Nilsson C.A., Carlson R., Blomquist G. and Andersson B. (1995d): Influence of temperature, oxygen and carbon dioxide levels on the production of volatile metabolites from *Streptomyces albidoflavus* cultivated on gypsum board and tryptone glucose extract agar. In: Sunesson A.L., Volatile metabolites from microorganisms in indoor environments–sampling, analysis and identification. Thesis, Umeå University, Sweden.

Sunesson A.L., Nilsson C.A., Andersson B. and Blomquist G. (1996): Volatile metabolites produced by two fungal species cultivated on building materials. Ann. Occup. Hygiene, *40*, 397–410.

Sydor R. and Pietrzyk D.J. (1978): Comparison of porous copolymers and related adsorbents for the stripping of low molecular weight compounds from a flowing air stream. Anal. Chem., *50*, 1842–1847.

Terziev N., Bjurman J. and Boutelje J.B. (1996): Effect of planing on mold susceptibility of kiln and air-dried scots pine (*Pinus sylvestris* L.) lumber. Mat. Org., *30*, 95–104.

Tressl R., Bahri D. and Engel K.H. (1982): Formation of eight-carbon and ten-carbon components in mushrooms (*Agaricus campestris*). J. Agric. Food. Chem., *10*, 89–93.

Unger D.R., Lam T.T., Schaefer C.E. and Kosson D.S. (1996): Predicting the effect of moisture on vapor-phase sorption of volatile organic compounds to soils. Environ. Sci. Technol., *30*, 1081–1091.

Viegas C.A. and Sá Correia I. (1995): Toxicity of octanoic acid in *Saccharomyces cerevisiae* at temperatures between 8.5 and 30 °C. Enzyme Microbiol. Technol., *17*, 826–831.

Viitanen H. and Bjurman J. (1994): Mold growth on wood under fluctuating humidity conditions. Mat. Org., *29*, 28–46.

Vining L.C. (1992): Role of secondary metabolites from microbes. In: Chadwick D.J. and Whelan J. (eds) Secondary Metabolites their Function and Evolution. Wiley, Chichester, 184–194.

Welsh F.W., Murray W.D. and Williams R.E. (1989): Microbial and enzymatic production of flavor and fragrance chemicals. Crit. Rev. Biotechnol., *9*, 105–169.

Whitfield F.B., Last J.H., Shaw K.J. and Mugford D.C. (1984): 2,4,6-Trichloroanisole and 2,3,4,6-tetrachloroanisole: important off-odor components of tainted jute sacks. Chem. Ind. (London), *20*, 744–745.

Wilkins K. (1996): Volatile metabolites from *actinomycetes*. Chemosphere, *32*, 1427–1434.

Wilkins K. and Larsen T.O. (1995): Variation of volatile organic compound patterns of mold species from damp buildings. Chemosphere, *31*, 3225–3236.

Wilkins K., Nielsen E.M. and Wolkoff P. (1997): Pattern in volatile compounds in dust from moldy buildings. Indoor Air, *7*, 128–134.

Wurzenberger M. and Grosch W. (1984): Origin of the oxygen in the products of the enzymatic cleavage reaction of linoleic acid to 1-octen-3-ol and 10-oxo-trans-8–decenoic acid in mushrooms (*Psalliota bispora*). Biochem. Biophys. Acta., *784*, 18–24.

Yong F.M. and Lim G. (1986): Effect of carbon source on aroma production by *Trichoderma viride*. MIRCEN Journal 2, 483–488.

Yong F.M., Wong H.A. and Lim G. (1985): Effect of nitrogen source on aroma production by *Trichoderma viride*. Appl. Microbiol. Biotechnol., *22*, 146–147.

Yong L. (1992): The effect of carbon and nitrogen sources on the growth and aroma production of *Penicillium italicum*. In: Charalambous G. (ed) Food Science and Human Nutrition. Elsevier, Amsterdam, 115–122.

Zang Y. and Ho C.T. (1991): Comparison of the volatile compounds formed from the thermal reaction of glucose with cysteine and glutathione. J. Agric. Food. Chem., *39*, 760–763.

Zechman J.M. and Labows J.N. Jr. (1985): Volatiles of *Pseudomonas aeruginosa* and related species by automated headspace concentration gas chromatography. Can. J. Microbiol., *31*, 232–237.

Zeringue H.J. Jr., Bhatnagar D. and Cleveland T.E. (1993): C15H24 volatile compounds unique to aflatoxigenic strains of *Aspergillus flavus*. Appl. Environ. Microbiol., *59*, 2264–2270.

3.8 Indoor Bioaerosols – Sources and Characteristics

Peter S. Thorne and Dick Heederik

3.8.1 Indoor Ecosystems and Bioaerosols

The indoor environment has changed dramatically since humans lived in caves or in huts made from grasses, mud and dung. As the construction materials of our domiciles have advanced, the array of chemical contaminants has increased. However, this complexity has not eclipsed that of the bioaerosols. Bioaerosols are airborne particulates of biological origin that arise from both indoor and outdoor sources. These include microorganisms and their products (see Table 3.8-1) and allergens from some arthropods, plants and vertebrate animals (see Table 3.8-2). Bioaerosols are a key contaminant in many problem buildings.

Table 3.8-1. Important indoor microbial bioaerosols.

Bacteria
- Infectious organisms
  - *Mycobacteria, Legionella, Streptococcus, Meningococcus, Nocardia*
- Non-infectious organisms*
  - *Pseudomonas, Bacillus, Thermoactinomyces, Enterobacter, Corynebacterium*
- Bacterial products or components
  - Endotoxins
  - Exotoxins
  - Peptidoglycans
  - Teichoic acids
  - Bacterial DNA bearing CpG motifs

Fungi (principally molds and yeasts)
- Infectious fungi
  - *Histoplasma capsulatum, Aspergillus fumigatus, A. flavus*
- Non-infectious fungi*
  - *Penicillium, Aspergillus, Cladosporium, Alternaria, Paecilomyces*
- Fungal products or components
  - Conidia, microconidia and hyphal fragments
  - Mycotoxins
  - Glucans

Infectious airborne viral particles
- Rhinoviruses (common cold)
- Orthomyxoviruses (influenza)
- Parainfluenza viruses (measles, mumps, rubella)

*Denotes that most species within the genus are non-pathogenic by inhalation in immunocompetent subjects.

Table 3.8-2. Important indoor aeroallergens.

Arthropods		
	Mite antigens	
		House dust mites – *Dermatophagoides pteronyssinus, D. farinae*
	Insect antigens	
		Cockroaches – *Blattella germanica, Periplanetta americana, Blatta orientalis*
Animals		
	Dermal and urine antigens from pets or vermin	
		Cat, Dog, Ferret, Guinea Pig, Hamster, Rabbit, Rat, Mouse
	Avian Proteins	
		Pigeon, Cockatoo, Cockateel
Plants		
	Pollens	
		Trees – Maple, Birch, Oak, Sycamore
		Grasses – Bluegrass, Timothy
		Weeds – Ragweed, Cocklebur, Plantain, Lamb's quarters
	Latex allergens – dust from latex gloves	
Low molecular weight chemical haptens		
	Perfumes, Preservatives, Drugs	

Indoor environments develop a complex ecology consisting of people, molds, mites, plants, and sometimes pets or vermin. The environment requires four basic elements: oxygen, heat, moisture and organic matter. This ecosystem is illustrated in Fig. 3.8-1. Abundant organic matter is supplied by humans, animals, ventilation systems and ex-

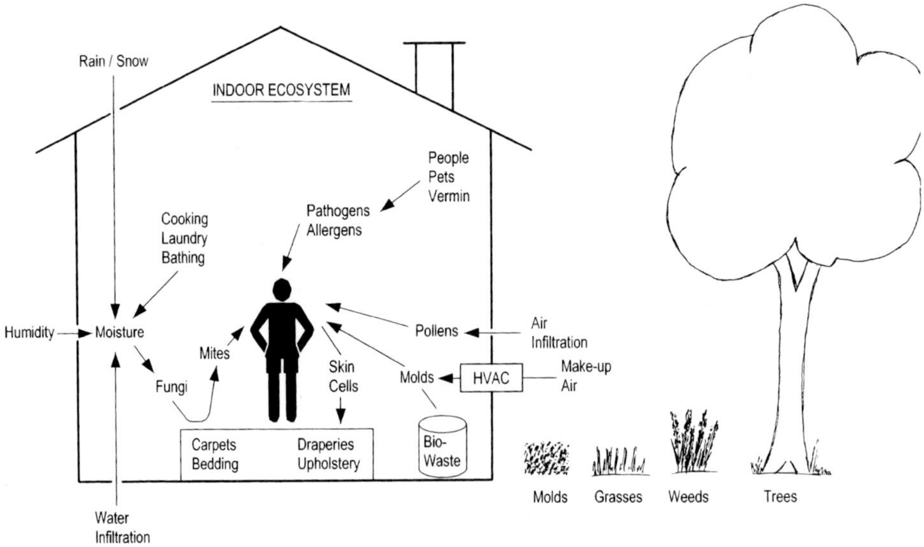

Figure 3.8-1. Bioaerosols in the indoor environment.

ternal sources. This material accumulates in furniture, draperies, bedding and carpets. An important component of this indoor ecosystem is found in the relationship between skin cells, fungi and house dust mites. As much as 1000 mg of skin cells are sloughed from each person every day. Fungi quickly colonize these skin cells in bedding, carpets and upholstery. The colonized skin cells then become food for mites that live optimally at 25 °C and at 70 to 80 % relative humidity. The mites produce gastric enzymes to aid in the digestion of the fungi and skin cells, and, unfortunately, the protease enzymes are potent human allergens. Excrement from the mites contains these allergenic molecules and when inhaled they complete the exposure pathway that leads to allergy and asthma. This example suggests that controlling temperature and humidity at lower levels while maintaining a clean indoor environment with minimal sites for sequestration of organic matter can disrupt the ecosystems that favor the growth of mites. In some cases the occupied space of a building may be clean and dry but local amplification sites for molds may develop. These may arise in ventilation systems, in utility closets, in subfloors or basements that serve as return air plenums, or in local sites of water damage. These can become microorganism and aeroallergen sources of sufficient magnitude to generate significant bioaerosol exposures throughout the environment. Point sources of bacteria and fungi such as diaper pails and household compostable waste containers (biowaste) can also act as significant contributors of microorganisms to the indoor environment.

Indoor bioaerosol exposures present a significant health burden to society. Airborne viruses, bacteria and fungi are responsible for a variety of building-related illnesses (BRIs) arising from organisms that are pathogenic to humans. Non-pathogenic microorganisms may induce symptoms or diseases through inflammatory processes, by stimulating the immune response, or by releasing noxious odors, allergenic compounds, or bioactive macromolecules. The result of exposures of inhabitants to microorganisms and their products may be manifest as outbreaks that affect a large number of the inhabitants, endemic problems affecting primarily susceptible individuals, or conditions that produce annoyance or subclinical effects on a large number of inhabitants. Outbreaks are especially important with infectious bioaerosols but also occur with widespread exposure to toxicants that most people will react to, such as bacterial endotoxins. There can be outbreaks that result from perennial problems such as influenza, from new organisms such as happened with Legionnaire's disease, or from returning old problems such as tuberculosis. Indeed, the "influenza season" results from people spending more time indoors during the winter months. A recent resurgence of tuberculosis and increases in organisms with resistance to multiple antimicrobial agents are frightening trends. Endemic problems of indoor air include allergic rhinitis and asthma. Asthma, in particular, is a disease that is becoming more widespread. Increases in new diagnoses of asthma, hospitalizations for asthma, sales of asthma medications, and mortality from asthma all attest to its increasing severity (Buist and Vollmer, 1994; CDC, 1995; Etzel, 1995; Malveaux and Fletcher-Vincent, 1995). The United States Center for Disease Control reported that from 1980 to 1993 the overall asthma death rate among children aged 5 to 14 years doubled, and in certain demographic groups the rate quadrupled (CDC, 1996). Rates of hospitalizations for asthma during this period nearly doubled for infants. Although some of this rise may be attributed to improved identification and surveillance, a greater proportion is attributable to the changing nature of the indoor envir-

Table 3.8-3. Sizes of bioaerosol components.*

Bioaerosols	Overall size range	Predominant size range
Tree pollens	20–150 μm	30–45 μm
Grass pollens	15–170 μm	30–50 μm
Fungi	5–500 μm	20–100 μm
Bacteria	0.3–30 μm	8–20 μm
Fungi conidia	1–50 μm	5–15 μm
Bacterial spores	0.3–10 μm	0.5–3 μm
Viruses	0.003–0.2 μm	0.01–0.05 μm
Droplet nuclei	2–15 μm	5–10 μm

* Sizes indicated were determined by light or electron microscopy and do not represent aerodynamic size. Fragmentation, agglomeration, or formation of droplet nuclei may alter the bioaerosol size of these components.

onment. Tighter buildings, higher humidities, and increased concentrations of bioaerosols and aeroallergens play an important role. Key sources of bioaerosols associated with asthma are house dust mites, cats, cockroaches, rodents, molds, bacteria and viruses.

Microbial bioaerosols can be live or dead and include bacteria, fungi, viruses, and occasionally algae and protozoa. Microorganisms, like higher organisms, are categorized by taxonomy. For bacteria (kingdom Monera) there are nearly 600 genera in 35 distinct phenotypic groups (Holt et al., 1994). The most important bacterial groups affecting indoor air are those that live well in environments rich in oxygen, the aerobic and facultative anaerobic Gram-negative (Gm –) rods and cocci, aerobic Gram-positive (Gm +) cocci, and actinomycetes. The kingdom Fungi includes nearly 100 000 species of molds and yeasts and other groups of fungi. The saprophytic fungi imperfecti are the most commonly represented fungi and include most of the organisms generally referred to as molds. Saprophytic organisms are those that live by digesting dead organic matter. Bioaerosols range in size from viruses that are as small as 0.003 μm to plant pollens that may be as large as 100 μm (see Table 3.8-3). Bioaerosols of primary importance in indoor environments are generally in the range 0.5 to 30 μm and may be inhaled and deposited throughout the lung and conducting airways depending upon their aerodynamic behavior. Spore-forming bacteria and fungi produce spores, the smallest of which may remain airborne for long periods and can transit the nasopharynx and reach the deeper airways. Bioaerosols are most conveniently divided into infectious bioaerosols, non-infectious bioaerosols, aeroallergens, and toxins. These are discussed sequentially below.

3.8.1.1 Infectious Microorganisms

A multitude of species of infectious bacteria, fungi, protozoa, and viruses have long challenged *Homo sapiens* for their place in the biosphere. While both non-infectious and infectious organisms represent problem bioaerosols (Table 3.8-1), infectious organisms may present more serious consequences with each exposure. Some bioaerosols are highly virulent. The great flu pandemic of 1918-19 killed 21 million people worldwide. This flu was caused by an influenza virus which is 0.11 μm in diameter, but is transmitted via virus-containing aerosols that are generally 5 to 10 μm in mass median aerodynamic diameter (Table 3.8-3) (Dowdle et al., 1977; Schlegel, 1995). Tuberculosis has been a recognized disease as far back as 1000 B.C. Hippocrates referred to the disease as phthisis, from the Greek *phthinein*, to waste away. Tuberculosis currently infects half the world's population and kills about 3 million people annually. This disease is transmitted by inhalation of droplet nuclei containing *Mycobacterium tuberculosis.* Droplet nuclei are infectious aerosols comprised of secretions that become airborne through talking, coughing and sneezing. They are small enough to dehydrate once airborne and may remain aloft for prolonged periods (Des Prez and Heim, 1990). Tuberculosis may be transmitted from an infected person by ventilation systems (Rao et al., 1980; Wales et al., 1985). Survival of infectious organisms improves with higher humidity and higher temperatures. Infectious organisms may be especially problematic in military installations, day care centers, dormitories, hospitals, nursing care facilities, shelters for homeless people, schools and hotels.

Whereas tuberculosis and influenza have been problematic as long as people have lived in communities, other pathogenic bioaerosols appear to have emerged more recently. Outbreaks of Legionnaire's Disease and Pontiac fever were discovered to have arisen from *Legionella pneumophilia* in the wake of the July 1976 American Legion convention in the Bellevue-Stratton Hotel in Philadelphia (Fraser et al., 1977; Terranova et al., 1978). Subsequent investigation of preserved tissue samples showed that *Legionella pneumophilia* was responsible for cases as early as 1947 (McDade et al., 1979). It is difficult to determine if an organism such as *Legionella pneumophilia* is newly emerging or if it simply went unrecognized. In some cases emerging technologies or new industrial processes can create a niche for particular organisms that have previously been rare. *Legionella* contamination of cooling water is now a recognized problem of ventilation systems.

Some animal-associated bioaerosols cause disease in humans through inhalation exposure. Psittacosis or Parrot (Avian) Fever arises from inhalation of aerosols of dried excreta from a variety of birds that contain the virus-like bacterium *Chlamydia psittaci*. Histoplasmosis is a low virulence opportunistic lung infection arising from the dimorphic fungus *Histoplasma capsulatum* that grows well in moist bird excrement. Products of the fungi are not toxins *per se* but are highly antigenic and inflammatory. *H. capsulatum* generally exists in a mycelial form, but it converts to a yeast phase at 37 °C in the body. The mycelial hyphae liberate macroconidia (8–14 μm) and microconidia (2–6 μm) that readily deposit in the conducting airways and the alveoli. *Aspergillus* is an example of a genus of fungus that can cause infectious disease in a suitable host but more commonly induces allergic responses. Aspergillosis is most commonly associated with *A. fumigatus*, but *A. flavus* and *A. niger* may also cause disease. Because this mold taxon is thermophilic (grows best at high temperatures), it grows well in compost buckets and on warm, damp surfaces. Inhalation of the 2–4 μm conidia can carry mycotoxins such as aflatoxin, ochratoxins and clavacin to the respiratory bronchioles and alveoli although it is not known whether this can result in toxic doses (Ciegler et al., 1983; Palmgren and Lee, 1986). Aspergillus species are frequently isolated from hospital air handling systems and are of particular concern for immunocompromised individuals. However, for most individuals, lung colonization is rare, and exposure to *Aspergillus* is thought to more frequently lead to allergy and airway inflammation. Yeasts may also be problems in indoor air. Rhodotorula has been found growing in humidifiers, chillers, and home mist vaporizers.

Community outbreaks of Streptococcal pneumonia (*Streptococcus pneumoniae*) and, less commonly, Meningococcal meningitis (*Neisseria meningitidis*) arise from aerial transmission of droplets of infected oropharyngeal secretions containing viable bacteria. In addition, many common respiratory viruses can be spread via bioaerosols. Since viruses are not independent organisms they must grow in living cells. Their reproduction occurs within host cells and is host and tissue specific. Airborne viral particles may transmit the common cold, influenza, measles, mumps, and rubella (Table 3.8-1). There are several protozoa that may be transmitted via the air. These include *Entamoeba histolytica*, which causes diarrheal disease, *Pneumocystis carinii*, which causes pneumonia, and *Toxoplasma gondii*, which is responsible for toxoplasmosis. However, it is important to recognize the role of immunocompetence as a determinant of the risk of bioaerosol exposure to the host. If host resistance is impaired due to poor nutritional status, from chemotherapeutic measures or from other disease conditions, airborne organisms that are not normally pathogenic may become so. Similarly, impaired host resistance serves to shift the exposure-response curve to the left such that levels of exposure that are insignificant for a normal individual may lead to serious disease in an immunosuppressed individual.

3.8.1.2 Non-Infectious Microorganisms

Potentially toxic bioaerosols include microorganisms and their products, aeroallergens, and a variety of other agents (Table 3.8-1). Non-infectious bioaerosols are responsible for a variety of upper airway and pulmonary conditions including acute airway inflammation, mucous membrane irritation, chronic bronchitis, organic dust toxic syndrome, occupational asthma, allergic rhinitis and hypersensitivity pneumonitis. Most of these toxicants are ubiquitous in the indoor environment. However, far higher concentrations than typical may occur in environments associated with increased risk such as buildings with new outbreaks or recognized problem buildings. Microorganisms, spores and pollens cover a range of sizes from the smallest viruses to large pollens and fungi (Table 3.8-3). The small organisms may agglomerate, attach to dust or droplets, and be suspended as a larger aerosol. Large organisms may fractionate and be suspended in air as respirable fragments. However, many bacteria, bacterial spores, fungal conidia, and pollens exist as free aerosols that can be readily inhaled.

A number of studies have examined the microorganism taxa most commonly occurring indoors. A list of the more common organisms is shown in Table 3.8-4. In general, the bacteria are common soil and plant microbes and those associated with humans and pets. Fungi are represented primarily by molds and yeasts. Some microorganisms are commonly found in air samples collected indoors. These include organisms of the following bacterial genera: *Pseudomonas, Enterobacter, Flavobacterium, Bacillus, Cory-*

Table 3.8-4. Principal genera of microorganisms in indoor bioaerosols.

Gm– bacteria	Gm+ bacteria	Fungi	Thermophilic bacteria
Pseudomonas	*Bacillus*	*Cladosporium*	*Thermoactinomyces*
Enterobacter	*Corynebacterium*	*Aspergillus*	*Saccharopolyspora*
Flavobacterium	*Staphylococcus*	*Penicillium*	*Bacillus* (some sp.)
Serratia	*Streptomyces*	*Alternaria*	
Klebsiella	*Micrococcus*	*Fusarium*	
Alkaligenes		*Paecilomyces*	
Achromobacter		*Rhizopus*	
Citrobacter		*Absidia*	**Yeast**
Escherichia		*Mucor*	*Candida*
		Histoplasma	*Rhodotorula*
		Blastomyces	*Cryptococcus*
		Coccidioides	*Trichosporon*
		Botrytis	*Torulopsis*
		Stachybotrys	
		Trichoderma	
		Epicoccum	
		Stemphylium	
		Sporobolomyces	
		Monosporium	

nebacterium and *Micrococcus* and the following mold genera: *Cladosporium, Penicillium, Aspergillus, Alternaria, Fusarium* and *Rhizopus* (Miller, 1992; DeKoster and Thorne, 1995; Verhoeff et al., 1992). Acute high-level exposures to some of these organisms can result in mucous membrane irritation and inflammation of the lung and airways. Exposure to Gm- and Gm+ bacteria exposes the occupant to the endotoxins and peptidoglycans that comprise the cell walls. Similarly, exposure to fungal conidia may carry mycotoxins to the lung. Inhalation of segments of fungal hyphae produces exposure to the glucans within those structures.

3.8.1.3 Microbial Products

3.8.1.3.1 Endotoxins

Inhaled endotoxins are an important component of bioaerosols in both occupational and non-occupational settings. They may produce fever, malaise, airway obstruction and acute pulmonary inflammation. Endotoxins are integral components of the outer membrane of Gm– bacteria and are composed of lipopolysaccharides associated with proteins and lipids. The term 'endotoxin' refers to the toxin as present in the bacterial cell wall and in cell wall fragments liberated upon bacterial cytolysis. Lipopolysaccharide, or LPS, refers to a class of water-soluble pure lipid carbohydrate molecules (free of protein and other cell wall components) that are particularly bioactive and are held responsible for the toxicity of bacterial endotoxins. LPS molecules are amphoteric, containing a lipid region (called Lipid A) and a long-chain polysaccharide moiety. Lipid A exhibits relatively little variation across genera. The polysaccharide is further subdivided into a low-diversity core portion and the O-specific chain that displays high diversity across genera and species of Gm- bacteria. The two innermost saccharide units are usually composed of components only found in LPS, (L-glycero-D-manno-heptose and 2-keto-3-deoxyoctonic acid, KDO). KDO is linked to the disaccharide of lipid A and is nearly always present in LPS. The fatty acids are 3-hydroxy-substituted, the most common being 3-hydroxymyristic acid. These are not found in Gm+ bacteria or in fungi and have been used as chemical markers of Gm- bacteria (Beije and Lundberg, 1991). The Lipid A moiety imparts the toxicity to the LPS macromolecule and to endotoxin, with the polysaccharide component aiding in conformational changes to facilitate association with receptors such as LPS-binding protein and cellular components (Rietschel and Brade, 1992; Rietschel et al., 1985). Even though most of the bioactivity of endotoxins can be reproduced by purified LPS, it is not correct to assume that this term is always preferable to the term endotoxin in defining biologic responses to this bacterial product. The terms *endotoxins* and *LPS* are synonymous in the majority of scientific literature.

Environmental airborne endotoxins are usually associated with dust particles or aqueous aerosol with a broad size distribution. Endotoxin exposure has been associated with a variety of pulmonary and systemic diseases in homes, office buildings and occupational environments. These diseases or conditions include chronic nose and throat irritation, humidifier fever, organic dust toxic syndrome, grain fever, byssinosis, asthma-like syndrome, exacerbation of asthma, and progressive irreversible airflow obstruction (Heederik et al., 1991; Michel et al., 1996; Muittari et al., 1980; Olenchock, 1990; Rylander, 1996; Rylander et al., 1978; Schenker et al., 1998; Schwartz et al., 1994; Schwartz et al., 1995). Recognition of the link between endotoxin exposures and these conditions has led to proposals to establish occupational health exposure limits. The Netherlands Health Council has recently published a criteria document for endotoxin that recommends a health-based occupational exposure limit of 50 EU/m^3 for full shift, personal, inhalable dust sampling (Douwes and Heederik, 1998).

3.8.1.3.2 Glucans

β(1-3)-glucans are high-molecular-weight glucose polymers that arise from the cell walls of fungi, bacteria and plants. These polymers exist as β-linked glucans with or without β(1-6) side-chains and linkages to other cellular components. β(1-3)-glucans predominate in fungi and are of greatest interest as potential causative agents of adverse health effects in indoor air. Recent evidence suggests that β(1-3)-glucans may be important respiratory immunomodulatory agents (Rylander et al., 1992; Rylander and Peterson, 1993) in indoor air and occupational environments. However, given the low potency of the glucans, it may be that these glucans serve primarily as environmental markers for exposure to fungi. Knowledge about respiratory health effects from airborne exposure to β(1-3)-glucan is currently limited by lack of suitable widely available methods for their measurement in field studies (Douwes et al., 1996).

3.8.1.3.3 Mycotoxins

Mycotoxins are biomolecules produced by fungi that are toxic to animals and humans. They have arisen as a means for a fungus to achieve a competitive advantage over other organisms. A given fungal species may produce a different mix of toxins depending on the substrate. The mycotoxins produced by a particular fungal species may work synergistically to inhibit the growth of other organisms (Miller, 1992). In the case of *Penicillium*, one such compound is penicillin, which has well-recognized antibiotic properties. Thousands of mycotoxins have been classified (Krough, 1984; Nelson et al., 1983); however, the health impact of inhalation exposure to most of these mycotoxins is not known.

Mycotoxin exposure indoors results from the handling of moldy materials, such as water-damaged books, paper currency, furniture and textiles. There are case reports where trichothecenes (Buck and Côté, 1991) and spores of *Stachybotrys atra* (Croft et al., 1986) have caused acute illness in problem buildings. *Fusarium* is a genus of mold that is occasionally a problem in buildings and produces a variety of mycotoxins, including fumonisins, fusarins, zearalenone, nivalenols, T-2 toxin and other trichothecenes, depending on its substrate and growth conditions (Bacon et al., 1989; Ramakrishna et al., 1989). Aflatoxin is perhaps the most well-characterized deleterious mycotoxin in humans and is a known human carcinogen (IARC, 1987; NTP, 1994). Aflatoxin has been the subject of epidemiologic studies (Hayes et al., 1984) and has been widely reviewed (Bray and Ryan, 1991). Ochratoxin A is considered a possible human carcinogen (NTP, 1991). The most relevant route of exposure to aflatoxin and ochratoxin is by ingestion, and respiratory risks associated with their inhalation are not established. Additional research is needed to establish the role of mycotoxins in building-related illness.

3.8.1.3.4 Other Microbial Products

Peptidoglycans have been postulated as a possible causative agent for pulmonary inflammation associated with inhalation of Gm+ bacteria. Peptidoglycans are components of the cell wall envelope of bacteria and are especially prevalent in the backbone of Gm+ bacteria. They may act as endotoxin-like molecules when inhaled. Muramic acid is an amino sugar component of peptidoglycans that does not appear elsewhere in nature and is a suitable marker for analytical assays (Black et al., 1994; Fox et al., 1993; Fox et al., 1995; Sonesson et al., 1988). Although peptidoglycans have been found in hospital and home air conditioner filters (Fox and Rosario, 1994), to date there have been no systematic dose-response studies to identify their potency.

Exotoxins are bioactive molecules (usually proteins) that are secreted during the growth of Gm- bacteria but are also released upon lysis of bacteria. Exotoxins are generally categorized as cytotoxins, neurotoxins, or enterotoxins. Although generally associated with infectious diseases such as botulism, cholera, and tetanus, exotoxins may arise in substrates that support bacterial growth and subsequently become airborne. One example is Exotoxin A produced by *Pseudomonas* sp. (Khan and Cerniglia, 1994). Aqueous environments in which Pseudomonads may be plentiful, such as in dehumidification systems, could act as reservoirs for Exotoxin A that could become airborne with the wash fluid aerosols. A number of bacteria also produce phytotoxins that have evolved as plant toxins. These may also be toxic to humans and be components of bioaerosols. Phytotoxins are generally polysaccharides, peptides, or peptidoglycans. Genera that produce phytotoxins include *Pseudomonas*, *Enterobacter*, and *Corynebacterium*. Fungi also produce phytotoxins that affect a variety of trees, fruits, vegetables, legumes and grasses. Genera of importance include *Alternaria, Helminthosporium* and *Stemphyllium*. These fungal and bacterial genera are common. Relatively little is known about these exotoxins and phytotoxins as potential respiratory hazards in the indoor environment.

Bioeffluents are volatile or semivolatile chemicals with strong odors that emanate from sites of active microbial growth. Petri dishes supporting the growth of organisms from air samples collected in problem buildings often smell like the buildings where the samples were taken. Genera of both fungi and bacteria give off odors readily detectable by the human olfactory. Among the most important odoriferous compounds are hydrogen sulfide, ammonia, mercaptans, indoles, skatoles and various other nitrogen and sulfur-bearing organics. Detection of bioeffluents is described in Chapter 1.7 of this book.

3.8.1.4 Aeroallergens

Airborne allergens are a significant problem in the indoor environment. Allergy can be defined as adverse reactions resulting from immunologic sensitization towards any specific agent or component, called the allergen. In principle all macromolecules of non-human origin, including those of animal, plant and microbial origin, can be immunogenic in humans. Each immunogenic compound may also act as an allergen in previously sensitized subjects, inducing adverse reactions upon re-exposure. Reports characterizing the symptoms associated with exposure to aeroallergens appeared as early as 1555 (Magnus, 1909). Aeroallergens include a variety of fungal spores (e.g. *Aspergillus*, *Penicillium*, *Alternaria*), spores of thermophilic bacteria (e.g. *Saccharopolyspora rectivirgula*, *Thermoactinomyces vulgaris*), arthropod proteins, animal danders and proteins in excreta, plant pollens, and Latex dust. These are listed in Table 3.8-2.

Type I allergens are usually considered those macromolecules with the ability to induce specific IgE immune responses and to provoke allergic reactions in sensitized subjects. Antibody-based immunoassays are widely used for the measurement of specific major allergens in the air. Antibodies can originate from sensitized humans, animals (rabbits producing polyclonal antibodies) or fusions of myeloma cells and mouse spleen cells in culture producing monoclonal antibodies. Examples have been published using monoclonal antibody-based assays to measure allergens such as house dust mites and cats (Platts-Mills and Chapman, 1987; Platts-Mills and de Weck, 1989) in settled dust as well as in airborne dust samples (Price et al., 1990; Sakaguchi et al., 1989; Yasueda et al., 1989). Allergy to rubber tree (*Hevea braziliensis*) proteins from Latex gloves has become a problem of great concern for laboratory workers, practitioners in the medical arts and patients who undergo multiple surgeries (Hunt et al., 1995; Vandenplas, 1995; Woods et al., 1997). The key allergens are thought to be *Hev b* 1 to *Hev b* 7 (Poole, 1997). Methods for quantifying Latex aeroallergen exposure have been developed (Miguel et al., 1996). In contrast, no such assays have become generally available for the assessment of many other major allergens present in indoor environments. Mold exposure is mainly known as the cause of allergic rhinitis (Type III allergies); however, many of the IgG-inducing molds can also induce specific IgE sensitization.

Most reports in the literature on Type I allergy from indoor allergens deal with IgE responses and resulting reactions to house dust mite antigens. House dust mites are microscopic (300 μm) arthropods from the family Pyroglyphidae. The most important species in indoor environments are *Dermatophagoides pteronyssinus* and *D. farinae.* Grain storage mites include distinct but related genuses, but these are most commonly found in agricultural environments. House dust mites predominate indoors and flourish in a humid, warm, and usually moldy environment. They are a major risk factor for asthma worldwide. Their presence may be quantitatively expressed as the number of (living or dead) mite bodies per gram of dust. Floor or mattress dust samples usually range from 100–1000 mites/g (Harving et al., 1993; Korsgaard, 1983; Platts-Mills and de Weck, 1989). Studies of people allergic to house dust mites have demonstrated immunologic sensitization particularly to two allergenic proteins called *Der p* I and *Der p* II from *D. pteronyssinus.* Similar and cross-reactive proteins are present in *D. farinae.* These are protease enzymes present in the gut and feces of the mites. Thus, it is common in asthma studies to report the settled dust or airborne concentrations of house dust mite aeroallergens: *Der p* I and *Der p* II (and/or *Der f* I and *Der f* II). Levels typically found in houses of asthmatic patients are 2 to 10 μg/g settled dust and 5 to 50 ng/m^3 in air samples depending upon activity during sampling (Pope et al., 1993).

Animal proteins can be very potent allergens, as shown by the usually high prevalences of Type I allergy in the general population to common pets like cats, dogs, guinea pigs, and rabbits. A high risk of occupational allergy is faced by laboratory animal workers exposed to rats or mice, in whom allergen exposure at less than nanogram per m^3 levels may cause sensitization and allergic symptoms (Fuortes et al., 1996; Hollander et al., 1997). Allergens can be associated with hair, skin scrapings, fecal particles, or dusts containing urine or salivary proteins. Data regarding airborne allergen exposure levels should be interpreted with caution, since, in the assays developed to date, the results are calculated by relating the activity of test samples to that of an internal standard allergen preparation. No common set of standard allergens is available at this time. While exposure-response relationships have been established for some allergens, more work is needed in order to identify baseline rates of allergy to indoor allergens and perform quantitative risk assessments.

3.8.2 Summary

Indoor bioaerosols are complex mixtures of biological materials that cause disease primarily through infection, inflammation, irritation or allergy. The health burden of these exposures is manifest as disease outbreaks, as problems prevalent in certain types of buildings in particular geographical areas, or as problems with lower prevalence that affect building occupants everywhere. Understanding the ecology of the indoor environment may help develop strategies to minimize exposures and the resulting morbidity associated with bioaerosols.

References

Bacon C.W., Marijanovic D.R., Norred W.P. and Hinton D.M. (1989): Production of fusarin C on cereal and soybean by *Fusarium moniliforme*. Appl. Environ. Microbiol., *55*, 2745–2748.

Beije B. and Lundberg P. (1991): Criteria documents from the Nordic expert group. Arbete och Hälsa, 50.

Black G.E., Fox A., Fox K., Snyder A.P. and Smith P.B. (1994): Electrospray tandem mass spectrometry for analysis of native muramic acid in whole bacterial cell hydrolysates. Anal. Chem., *66*, 4181–4176.

Bray G.A. and Ryan D.H.(eds) (1991): Mycotoxins, cancer, and health. Louisiana State Univ. Press, Baton Rouge, La.

Buck W.B. and Côté L.M. (1991): Trichothecene mycotoxins. In: Keeler R.F. and Tu A.T.(eds) Toxicology of plant and fungal compounds. Dekker, New York, 523–555.

Buist S.A. and Vollmer W.M. (1994): Preventing deaths from asthma. New Engl. J. Med., *331*, 1584–1585.

CDC. (1995): Asthma–United States, 1982–1992. MMWR, *43*, 952–955.

CDC. (1996): Asthma mortality and hospitalization among children and young adults–United States, 1980–1993. MMWR, *45*, 350–353.

Ciegler A., Burmeister H.R. and Vesonder R.F. (1983): Poisonous fungi: mycotoxins and mycotoxicoses. In: Howard D.H. (ed) Fungi pathogenic for humans and animals, Vol. 3, Part B. Dekker, New York, 413.

Croft W.A., Jarvis B.B. and Yatawara C.S. (1986): Airborne outbreak of trichothecene toxicosis. Atmos. Environ., *20*, 549–552.

DeKoster J.A. and Thorne P.S. (1995): Bioaerosol concentrations in non-complaint, complaint and intervention homes in the midwest. Am. Ind. Hyg. Assoc. J., *56*, 573–580.

Des Prez R.M. and Heim C.R. (1990): Mycobacterial diseases. In: Mandell G.L., Douglas R.G., Jr. and Bennett J.E. (eds) Principles and practice of infectious diseases, 3rd edn. Churchill Livingstone, New York, 1879.

Douwes J. and Heederik D. (1998): Endotoxin criteria document: Environmental levels, human exposure and epidemiological studies. Dutch Expert Committee on Occupational Standards (DECOS). Netherlands Health Council, Voorburg, 1998.

Douwes J., Doekes G., Montijn R., Heederik D. and Brunekreef B. (1996): Measurement of $\beta(1\rightarrow3)$-glucans in occupational and home environments with an inhibition enzyme immunoassay. Appl. Environ. Microbiol., *62*, 3176–3182.

Dowdle W.R., Noble G.R. and Kendal A.P. (1977): Orthomyxovirus-influenza: Comparative diagnosis unifying concept. In: Krustac E. and Krustac C. (eds) Comparative diagnosis of viral diseases. Academic, New York, 447.

Etzel R.A. (1995): Indoor air pollution and childhood asthma: effective environmental interventions. Environ. Health Perspec., *103*, 55–58.

Fox A. and Rosario R.M.T. (1994): Quantitation of muramic acid, a marker for bacterial peptidoglycan, in dust collected from hospital and home air-conditioning filters using gas chromatography-mass spectrometry. Indoor Air, *4*, 239–247.

Fox A., Rosario R.M.T. and Larsson L. (1993): Monitoring of bacterial sugars and hydroxy fatty acids in dust from air conditioners by gas chromatography-mass spectrometry. Appl. Environ. Microbiol., *59*, 4354–4360.

Fox A., Wright L. and Fox K. (1995): Gas chromatography-tandem mass spectrometry for trace detection of muramic acid, a peptidoglycan chemical marker, in organic dust. J. Microbiol. Meth., *22*, 11–26.
Fraser D.W., Tsai T.R., Orenstein W., Parkin W.E., Beecham H.J., Sharrar R.G., Harris J., Mallison G.F., Martin S.M., McDade J.E., Shepard C.C. and Brachman P.S. (1977): Legionnaire's disease: description of an epidemic of pneumonia. New Engl. J. Med., *297*, 1183–1197.
Fuortes L.J., Weih L.A., Jones M., Burmeister L., Thorne P.S., Pollen S. and Merchant J.A. (1996): Epidemiologic assessment of laboratory animal allergy among university employees. Am. J. Ind. Med., *29*, 67–74.
Harving H., Korsgaard J. and Dahl R. (1993): House dust mites and associated environmental conditions in Danish homes. Allergy, *48*, 106–109.
Hayes R.B., Van Nieuwenhuize J.P., Raatgever J.W. and Ten Kate F.J.W. (1984): Aflatoxin exposures in the industrial setting: An epidemiological study of mortality. Food Chem. Toxicol., *22*, 39–43.
Heederik D., Brouwer R., Biersteker K. and Boleij J.S.M. (1991): Relationship of airborne endotoxin and bacteria levels in pig farms with the lung function and respiratory symptoms of farmers. Int. Arch. Occup. Environ. Health, *62*, 595–601.
Hollander A., Heederik D. and Doekes G. (1997): Respiratory allergy to rats: exposure-response relationships in laboratory animal workers. Am. J. Crit. Care Med., *155*, 562–567.
Holt J.G., Krieg N.R., Sneath P.H.A., Staley J.T. and Williams S.T., (eds) (1994): Bergey's Manual of Determinative Bacteriology, 9th edn. Williams & Wilkins, Baltimore.
Hunt L.W., Fransway A.F., Reed C.E., Miller L.K., Jones R.T., Swanson M.C. and Yunginger J.W. (1995): An epidemic of occupational allergy to latex involving health care workers. J. Occup. Environ. Med., *37*, 1204–1209.
IARC (1987): International Agency for Research on Cancer. IARC Monographs on the evaluation of carcinogenic risks to humans. Overall Evaluations of carcinogenicity. Supplement 7, Lyon, France.
Khan A.A. and Cerniglia C.E. (1994): Detection of *Pseudomonas aeruginosa* from clinical and environmental samples by amplification of the exotoxin A gene using PCR. Appl. Environ. Microbiol., *60*, 3739–3745.
Korsgaard J. (1983): Mite asthma and residency: a case-control study on the impact of exposure to house dust mites in dwellings. Am. Rev. Respir. Dis., *128*, 231–235.
Krough P. (ed) (1984): Mycotoxins in food. Academic, San Diego, Calif.
Magnus O. (1909): In: Historia om de nordiska folken, Almquist and Wiksell. Uppsala-Stockholm, *13*, 41.
Malveaux F.J.and Fletcher-Vincent S.A. (1995): Environmental risk factors of childhood asthma in urban centers. Environ. Health Perspec., *103*, 59–62.
McDade J.E., Brenner D.J. and Bozeman F.M. (1979): Legionnaire's disease bacterium isolated in 1947. Ann. Intern. Med., *90*, 659–661.
Michel O., Kips J., Duchateau J., Vertongen F., Robert L., Collet H., Pauwels R. and Sergysels R. (1996): Severity of asthma is related to endotoxin in house dust. Am. J. Respir. Crit. Care Med., *154*, 1641–1646.
Miguel A.G., Cass G.R., Weiss J. and Glovsky M.M. (1996): Latex allergens in tire dust and airborne particles. Environ. Health Perspec., *104*, 1180–1186.
Miller J.D. (1992): Fungi as contaminants in indoor air. Atmos. Environ., *26A*, 2163–2172.
Muittari A.R., Rylander R. and Salkinoja-Salonen M. (1980): Endotoxin and bathwater fever. Lancet, *ii*, 89.
Nelson P.E., Toussoun T.A. and Marassas W.F.O. (1983): *Fusarium* species: An illustrated manual for identification. Penn State Univ. Press, University Park, Pa.
NTP (1994): Seventh Annual Report on Carcinogens: 1994 Summary. National Toxicology Program, U.S. Public Health Service.
Olenchock S.A. (1990): Endotoxins. In: Morey P.R., Feeley J.C., Sr. and Otten J.A. (eds) Biological contaminants in indoor environments, ASTM STP 1071. ASTM, Philadelphia, 190–200.
Palmgren M.S. and Lee L.S. (1986): Separation of mycotoxin-containing sources in grain dust and determination of their mycotoxin potential. Environ. Health Perspec., *66*, 105–108.
Platts-Mills T.A.E. and Chapman M.D. (1987): Dust mites: immunology, allergic diseas, and environmental control. J. Allergy Clin. Immunol., *80*, 755–775.
Platts-Mills T.A.E. and de Weck A.L. (1989): Dust mite allergens and asthma–a worldwide problem. J. Allergy Clin. Immunol., *83*, 416–427.
Poole C.J.M. (1997): Hazards of powdered surgical gloves. Lancet, *350*, 973–974.

Pope A.M., Patterson R. and Burge H. (eds) (1993): Indoor allergens: assessing and controlling adverse health effects. National Academy Press, Washington, D.C., 1993.
Price J.A., Pollock I., Little S.A., Longbottom J.L. and Warner J.O. (1990): Measurement of airborne mite antigen in homes of asthmatic children. Lancet, *336*, 895–897.
Ramakrishna Y., Bhat R.V. and Ravindranath V. (1989): Production of deoxynivalenol by *Fusarium* isolates from samples of wheat associated with a human mycotoxicosis outbreak and from sorghum cultivars. Appl. Environ. Microbiol., *55*, 2619–2620.
Rao V.R., Joanes R.F., Kilbane P. and Galbraith N.S. (1980): Outbreak of tuberculosis after minimal exposure to infection. Br. Med. J., *281*, 187–189.
Rietschel E.T. and Brade H. (1992): Bacterial endotoxins. Sci. Amer., *Aug92*, 54–61.
Rietschel E.T., Brade H., Brade L., Kaca W., Kawahara K., Lindner B., Lüderitz T., Tomita T., Schade U., Seydel U. and Zähringer U. (1985): Newer aspects of the chemical structure and biological activity of bacterial endotoxins. In: ten Cate J.W., Büller H.R., Sturk A. and Levin J. (eds) Bacterial endotoxins; structure, biomedical significance, and detection with the *Limulus* amebocyte lysate test. Alan R. Liss, New York, 31–50.
Rylander R. (1996): Airway responsiveness and chest symptoms after inhalation of endotoxin or (13)-β-D-glucan. Indoor Built Environ., *5*, 106–111.
Rylander R. and Peterson Y. (1993): Second glucan inhalation toxicology workshop. ICOH Committee on Organic Dusts Report.
Rylander R., Haglind P., Lundholm M., Mattsby I. and Stenqvistn K. (1978): Humidifier fever and endotoxin exposure. Clin. Allergy, *8*, 511–516.
Rylander R., Persson K., Goto H., Yuasa K. and Tanaka S. (1992): Airborne β,1–3-glucan may be related to symptoms in sick buildings. Indoor Environ., *1*, 263–267.
Sakaguchi M., Inouye S., Yasueda H., Irie T., Yoshizawa S. and Shida T. (1989) Measurement of allergens associated with dust mite allergy, II. Concentrations of airborne mite allergens (Der I and Der II) in the house. Int. Arch. Allergy Appl. Immunol., *90*, 190–193.
Schenker M.B., Christiani D., Cormier Y., et al. (1998): Respiratory health hazards in agriculture: Amer. Thoracic Soc. Workshop Report. Am. J. Respir. Crit. Care Med. *158*, 1–76.
Schlegel H.G. (1993): General microbiology, 7th edn, Translated by M Kogut. Cambridge Univ. Press, Cambridge UK, 447.
Schwartz D.A., Thorne P.S., Jagielo P.J., White G.E., Bleuer S.A. and Frees K.L. (1994): Endotoxin responsiveness and grain dust-induced inflammation in the lower respiratory tract. J. Appl. Physiol. *267* (Lung Cell. Mol. Physiol. *11*), L609–L617.
Schwartz D.A., Thorne P.S., Yagla S.J., Burmeister L.F., Olenchock S.A., Watt J.L. and Quinn T.J. (1995): The role of endotoxin in grain dust-induced lung disease. Am. J. Respir. Crit. Care Med., *152*, 603–608.
Sonesson A., Larsson L., Fox A., Westerdahl G. and Odham G. (1988): Determination of environmental levels of peptidoglycan and lipopolysaccharide using gas chromatography-mass spectrometry utilizing bacterial amino acids and hydroxy fatty acids as biomarkers. J. Chromatogr. Biomed. Appl., *431*, 1–15.
Terranova W., Cohen M.L. and Fraser D.W. (1978): 1974 outbreak of Legionnaire's disease diagnosed in 1977. Lancet, *2*, 122–124.
Vandenplas O. (1995): Occupational asthma caused by natural rubber latex. Eur. Respir. J., *8*, 1957–1965.
Verhoeff A.P., van Wijnen J.H., Brunekreef B., Fischer P, van Reenen-Hoekstra E.S. and Samson R.A. (1992): Presence of viable mould propagules in indoor air in relation to house dampness and outdoor air. Allergy, *47*, 83–91.
Wales J.M., Buchan A.R., Cookson J.B., Jones D.A. and Marshall B.S.M. (1985): Tuberculosis in a primary school: the Uppingham outbreak. Br. Med. J., *291*, 1039–1040.
Woods J.A., Lambert S., Platts-Mills T.A.E., Drake D.B. and Edlich R.F. (1997): Natural rubber latex allergy: spectrum, diagnostic approach, and therapy. J. Emerg. Med., *15*, 71–85.
Yasueda H., Mita H., Yui Y. and Shida T. (1989): Measurement of allergens associated with dust mite allergy, I. Development of sensitive radioimmunoassays for the two groups of *Dermatophagoides* mite allergens, Der I and Der II. Int. Arch. Allergy Appl. Immunol., *90*, 182–189.

4. Investigation concepts and quality guidelines for organic air pollutants

4.1 Indoor Air Quality Guidelines

Peter Pluschke

4.1.1 Introduction

Indoor air quality has become a matter of concern only fairly recently, even though hygienists as early as in the 18th century discussed and studied the consequences of inadequate ventilation in the indoor environment. John Arbuthnot (1751) studied the effects of air on human bodies and concluded: *"XVIII. The Air of Cities is not so friendly to the Lungs as that of the Country, for it is replete with sulphureous Steams of Fuel, and the perspirable Matter of Animals; therefore the Consumptive and Asthmatic are better in the Country. ...XXI. Private Houses ought to be perflated once a Day, by opening Doors and Windows, to blow off the Animal Steams. XXII. Houses, for the sake of Warmth fenc'd from Wind, and where the Carpenters Work is so nice as to exclude all outward Air, are not the most wholsom."* – a brief, scientifically justified guideline on indoor air quality, already taking into consideration the effects of improved building insulation.

Modern guidelines still have a lot in common with Arbuthnot's recommendations but are facing a more complex situation in the buildings. Basically there are two different areas of regulation:

(a) Guidelines and standards for air quality in the **workplace environment** are well developed, and a large number of chemical parameters have been set. In the USA the American Conference of Governmental Industrial Hygienists (ACGIH) proposed in 1946 Threshold Limit Values (TLV) for 146 substances, and their number grew to nearly 700 in the 1990s (Gammage et al., 1993). Similar regulations exist in most industrialized countries. For example, in Germany the number of published MAK (Maximale Arbeitsplatzkonzentration) and BAT (Biologische Arbeitsstoff Toleranz) values is about the same as in the USA. These values address the situation of a healthy worker who is to be protected against individual compounds predominating in industrial or manual work processes. The exposure in this case is limited to the regular working hours.

Occupational exposure limits (OEL) – like the TLV or the MAK – have frequently been used to derive tentative values for air quality in non-industrial indoor environments. As a rule of thumb it has been proposed to take

- 1/10 of the TLV (ASHRAE, 1990; Levin, 1998),
- 1/20 of the MAK (Kunde, 1982; VDI, 1974) or respectively 5 % of the TLV (US-EPA, 1996),
- or 1/40 of the applicable OEL

as indoor air quality standards. In ASHRAE Standard 62–1989 it is stated that a concentration of 1/10 TLV *"would not produce complaints in a non-industrial population"* but *"may not provide an environment satisfactory to individuals who are extremely sensitive to an irritant"*. The German guideline which referred to the 1/20 MAK as tentative immission standards has been withdrawn meanwhile, whereas Nielsen et al.

(1997) presented their concept for indoor air quality standards based on 1/40 of the OEL as a first approximation only fairly recently. It has been recognized as a useful tool to evaluate potential health effects in indoor environments if no specific indoor air quality standard or similar guidance are available.

(b) Indoor air quality guidelines for **the non-industrial indoor environment**, on the other hand, are intended to provide a comfortable and healthy atmosphere in buildings which are used not only by adult and healthy people but also by minors and aged or even sick people who may be more sensitive to chemical stress. It is well known from time-budget surveys that most people spend more than 16 hours per day in indoor environments. On average, in most countries mean time spent in indoor environments exceeds 20 hours (Szalai, 1972; a summary of data is given by Pluschke, 1996). Total exposure to airborne contaminants – especially to organic compounds – is in many cases dominated by air pollution in residences and other non-industrial environments, but regulations and standards for these are still fragmentary.

Indoor air contaminants have become even more important in recent years as the comparative analysis of data on ambient and indoor air pollution show, because – at least in advanced industrialized countries – ambient air pollution has been cut significantly thanks to air pollution control regulations which were successfully implemented for industrial and energy conversion processes and vehicular traffic. Ambient air quality up to the 1980s could be characterized by the fact that *"the organic and inorganic pollution load in the atmosphere continues to increase"* (Suess, 1988), whereas over the last one or two decades pollution loads decreased significantly for quite a number of parameters, especially sulfur dioxide, carbon monoxide, suspended particles, lead and cadmium. As a consequence of this development, ambient air quality in most places is in compliance with air quality standards for most parameters, and indoor air tends to be more polluted than ambient air with certain regional and temporal exceptions, especially during photo-smog episodes and in the vicinity of major traffic arteries. Thus, from a public health point of view, air quality management requires new regulatory and intervention strategies to deal with this situation. As Spengler and Sexton (1983) have already pointed out, there exists a fundamental difference between indoor and outdoor air in so far as outdoor air *"is a public good in the sense that members of a community breathe basically the same ambient air. The rationale for Government regulation of outdoor air pollution has focused on the issue that those who suffer the effects are not compensated, nor is their interest in cleaner air readily effective in influencing polluters"*.

Exposure to indoor air pollution, especially in private residences and other non-public environments, may be caused by the persons occupying a building, and the decision to improve indoor air quality is to be made by them – bearing the cost and enjoying the benefits. To what extent indoor air quality in private residences and premises is to be regulated and what is the adequate public response to this problem are matters under discussion.

Personal habits, cultural practices and individual perceptions strongly influence the acceptance of a building and its indoor air quality. Public intervention in non-public buildings and the enforcement of strict indoor air quality regulations beyond the requirements of building codes, fire ordinances and safety rules face serious impediments.

Indoor Air Quality guidelines may set the standards

- which have to be met in non-industrial, public buildings,
- which are to be taken into consideration in building design and management and
- which may be imposed upon the owner of a building for the comfort and health of its tenants or users.

4.1.2 Legal Status and Basic Concepts for Indoor Air Quality Guidelines

Indoor air quality is influenced by many parameters such as:

- emissions from building materials and technical equipment,
- permeability of the wall structures,
- ventilation practices and ventilation rate,
- building maintenance and cleaning habits,
- emissions of products which are used indoors,
- cultivation of plants and flowers in the indoor environment,
- body effluents,
- ambient air quality.

It follows from the variety of factors that indoor air quality regulation and management have to be organized and guided in a multidisciplinary way.

In many countries the building codes include certain general regulations to guarantee proper and sanitary conditions in buildings. Generally it is required that the indoor environment should not be injurious to health and that indoor air quality should be at least the same as outdoor air quality, e. g., as is stated in a German directive on non-industrial workplaces (Bundesanstalt für Arbeitsschutz, 1988).

From nation to nation the legal status of regulations on sanitary conditions within buildings varies, just as it does with respect to indoor air quality, and in most federal political systems both state and local authorities have a significant share in regulating indoor air quality.

In Japan the government enacted a *"Law for Maintenance of Sanitation in Buildings"* in 1970 which has been revised and amended in the 1980s and specified by a number of ordinances (Kagawa, 1993) requiring regular on-the-spot inspections for specified buildings larger than a certain size.

In Europe the national building codes often include some general formula on sanitary conditions and ventilation in the buildings, whereas more detailed regulations on contaminants and indoor air pollution are rare and often legally defined in other sectors of law (Türck and Böttger, 1993; Lemaire, 1994), e.g. in environmental regulations, regulations for chemical and consumer products or in public health legislation.

In some countries legal regulations have been supplemented by technical rules which serve as recommendations and do not overrule official building codes or interpretations of these. The Finnish *"Classification of Indoor Climate, Construction, and Finishing Materials"* (FiSIAQ et al., 1995) gives a good and elaborate example for this type of advisory guidelines.

Regulations and guidelines for indoor air quality may be grouped into three categories, dealing with

1. the physical environment, providing **standards for ventilation** (including guidelines for operation and maintenance of a building and its technical equipment) such as the European guidelines (Commission of the European Communities, 1992) and national standards such as the ASHRAE Standard 62–1989 (1989) for the USA, the German DIN 1946, the British BS 5925 (1991) or the Australian Standard 1668.2–1992;
2. indoor air quality, stating **exposure standards,** i. e. (numerically) defined levels of concentration for individual compounds as well as for complex mixtures, so guaranteeing a comfortable and healthy indoor environment with negligible health risk, but keeping in mind that people who have developed hypersensitivity to chemicals are not protected by these standards (Seifert et al., 1993). In some cases exposure standards are integrated in ventilation standards (ASHRAE, 1990; Janssen 1989), but in most cases exposure standards are treated separately, as in Canada (Health and Welfare Canada, 1989), Germany (BGA, 1993; IRK, 1996) or Norway (Helsedirektoratet, 1990);
3. **product quality standards** in order to minimize the exposure to volatile compounds emitted from consumer products. Ecolabelling regulations may include indoor climate aspects, and can result in considerable improvements in building and consumer products, as, for example, was documented by Salthammer et al. (1996) for wallcoverings.

Even though volatile organic compounds (VOCs) play a prominent role in indoor air pollution, few exposure standards and guideline values have been defined for them. Building code regulations and other technical recommendations focus mostly on inorganic parameters such as CO_2, CO and NO_2, which may serve as indicators of a good and healthy indoor climate. Regulating indoor air pollution by volatile organic compounds is more difficult because of their diversity and chemical reactivity. Samet and Speizer (1993) commented that this *"restricted focus undoubtedly reflects, in part, the difficulty of accurately estimating personal exposures to multiple pollutants and assessing mutiple health outcomes"*.

4.1.3 Exposure Standards and Guidelines for Organic Compounds

In the 1990s new concepts for regulating indoor air exposure to organic compounds have been developed. Two complementary strategies have been followed:

1. **Guideline values for individual compounds** define an upper level which requires remedial action and a lower level which is recommended as a target value to give prevention. In a few cases, regulatory indoor air quality standards have been set (e.g. for formaldehyde and, in Germany, for tetrachloroethylene). Ambient air quality standards (such as NAAQS in the USA or MIK values in Germany) may serve as well as indoor air quality standards because they address the same situation from a public health point of view, as was stated by WHO (1987). The US Environmental Protection Agency (EPA) claimed for their own buildings that the Maximum Allowable Air Concentration for all regulated pollutants should be below the National Ambient Air Quality Standard (US-EPA, 1996).

2. The **TVOC (Total Volatile Organic Compounds) guideline value for the complex mixture** is defined as the sum of the individual compounds separated and quantified by a gas-chromatographic technique (Seifert, 1990).

The proposed and implemented guideline values from a number of national and international regulatory or advisory bodies have been summarized:

- in Table 4.1-1 a for CO_2 and CO,
- in Tables 4.1-1 b and c for a number of volatile and semi-volatile compounds; additionally, in Table 4.1-2 the guideline values derived by Nielsen et al. (1997; 1998 a, b, c) which are based on toxicological considerations and reflect *"action limits where something may need to be done"* are presented,
- in Table 4.1-3 for TVOC.

Even though considerable progress has been made during the last decade towards defining chemical standards for indoor air quality with respect to a number of parameters, a general procedure for evaluating the health consequences of indoor exposure has not yet been developed and accepted on an international level. Recently basic approaches for setting indoor air quality standards have been published (IRK, 1996, Nielsen et al., 1997) and applied to a number of indoor pollutants. The resulting guideline values have been summarized in Tables 4.1-1 b, c and 4.1-2.

Risk assessment should be based on a careful analytical study, for which Mølhave (1998) defined the basic requirements and results:

„A list of identified compounds. Concentrations of each compound reported as:

- *lifetime integrated absorbed dose,*
- *average concentration (e.g. 24 h),*
- *peak exposures (e.g. 10–30 s),*
- *peak exposure frequency (e.g. 90 % fractile),*
- *variation in space,*
- *TVOC according to a standardized procedure,*
- *timing of the sampling with respect to measurements of health effects."*

Such a complete set of information is available only in a limited number of cases. In order to develop a simple and fairly straightforward instrument for assessing indoor air quality, there is a strong interest in the TVOC concept. Recently, analytical procedures and a standardized set of parameters have been proposed (Mølhave et al., 1997) for indoor air quality examinations. The most detailed guideline for the evaluation of analytical results based on the TVOC concept was published by Seifert in 1990 (Seifert, 1990; see Table 4.1-3 (Germany)). It is not based on toxicological considerations but describes a standard for a good (chemical) indoor climate which can be achieved if a building is properly constructed and maintained and products used in the building are carefully chosen.

The target guideline value for TVOC has been set by Seifert at 300 $\mu g/m^3$ (see Table 4.1-3). According to his concept *„the VOCs belonging to each of the following chemical classes are ranked according to their measured concentration: alkanes, aromatic hydrocarbons, terpenes, halocarbons, esters, carbonyl compounds (excluding formaldehyde)*

Table 4.1-1a. Exposure standards and guideline values for carbon dioxide and carbon monoxide

Substance	Averaging time	WHO	USA	Canada	Finland	Germany
Carbon dioxide, CO_2	Continuous	No concern: 1800 µg/m³ Concern: 12 000 µg/m³	a) 1800 mg/m³	ALTER: 6300 mg/m³	S3: 2700 mg/m³ S2: 2250 mg/m³ S1: 1800 mg/m³	200 mg/m³
Carbon monoxide, CO	8 h	10 mg/m³	a) 10 mg/m³	ASTER: 12.6 mg/m³	S3: 8 mg/m³ S2: 5 mg/m³ S1: 2 mg/m³	RW II L: 5 mg/m³ RW I L: 1.5 mg/m³
	1 h	30 mg/m³	a) 40 mg/m³	ASTER: 28.6 mg/m³	–	RW II K: 60 mg/m³ RW I K: 6 mg/m³
Remarks				ALTER: acceptable long-term exposure range ASTER: acceptable short-term exposure range	S 1, S 2, S 3: target values for indoor air quality S 1 – best category (> 90 % of occupants satisfied), S 3 – minimum requirements according to building codes	RW (Richtwert): • guide value for intervention: RW II • target value: RW I for short term (K) and long-term (L) exposure
Ref.		WHO (1987)	a) ASHRAE (1990)	Health and Welfare Canada (1989)	FiSIAQ (1995)	BGA (1993) Englert (1997)

Table 4.1-1b. Exposure standards and guideline values for volatile organic pollutants in the indoor environment.

Substance	Averaging time	WHO	USA	Canada	Finland	Germany
Acrolein	8 h	–	–	50 µg/m³	–	–
Formaldehyde		100 µg/m³	b) 20 µg/m³	Action level: 120 µg/m³ Target level: 60 µg/m³	S3: 150 µg/m³ S2: 70 µg/m³ S1: 30 µg/m³	120 µg/m³
Methylenechloride	Continuous	–	–	–	–	RW II: 2 mg/m³ RW I: 0.2mg/m³
Styrene	24 h 8 h	800 µg/m³ 70 µg/m³	– –	–	– –	RW II: 300 µg/m³ RW I: 30 µg/m³
Tetrachloroethylene	24 h ½ h	a) 5000 µg/m³ b) 8000 µg/m³	– –	a)	– –	7-d-average: 100 µg/m³ –
Toluene	24 h ½ h	a) 700 µg/m³ b) 1000 µg/m³	c) 400 µg/m³ –	– –	– –	RW II: 3000 µg/m³ RW I: 300 µg/m³ –
Trichloroethylene	24 h	1000 µg/m³	–	a)	–	–
Remarks		a) Guideline level based on toxicological considerations b) Air quality guideline based on odor threshold	For non-regulated pollutants: < 5 % of TLV and time-weighted average, US-EPA (1996)	a) Potential for adverse health effects		RW (Richtwert): • guide value for intervention: RW II • target value: RW I
Ref.		WHO (1987)	a) ASHRAE (1990) b) US-EPA (1996) c) RfC: Inhalation Reference Concentration, Greenberg (1997)	Health and Welfare Canada (1989)		BGA (1993) Sagunski (1996; 1998) IRK (1997; 1998) Lederer (1997)

Table 4.1-1c. Exposure standards and guideline values for semi-volatile organic pollutants in the indoor environment.

Substance	Averaging time	WHO	USA	Canada	Finland	Germany
Chlordane	Continuous 8 h	– –	a) 5 µg/m³ a) 500 µg/m³	–	– –	–
DDT	Continuous	–	–	b)	–	Proposed target value: 0.1 µg/m³
Lindane	Continuous	–	–	b)	–	1.0 µg/m³
Pentachlorophenol (PCP)	Continuous	–	–	b)	– –	RW II: 1.0 µg/m³ RW I: 0.1 µg/m³
Polychlorinated biphenyls (PCBs)	Continuous	–	–	–	– –	RW II: 3000 ng/m³ RW I: 300 ng/m³
4-Phenylcyclohexene (4-PCH)		–	b) 3 µg/m³	–	–	–
Remarks			For non-regulated pollutants: < 5 % of TLV and time-weighted average, US-EPA (1996)	b) Misuse may create a health hazard		RW (Richtwert): • guide value for intervention: RW II • target value: RW I
Ref.			a) ASHRAE (1990) b) US-EPA (1996)	Health and Welfare Canada (1989)		BGA (1993) Sagunski (1996) IRK (1997; 1998) Lederer (1997)

Table 4.1-2. Exposure standards and guideline values for organic pollutants in the indoor environment according to the approach of Nielsen et al. (1997; 1998a, 1998b, 1998c)

Substance	Odor threshold [µg/m^3]	Sensory irritation exposure limit estimate [µg/m^3]	Health-based indoor air exposure limit estimate [µg/m^3]	1/40 TLV [µg/m^3]
Toluene	1000	8000	8000	–
Ammonia	4000	4000	4000	–
Benzyl alcohol	100 – <670000	–	1000	–
2–Butanone oxime	<4000 – 18000	4000 – 18000	100	–
Propylene glycol	–	–	4000	–
2,2,4-Trimethyl-1,3-pentanediol monoisobutyrate	<150000	–	1000	–
Formic acid	53000	2000	300	200
Acetic acid	400	2500	1000	600
Propionic acid	100	3000	7000	800
Butyric acid	14	4000	1000	(900)*
Phenol	420	4000	400	500
Butylated hydroxytoluene (2,6–bis(1,1-dimethylethyl)-4-methylphenol)/BHT	–	–	500	–
2-Ethoxyethanol	4600	10000	400	500
2-(2-Ethoxyethoxy)ethanol	4000	2500	6000	–
2-(2-Butoxyethoxy)ethanol	9.2	–	9000	–
1-Methoxy-2-propanol	700	10000	10000	–

*No TLV established for butyric acid. The value given in parenthesis is based on the hypothesis that it should not be far from the value for propionic acid.

Table 4.1-3. Guideline values for total volatile organic compounds (TVOC) in the indoor environment

Substance	USA	Finland	Germany
TVOC	200 μg/m^3	S3: 600 μg/m^3 S2: 300 μg/m^3 S1: 200 μg/m^3	Proposed target value: 300 μg/m^3
Remarks	Maximum allowable air concentration standard for EPA buildings/internal standard	S 1, S 2, S 3: target values for indoor air quality S 1 – best category (> 90 % of occupants satisfied), S 3 – minimum requirements according to building codes	To obtain the sum, the VOCs belonging to one of the following chemical classes are ranked according to their measured concentration: alkanes, aromatic hydrocarbons, terpenes, halocarbons, esters, carbonyl compounds (excluding formaldehyde) and "other". The concentrations of the first ten VOCs in each class are summed
Ref.	US-EPA (1996)	FiSIAQ (1995)	Seifert (1990)

and "other". The concentration of the first ten VOCs in each class are summed" to obtain the TVOC with the following additional conditions: „*no individual VOC should exceed 50 % of the concentration allotted to its class or 10 % of the TVOC concentration*", i.e. no individual VOC concentration should exceed a concentration of 30 μg/m^3. The complete set of proposed target guideline values is defined in Table 4.1-4.

Table 4.1-4. Target guideline values for compound classes and TVOC (Seifert, 1990)

Chemical class of VOC	Concentration (μg/m^3)
Alkanes	100
Aromatic hydrocarbons	50
Terpenes	30
Halocarbons	30
Esters	20
Aldehydes and ketones (formaldehyde excluded)	20
Other	50
TVOC (sum of VOCs)	300

It is expected that under "normal" long-term conditions indoor air concentrations of VOCs should not exceed these target values if a building is properly managed. In new or renovated buildings the proposed concentration levels will be exceeded for some time due to emissions from various building and finishing materials. According to Seifert (1990), during the first week after renovation 50 times higher concentrations are acceptable and for another 6 weeks the concentrations may be 10 times higher than the proposed target values.

The various proposed TVOC target values in Table 4.1-3 are all in good agreement with the concept outlined by Seifert and cover a range from 200 μg/m^3 to 600 μg/m^3. Recently Seifert (1999) published a revised version of his TVOC concept recommending officially intervention if the TVOC concentration exceeds a range of 1000 μg/m^3 to

3000 μg/m^3 and defining a target value in the range of 200 μg/m^3 to 300 μg/m^3. These TVOC standards are fairly strict if compared to real world TVOC concentrations as published in a number of reviews (Krause et al., 1991; Brown et al., 1994; Lindstrom et al., 1995; see also Capter 3.1)

It has been reported (Krause et al., 1987; 1991) that in German established buildings (private homes,) the arithmetic mean TVOC concentration is in the order of 400 μg/m^3, whereas the 50 % fractile is in the order of 320 μg/m^3.

Brown et al. (1994) evaluated data from 50 studies which provided measurements of the indoor air concentration of VOCs in American and European buildings. The reported data for 1081 established dwellings based on 2478 measurements resulted in an overall weighted average geometric mean (WAGM) concentration for the TVOC of 1130 μg/m^3 and significantly lower concentrations in office buildings (WAGM: 180 μg/m^3), school buildings (WAGM: 70 μg/m^3) and hospitals (WAGM: 410 μg/m^3).

It is interesting to compare these results with the TVOC concentrations reported by Lindstrom et al. (1995) for a number of experimental dwellings "*that had been built with materials chosen for low pollutant emission and other modified design features to provide enhanced residential indoor air quality*". These workers studied indoor air quality in the preoccupancy phase as well as some time later in the occupancy phase and observed a range of TVOC concentrations well below 1000 μg/m^3 (Table 4.1-5).

This study from the USA shows that the largest differences in air quality during the occupancy phase were related to automobiles and the use of an attached garage significantly influencing indoor air quality (introducing fairly high concentrations of aromatic hydrocarbons).

Our own studies of indoor air quality in non-conventional buildings in Germany showed three major differences to conventional buildings:

- generally a reduced TVOC level,
- a certain temporal delay of emissions and, as a consequence, of immissions,
- a different characteristic of the TVOC with higher proportions of terpenes (and related compounds) and alkanes, no halogenated compounds, but considerable concentrations of semi-volatile compounds.

For conventional and non conventional homes the data prove that the TVOC concentrations are dropping – on a different time-scale – in a way which is well in agreement with the TVOC guideline proposed by Seifert (1990).

Even though the association between TVOC and complaints about indoor air quality or signs of a sick building syndrome are not yet well understood (Wolkoff, 1995) and the relationship between perceived air quality and TVOC concentration is in question

Table 4.1-5. Range of TVOC concentrations in conventional and experimental homes in the USA (Lindstrom et al., 1995).

	Conventional homes ($n = 3$) (μg/m^3)	Experimental homes ($n = 6$) (μg/m^3)
Preoccupancy phase	339 – 669	7 – 105
Occupancy phase	50 – 162	30 – 333

(Fitzner and Finke, 1996), nonetheless the TVOC concept seems to be appropriate to establish a general indicator for an acceptable (chemical) indoor climate.

Regulations for individual compounds with odorous, irritative, carcinogenic or other annoying properties will supplement any regulation which is based on the TVOC concentration. The definition of specific guideline values for individual compounds will be an ongoing process following the technical development in construction, building maintenance and in the consumer product sphere.

Indoor air quality guidelines and standards are in the making and have not yet reached a mature state, but it seems that all provisional standards which have been put into force in the past have served fairly well as an initial point for improvements. This can be demonstrated with reference to indoor air pollution by formaldehyde, wood preservers or halocarbons, which decreased significantly after introducing and enforcing the respective standards.

With this in mind, scientific endeavors and regulatory efforts in the development of more sophisticated and conclusive indoor air quality guidelines will undoubtedly provide support for practical activities aimed at preserving and improving comfortable and healthy conditions in the indoor environment.

References

Arbuthnot J. (1751): An Essay concerning the Effects of Air on Human Bodies. J. and R. Tonson and S. Draper, London.

ASHRAE/The American Society of Heating, Refrigerating, and Air Conditioning Engineers (1990): Ventilation for Acceptable Indoor Air Quality, ANSI/ASHRAE 62–1989 (including ANSI/ASHRAE Addendum 62a–1990), Atlanta.

Australian Standard 1668.2–1992 (1992): The use of mechanical ventilation and air conditioning in buildings. Part 2: Mechanical ventilation for acceptable indoor air quality. Standards Australia.

BGA/Bundesgesundheitsamt (1993): Bekanntmachungen des BGA. Bewertung der Luftqualität in Innenräumen. Bundesgesundhbl., *36*, 117–118.

British Standard 5925 (1991): Code of Practice for Ventilation Principles and Designing for Natural Ventilation. BSI Linford Wood, Milton Keynes.

Brown S.K., Sim M.R., Abramson M.J. and Gray C.N. (1994): Concentrations of volatile organic compounds in indoor air – a review. Indoor Air, *4*, 123–134.

Bundesanstalt für Arbeitsschutz (1988): Amtliche Mitteilungen der Bundesanstalt für Arbeitsschutz, No. 4, 14.

Commission of the European Communities (ed) (1992): Guidelines for ventilation requirements in buildings. European Concerted Action Indoor Air Quality & Its Impact on Man, Report No. 11. ECSC/EEC/EAEC, Brussels/Luxembourg.

DIN/Deutsches Institut für Normung (1992): DIN 1946, Teil 2: Raumlufttechnik–Gesundheitstechnische Anforderungen. Berlin, (draft).

Englert N. (1997): Richtwerte für die Innenraumluft: Kohlenmonoxid. Bundesgesundhbl., *40*, 425–428.

FiSIAQ/Finnish Society of Indoor Air Quality and Climate, Finnish Association of Construction Clients/RAKLI, Finnish Association of Architects/SAFA, Finnish Association of Consulting Firms/SKOL (1995): Classification of Indoor Climate, Construction, and Finishing Materials. Helsinki.

Fitzner K. and Finke U. (1996): Bestimmung der empfundenen Luftqualität in Bürogebäuden – Ergebnisse und Wertungen. Gesundheits-Ing.–Haustechnik–Bauphysik–Umwelttechnik, *117*, 192–201.

Gammage R.B., Tiffany J.A., Hodgson M., Light E. and Miller C. (1993): TLVs and IAQ guidelines: resolving the ambiguities. In: Indoor Air "93. Workshop Summaries, H. Levin (ed) Indoor Air '93. The 6th International Conference on Indoor Air Quality and Climate, Helsinki, 11–12.

Greenberg M.M. (1997): The central nervous system and exposure to toluene: a risk characterization. Environ. Res., *72*, 1–7.
Health and Welfare Canada (1989): Exposure guidelines for residential indoor air quality. Ottawa (revised edition).
Helsedirektoratet (1990): Retningslinjer for inneluft-kvalitet. Helsedirektoratets utredningsserie No. 6–90, IK-2322.
IRK/Innenraumlufthygiene-Kommission des Umweltbundesamtes und Ausschuß für Umwelthygiene der AGLMB (1996): Richtwerte für die Innenraumluft: Basisschema. Bundesundhbl., *39*, 422–426.
IRK/Innenraumlufthygiene-Kommission des Umweltbundesamtes und Ausschuß für Umwelthygiene der AGLMB (1997): Richtwerte für die Innenraumluft: Pentachlorphenol. Bundesundhbl., *40*, 234–236.
IRK/Innenraumlufthygiene-Kommission des Umweltbundesamtes und Ausschuß für Umwelthygiene der AGLMB (1998): Richtwerte für die Innenraumluft: Styrol. Bundesgesundhbl., *41*, 392–398.
Janssen J.E. (1989): Ventilation for acceptable indoor air quality. ASHRAE J., *31*, 40–48.
Kagawa J. (1993): Indoor air quality standards and regulations in Japan. Indoor Environ., *2*, 223–231.
Krause C., Mailahn W., Nagel R., Schulz C., Seifert B. and Ullrich D. (1987): Occurrence of volatile organic compounds in the air of 500 homes in the Federal Republic of Germany. Indoor Air '87. The 4th International Conference on Indoor Air Quality and Climate, Berlin, Vol. 1, 102–106.
Krause C., Chutsch M., Henke M., Kliem C., Leiske M., Schulz C. and Schwarz E. (1991): Umwelt-Survey. Messung und Analyse von Umweltbelastungsfaktoren in der Bundesrepublik Deutschland – Umwelt und Gesundheit. Band IIIc – Wohn-Innenraum: Raumluft, WaBoLu-Hefte 4/1991, Berlin.
Kunde M. (1982): Erfahrungen bei der Bewertung von Holzschutzmitteln. In: Aurand K., Seifert B., Wegner J. (ed) Luftqualität in Innenräumen, G. Fischer, Stuttgart/New York, 317–325.
Lederer P. and Angerer J. (1997): Raumluftbelastung durch DDT als Holzschutzmittel. Umweltmed. Forsch. Prax., *2*, 32.
Lemaire, M.-C. (1994): Normes dans l'environment et les milieux intérieurs. Secteur bâtiment. Poll. Atmosphérique, Oct.-Dec. 1994, 90–94.
Levin H. (1998): Introduction (to Indoor Air Guideline Values for Organic Acids, Phenols, and Glycol Ethers). Indoor Air Suppl. *5*, 5–7.
Lindstrom A. B., Proffitt D. and Fortune C.R. (1995): Effects of modified residential construction on indoor air quality. Indoor Air, *5*, 258–269.
Mølhave L. (1998): Principles for evaluation of health and comfort hazards caused by indoor air pollution. Indoor Air Suppl., 4, 17–25.
Mølhave L., Clausen G., Berglund B., De Ceaurriz J., Kettrup A., Lindvall T., Maroni M., Pickering A.C., Risse U., Rothweiler H., Seifert B. and Younes M. (1997): Total volatile organic compounds (TVOC) in indoor air quality investigations. Indoor Air, *7*, 225–240.
Nielsen G.D., Hansen L.F., Wolkoff P. (1997): Chemical and Biological Evaluation of Building Material Emissions. II. Approaches for Setting Indoor Air Standards or Guidelines for Chemicals. Indoor Air *7*, 17–32.
Nielsen G.D., Hansen L.F., Nexø B.A., Poulsen O.M. (1998 a): Indoor Air Guideline Levels for Formic, Acetic, Propionic and Butyric Acids. Indoor Air Suppl. *5*, 8–24.
Nielsen G.D., Hansen L.F., Nexø B.A., Poulsen O.M. (1998 b): Indoor Air Guideline Levels for Phenol and Butylated Hydroxytoluene (BHT). Indoor Air Suppl. *5*, 25–36.
Nielsen G.D., Hansen L.F., Nexø B.A., Poulsen O.M. (1998 c): Indoor Air Guideline Levels for 2-ethoxyethanol, 2-(2-ethoxyethoxy)ethanol, 2-(2-butoxyethoxy)ethanol and 1-methoxy-2-propanol. Indoor Air Suppl. *5*, 37–54.
Pluschke P. (1996): Luftschadstoffe in Innenräumen. Springer, Berlin Heidelberg New York.
Sagunski H. (1996): Richtwerte für die Innenraumluft: Toluol. Bundesgesundhbl., *39*, 416–421.
Sagunski H. (1998): Richtwerte für die Innenraumluft: Zum derzeitigen Stand in der Ad-hoc-Arbeitsgruppe der Innenraumlufthygiene-Kommission des Umweltbundesamtes und der Arbeitsgemeinschaft der Obersten Landesgesundheitsbehörden (Ad-hoc AG IRK/AOLG). Umweltmed. Forsch. Prax. *3*, 230.
Salthammer T., Fuhrmann F., Meininghaus R., Miertzsch H. and Wismach C. (1996): The German RAL wallcovering label: experience and progress. In: Indoor Air '96. Proceedings of the 7th International Conference on Indoor Air Quality and Climate, Nagoya, Vol. 3, 543–548.
Samet J.M. and Speizer F.E. (1993): Introduction and recommendations: working group on indoor air and other complex mixtures. Environ. Health Persp, 101, Suppl. 4, 143–147.
Seifert B. (1990): Regulating indoor air. In: Indoor Air '90. Proceedings of the 5th International Conference on Indoor Air Quality and Climate, Toronto/Ottawa, Vol.5, 35–49.

Seifert B., Levin H., Lindvall T. and Moschandreas D. (1993): A critical review of criteria and procedures for developing indoor air quality guidelines and standards. In: Indoor Air '93. Proceedings of the 6th International Conference on Indoor Air Quality and Climate, Helsinki, Vol. 3, 465–470.

Seifert, B. (1999): Richtwerte für die Innenraumluft. Die Beurteilung der Innenraumluftqualität mit Hilfe der Summe der flüchtigen organischen Verbindungen (TVOC-Wert) Bundesgesundhblatt, 42, 270–278.

Spengler J.D. and Sexton K. (1983): Indoor air pollution: a public health perspective. Science, *221*, 9–17.

Suess M. (1988): Comparative evaluation of indoor and ambient air quality–health guidelines. Environ. Technol. Lett., *9*, 461–470.

Szalai A. (1972): The use of time: daily activities of urban and suburban population in twelve countries. Mouton, Den Haag Paris.

Türck R. and Böttger A. (1993): Concept of the Federal Government for improving indoor air quality in Germany. In: Indoor Air '93. Proceedings of the 6th International Conference on Indoor Air Quality and Climate. Helsinki, Vol. 3, 561–566. This report is based on a policy paper of the German government: Bundesminister für Umwelt, Naturschutz und Reaktorsicherheit /BMU (ed) (1992): Konzeption der Bundesregierung zur Verbesserung der Luftqualität in Innenräumen, Bonn.

US-EPA (1996): IAQ a Concern in EPA's Planned 232-Million Research Facility. Indoor Air Quality Update *9*, no. 10, 8–13.

VDI 2310 (1974): Maximale Immissions-Werte. Beuth, Berlin.

WHO/World Health Organization–Regional Office for Europe (1987): Air quality guidelines for europe. WHO Regional Publications, European Series No. 23, Copenhagen.

Wolkoff P. (1995): Volatile organic compounds. Sources, measurements, emissions and the impact on indoor air quality. Indoor Air Suppl., *3*, 1–73.

4.2 The TVOC Concept

Lars Mølhave

4.2.1 Introduction

In the past, when human bioeffluents were considered to be the dominating pollutants of indoor air in non-industrial buildings, carbon dioxide (CO_2) was generally accepted as an indicator for indoor air quality (IAQ). This function of CO_2 is less important now, partly because today many more sources than human beings emit pollutants into indoor air. In fact, the widespread use of new products, processes and materials in our days has resulted in increased concentrations of other indoor pollutants, especially of volatile organic compounds (VOCs) which may affect human health.

A classification of organic air pollutants was given by a WHO working group on organic indoor air pollutants (WHO, 1989). This group initiated the common practice of dividing organic chemicals according to boiling point ranges and to discriminate between Very Volatile Organic Compounds (VVOCs), Volatile Organic Compounds (VOCs), Semivolatile Organic Compounds (SVOCs) and Particulate Organic Compounds (POMs). The VOC class includes compounds with boiling points between 50–100°C and 240–260°C, where the higher values refer to polar compounds (Table 4.2-1).

In Western Europe human beings are usually exposed more than 20 h per day to indoor air pollutants. The concentrations of VOCs in indoor air have been associated with odors,

Table 4.2-1. Classification of indoor organic pollutants (WHO, 1989).

Category	Description	Abbreviation	Boiling point range* [°C]	Sampling media typically used in field studies
1	Very volatile (gaseous) organic compounds	VVOC	< 0 to 50–100	Batch sampling; adsorption on charcoal
2	Volatile organic compounds	VOC	50–100 to 240–260	Adsorption on Tenax, graphitized carbon black or charcoal
3	Semivolatile organic compounds	SVOC	240–260 to 380–400	Adsorption on polyurethane foam or XAD-2
4	Organic compounds associated with particulate matter or particulate organic matter	POM	> 380	Collection on filters

* Polar compounds appear at the higher end of the range.

irritation of mucous membranes of eyes, nose and mouth, Sick Building Syndrome (SBS) (WHO, 1986) and neurotoxicity (Berglund et al., 1992; EU-ECA-IAQ, 1991).

This association between increasing concentrations of air pollutants and symptoms prevalence in non-industrial buildings is the reason for the growing interest in ways of characterizing the air in enclosed spaces. As a result, the air of all kinds of indoor spaces is frequently analyzed for VOCs (Brown et al., 1994).

As generally accepted approaches are lacking, the assessment of exposures to VOC and the evaluation of the associated health and discomfort effects are often based on some sort of integrating or summarizing measure. An often-used measure is the Total Volatile Organic Compounds (TVOC). This means that one single value represents the mixture of many VOCs (Mølhave and Nielsen, 1992). Usually, TVOC values represent a summation of some or (rarely) all of the VOCs present in air indoors.

The TVOC value, or other measures of volatile organic compounds, may be used for a number of applications other than evaluations of health risks. Examples are: testing of materials, indication of insufficient or poorly designed ventilation in a building, and identification of highly polluting activities. Such applications are not discussed here (see EU-ECA WG 13, 1996).

The TVOC indicator is used by different authors both for reporting exposures (as substitute or indicator measure of exposure and of air quality) and as a predictor of the probability of health and comfort effects (Mølhave and Nielsen, 1992). The justification for the use of TVOC in the literature to describe indoor air exposures and to estimate health consequences and risks is mostly derived from the work of Mølhave and his group (Mølhave, 1986; Mølhave et al., 1986, 1991, 1993; Kjærgaard et al., 1991), who studied the health and comfort effects of a mixture of 22 VOCs, and the subsequent complementary work carried out at the laboratories of the US-EPA using almost the same mixture (Otto et al., 1990a, 1990b, 1992; Hudnell et al., 1992; Koren et al., 1992).

However, given the relatively small number of VOCs used in these studies and the specific composition of the mixture used, it cannot be expected a priori that the observed increased subjective ratings of general discomfort and CNS-mediated symptoms would also occur with another mixture even if the TVOC levels of the two mixtures were close to each other.

Literature shows that there is a large variety of ways to calculate a TVOC value from the results of an analysis (e.g., De Bortoli et al., 1986; Gammage et al., 1986; Krause et al., 1987; Mølhave, 1992; Rothweiler et al., 1992; Seifert, 1990; Wallace et al., 1991).

Different authors use different procedures for calculating the TVOC and for interpreting the probability of health and comfort consequences from the calculated TVOC exposure levels. In addition, differences may arise from the influence of the analytical system, including the adsorbent used for sampling, the sampling rate and volume, and the separation and detection system.

There are four main causes of differences between reported measures:

- Different ranges of compounds are included in the TVOC values and the range of compounds are not clearly defined.
- The specific compounds in the TVOC value are not identified.
- The TVOC value does not represent the total concentration of VOCs in the air sample as closely as possible.

- The TVOC value is not constructed in a way that favors as much as possible its usefulness in the evaluation of indoor air quality.

Hence, the number and nature of the VOCs on which a TVOC value is based can vary in published studies. Therefore the results found in literature are hardly comparable.

For all these reasons, TVOC data from the literature must be interpreted very cautiously and it would be beneficial to have a standardized method for reporting TVOC. This should include specific sampling and analytical protocol and should provide some validation of the methods used, including a specification of the simplifications made and limitations of the accuracy following from these simplifications.

This was the reason for the work of the EU-ECA working group 13 on: "The use of the TVOC as an indicator in IAQ investigations", EU-ECA-WG 13 (1996). The EU-ECA acknowledged that the acceptance of TVOC as a general principle faces serious barriers due to the enormous amount of existing data based on different methods and the investment represented in the sampling and analytical equipments already in use.

The EU-ECA-WG13 found, however, that the TVOC measure is a suitable way of reporting exposures to VOCs indoors if it is measured in the way proposed by the group. Although many uncertainties remain with the use of a TVOC value, especially with regard to the type and quantity of potentially associated health and comfort effects, the working group felt that such use can be beneficial if the limits of application are outlined and respected.

In their report, the EU-ECA-WG 13 (1996) discusses the advantages and disadvantages of the TVOC concept in respect of exposure assessment and prediction of health effects. The report has also been published as Mølhave et al., 1997. The recommendations cited from their report in this Chapter are in italics.

The conclusions of the working group are also supported by the conclusion of a TVOC workshop at the ASTM international conference on "The TVOC as an indoor climate metrics" in Washington, 1997. A substantial number of the participants indicated that they would continue to use the TVOC in the future, despite its imperfections, and that TVOC measurements can aid in limiting the concentrations of pollutants in the air breathed by the occupants.

The ASTM working group agreed that a standard definition of TVOC is needed. The EU-ECA definition was found to be useful in the context of evaluating the potential effects on IAQ of VOC. TVOC should only be used as a screening tool and should not be used alone as a basis for definite conclusions. The ASTM group agreed also that, while elaborated TVOC values can alert practitioners to potential problems, low values do not necessarily guarantee that a building is free of problems.

This Chapter summarizes and discusses the EU-ECA definition of TVOC and indicates the ways in which the TVOC could be used in air pollution control in the future.

4.2.2 Instruments for Measuring Exposure

There are two basic principles for the analysis of VOCs in indoor air. They differ in the degree of information they provide. The simplest procedure does not separate the mixture into its individual components, but instantly reports the total concentration. This is done by direct-reading instruments. In a more elaborate integrating procedure, the components of a chemical mixture are separated and some or all of the compounds (or possibly none of them) are identified. The concentrations of individual compounds are then added together.

4.2.2.1 Direct-Reading Instruments

The detectors used in gas chromatography for the detection of individual VOCs can also provide information about a mixture without a separation step. These VOC detectors include, for example, the flame-ionization detector (FID) and the photo-ionization detector (PID). Other direct-reading instruments have also been used for determining VOCs, such as photoacoustic spectroscopy (PAS) (see Chapter 1.6). Other types of sensors (e.g. "electronic noses") may become important in the future.

Direct-reading detectors are generally calibrated with one single compound, e.g. a hydrocarbon such as *n*-hexane or toluene. Consequently, the signal obtained from a mixture of VOCs is expressed in terms of concentration equivalents of the reference compound regardless of the composition of the mixture.

Direct-reading instruments are easy to use. They are portable and provide a real-time signal which makes it possible to detect rapid concentration changes also. Direct-reading monitors respond not only to VOCs but also to other organic compounds, especially VVOCs (often including methane). The output signal does not give information about the qualitative composition of the mixture.

Since the TVOC values measured by different direct-reading instruments differ from one another and also from the new TVOC value defined later in this chapter, EU-ECA WG13 (1996) suggests that, to avoid confusion in the future, *these integrating measurements are marked with a suffix indicating the type of direct-reading instrument used, such as* $TVOC_{FID}$, $TVOC_{PID}$ or $TVOC_{PAS}$.

The EU-ECA working group recommended that *the use of simple integrating instruments (e.g. FID or PID) for measurements of TVOC indicators should be restricted to samples of slightly varying composition (e.g. from the same source) and to situations where an acceptable correlation between the TVOC indicator values based on the simple measures and those obtained with the recommended procedure has been established.*

The TVOC assessment procedure as described later may start with a simple integrating detector reporting the concentration in toluene equivalents and may be followed by more detailed analyses in which individual compounds are identified and quantified. *If the value obtained with a simple integrating detector is above 0.3 mg/m^3, detailed analysis should be made using the recommended procedure.*

4.2.2.2 VOC Separation Methods

In most evaluations of IAQ, details of individual organic compounds are needed, and the chemical mixture of air pollutants is, therefore, separated into its constituents.

Most VOC analysis of indoor air are carried out using sampling on a sorbent and subsequent separation by gas chromatography (GC). The number of possible GC procedures used is large (Otson and Fellin, 1992), and no single procedure can be recommended as the only possible one. In a compilation of analytical procedures for indoor air analysis (Seifert, 1990), examples are given of GC procedures including short-term and long-term sampling.

The result of the separation step is usually a chromatogram containing a large number of peaks, each representing one or more compounds. In most systems, the integration of the peak areas is performed automatically by a computer.

The EU-ECA working group recommended that *the best way of creating a TVOC value is to base it on the identification of each individual VOC in the mixture and to use its own response factor (i.e. proportionality factor between instrumental response and air concentration) to calculate a concentration.* Although tedious, this approach has been used in practice (Krause et al., 1987). Only where *correct reference compounds* are not available can a reference compound such as toluene be used.

Rather than taking one single response factor, Wallace et al. (1991) used the average of the response factors of 17 target VOCs. The EU-ECA working group recommended *not to use average response factors of a selected group of reference compounds for the quantification of identified or unidentified compounds.*

To reduce the analytical work, some researchers have suggested to use only a limited number of marker compounds as representative of the entire range of VOCs which may be present in any atmosphere. Wallace et al. (1991) used the average of 17 target VOCs. The EU-ECA working group recommended *not to use TVOC summations based on the identification and quantification of a selected and limited group of marker compounds.*

However, often not all peaks can be identified. If some or all of the individual compounds have not been identified, the response factor of one single reference compound, e.g. *n*-hexane or toluene, is used to estimate the concentrations of unidentified compounds. The reference compound used must be published together with the results. *The group recommended the use of toluene* as a compromise between frequently appearing air pollutants, low vapor pressure (to reduce evaporation in the analytical process) and a response factor close to the average of VOCs in general.

Taking into account that a certain percentage of VOCs cannot usually be identified, Clausen et al. (1991) have combined the "individual calibration" approach and the "average response factor" approach. They defined the TVOC value as the sum of the correctly measured masses of identified VOCs plus the amount of non-identified peaks in the chromatogram, using the response factor of toluene. The EU-ECA working group *recommend this approach.*

4.2.2.3 Special VOCs

On a routine basis, the currently available GC/MS-based analytical procedures have a detection threshold that is too high for detecting a number of VOCs at indoor air concentrations.

Examples of such compounds of special interest are: formaldehyde, acetaldehyde, acetic acid, amines, diisocyanates, some polycyclic aromatic hydrocarbon, and many biocides (some of these may not be VOCs). Additional examples of such compounds are β-glucan, which is believed to cause symptoms in sick buildings (Rylander and Jacobs, 1994), and aldehydes, which are frequently determined using high-performance liquid chromatography following derivatization with 2,4-dinitrophenylhydrazine (DNPH).

Another reason for special interest in a particular VOC may be an extremely low threshold of effects of the compounds. Many odorous VOCs are perceived by some individuals at concentrations below the analytical detection limit, which frequently is of the order of 1 fg/m^3 (Devos et al., 1990).

If compounds with such special analytical or toxicological properties appear indoors, complaints may be reported even if the GC/MS-measured TVOC value in indoor air is found to be low. The EU-ECA working group recommended that *the GC/MS procedure is supplemented with additional methods if such special compounds can be expected.* It follows that *the investigator must consider this possibility and report the conclusion of such considerations together with the TVOC results.*

4.2.3 TVOC – a Summary of the EU-ECA Definition

In the following, the proposed new definition of TVOC from the EU-ECA working group (EU-ECA-WG 13, 1996) is summarized. It is important to underline that the TVOC value determined according to the procedure mentioned below does not include all organic compounds in indoor air. There are non-VOC organic air pollutants highly relevant for IAQ that are not reflected in the TVOC value.

4.2.3.1 Recommended Procedure for the Measurement of TVOC

The procedure starts with a consideration of the possible occurrence of relevant compounds which may not be identified by the procedure proposed below. If such compounds can be expected, additional analytical procedures must be applied.

Instrumentation

(1) Use Tenax GC or Tenax TA for sampling.
(2) Use a non-polar GC column for analysis (column polarity index of <10). The system must have a detection limit (three times the noise level) for toluene and 2-butoxyethanol of less than 0.5 and less than 1 μg/m^3, respectively.
(3) The analytical procedure should allow the identification and quantification of a sample of compounds representing the most frequently seen chemical classes of VOCs encountered in indoor air (Table 4.2-2).

Analytical window

(4) The selected analytical procedure should allow all components of the list of compounds of interest (Table 4.2-2) to have retention times between those of *n*-hexane and *n*-hexadecane.
(5) In calculating TVOC, consider all the compounds (including other compounds than those shown in Table 4.2-2) found in the part of the chromatogram from *n*-hexane to

n-hexadecane. In this procedure, the WHO definition (WHO 1989) has been slightly modified by replacing the range of boiling points by a definition of the "analytical window" in terms of the two specific reference compounds.

Quantification

(6) Based on individual calibration, quantify as many of the VOCs as possible. As a minimum, at least those contained in the list of known VOC pollutants (Table 4.2-2) and in addition, the ten highest peaks should be quantified. The list of compounds and concentrations (identified as well as unidentified by name) must be published together with the TVOC result.

Calculation of TVOC

(7) Calculate the combined concentration S_{id} (mg/m^3) of these identified and quantified compounds.
(8) Calculate S_{un} (mg/m^3) of unidentified VOCs using the response factor of toluene.
(9) An acceptable level of identification has been achieved if S_{id} exceeds three times S_{un}. (If the sum is lower than 1 mg/m^3, it may be sufficient if S_{id} equals S_{un}).
(10) The sum of S_{id} and S_{un} is called the TVOC concentration or TVOC value.
(11) If many compounds are observed outside the VOC range as defined at point 5) this information should be added to the TVOC value.

The use of other techniques than those described in (1), (2) and (3) is allowed provided that their results are documented to be comparable to the procedure described here.

4.2.4 TVOC as a Health Effect Indicator

4.2.4.1 Toxicity of VOCs Indoors

The health effects relevant to TVOC and low-level VOC exposures indoors are those which are likely to be common for most organic compounds (although appearing at different concentrations for different compounds). At the concentrations typically found indoors, sensory effects and hypersensitivity effects are considered to be the most relevant. These include sensory irritation, a feeling of dryness, weak inflammatory irritation in eyes, nose, airways and skin. The effects may be both immediate (acute) or delayed (subacute) (Berglund et al., 1992; EU-ECA-IAQ, 1991).

As VOCs belong to different chemical classes, the severity of the potential effects at one and the same concentration level of different compounds may differ by orders of magnitude. Dose-response relationships are expected to exist between exposure to any given VOC mixture in air and individual health or discomfort effects. However, these relationships, of which the exact forms are mostly unknown, are expected to be complex and much affected by other factors than the total amount of VOCs present and will vary much between different mixtures.

Table 4.2-2. Minimum number of compounds to include in TVOC analysis.

Chemical compound	CAS No.	Boiling point [°C]
AROMATIC HYDROCARBONS		
Benzene	71-43-2	80.1
Toluene	108-88-3	111
Ethylbenzene	100-41-4	136.2
m/*p*-Xylene	108-38-3 / 106-42-3	139.1 / 138.3
o-Xylene	95-47-6	144
n-Propylbenzene	103-65-1	159
1,2,4-Trimethylbenzene	95-63-6	169.4
1,3,5-Trimethylbenzene	108-67-8	165
2-Ethyltoluene	611-14-3	165.2
Styrene	100-42-5	145.2
Naphthalene	91-20-3	218
4-Phenylcyclohexene	4994-16-5	251-3[1]

[1]Value for 1-phenyl-cyclohexene.

Chemical compound	CAS No.	Boiling point [°C]
ALIPHATIC HYDROCARBONS		
n-C6 to *n*-C16		
n-Hexane	110-54-3	69
n-Heptane	142-82-5	98.4
n-Octane	111-65-9	125.7
n-Nonane	111-84-2	150.8
n-Decane	124-18-5	174.1
n-Undecane	1120-21-4	196
n-Dodecane	112-40-3	216.3
n-Tridecane	629-50-5	235.4
n-Tetradecane	629-59-4	253.7
n-Pentadecane	629-62-9	270.6
n-Hexadecane	544-76-3	287
2-Methylpentane	107-83-5	60.3
3-Methylpentane	96-14-0	63.3
1-Octene	111-66-0	121.3
1-Decene	872-05-9	170.5
Methylcyclopentane	96-37-7	71.8
Cyclohexane	110-82-7	81
Methylcyclohexane	108-87-2	101
TERPENES		
Δ^3-Carene	13466-78-9	167
α-Pinene	80-56-8	156
β-Pinene	18172-67-3	164
Limonene	138-86-3	170

Table 4.2-2. (Continued).

Chemical compound	CAS No.	Boiling point [°C]
ALCOHOLS		
2-Propanol	67-63-0	82.4
1-Butanol	71-36-3	118
2-Ethyl-1-hexanol	104-76-7	182
GLYCOLS/GLYCOL ETHERS		
2-Methoxyethanol	109-86-4	124-25
2-Ethoxyethanol	110-80-5	135
2-Butoxyethanol	111-76-2	171
1-Methoxy-2-propanol	107-98-2	118
2-Butoxyethoxyethanol	112-34-5	231
ALDEHYDES		
Butanal	123-72-8	76
Pentanal	110-62-3	103
Hexanal	66-25-1	129
Nonanal	124-19-6	190-2
Benzaldehyde	100-52-7	179
KETONES		
Methyl ethyl ketone	78-93-3	80
Methylisobutylketone	108-10-1	116.8
Cyclohexanone	108-94-1	155.6
Acetophenone	98-86-2	202
HALOCARBONS		
Trichloroethene	79-01-6	87
Tetrachloroethene	127-18-4	121
1,1,1-Trichloroethane	71-55-6	74.1
1,4-Dichlorobenzene	106-46-7	173
ACIDS		
Hexanoic acid	142-62-1	202-3
ESTERS		
Ethyl acetate	141-78-6	77
Butyl acetate	123-86-4	126.5
iso-Propyl acetate	108-21-4	85
2-Ethoxyethyl acetate	111-15-9	156.4
TXIB (Texanol *iso*-butyrate)	6846-50-0	
OTHER		
2-Pentylfuran	3777-69-3	>120 (2-*tert*-Butylfuran)
THF (Tetrahydrofuran)	109-99-9	67

From a toxicological point of view, an indicator of the potential of a mixture to cause one of health and comfort effects should ideally be based on a compound-by-compound procedure, correctly quantifying the contributions from all relevant single compounds and their interactions. Such a procedure is discussed by Mølhave (1998). Presently, no generally accepted toxicological principle exists for such evaluations.

According to basic toxicological knowledge, the effects of pollutants in a mixed exposure may be additive ($Effect_{Mix} = Effect_A + Effect_B + ...$), synergistic ($Effect_{Mix} > Effect_A + Effect_B + ...$), antagonistic ($Effect_{Mix} < Effect_A + Effect_B + ...$), or even independent of each other. For sensory reactions, the interaction mechanisms are known only for a small group of VOCs with strong odors for which hypoadditive behavior has been demonstrated (Berglund and Olsson, 1993).

Time and lack of data often prevent such a procedure from being adopted, and addition is therefore often assumed in practice. In general terms such "summation of effects of individual compounds in a mixture" can only be expected to predict the effects following exposure to a mixture if these compounds exert the same type of toxic effect(s) in the same organ(s), and if their mechanism of action is the same. Under these circumstances, additivity of effects may be assumed to be an acceptable estimate of the upper-bound risk level, and the sum of the individual (or "corrected") concentrations can be used to calculate a total level of exposure. However, for the exposure estimate, addition of mass (concentration) may not be the mathematical procedure which best reflects the biological principles involved in the comfort and health effects. Better models may be established in the future.

In the absence of a generally agreed procedure, and because of the obvious need resulting from indoor climate complaints, TVOC has been widely adopted as a simple indicator. Because the theoretical framework and health data are missing and because of problems with the addition of contributions from very different VOC compounds, many toxicological experts argue that TVOC may never be used to evaluate health risks.

4.2.4.2 TVOC and Health Evaluations

The EU-ECA working group found that at present, *the complexity of available alternative mathematical and toxicological models for construction of an indicator for health risks associated with VOC exposures and the hypothetical gains in accuracy which they may provide does not at this level of knowledge justify introduction of more complex procedures than those specified by the group for TVOC.* The EU-ECA working group, therefore, found that for practical use *TVOC as defined by the group presently is the best available indicator for VOC exposures. Further, they concluded that if TVOC is measured in the proposed way then for screening purposes it gives a first estimate of the consequences of the exposures.*

The working group recommend that *the use of the TVOC-indicator be restricted to estimations of the probability of non-specific immediate and delayed sensory irritation.*

As illustrated through several field studies, sensory effects can be caused by other environmental factors than VOC. *A high TVOC value therefore does not by itself establish causality.*

No simple relationship exists between VOC concentrations of mixtures and odor intensity. However, as TVOC concentration increases, the likelihood of a multicomponent mixture being odorous increases also. Odor thresholds are generally lower than irritative thresholds. *The consensus of the group is that although TVOC is expected to be appropriate for irritation effects it may also turn out to be useful for acute and long-term odors (as perceived by the sense of olfaction).*

Evaluations and ratings of, e.g., IAQ includes non-additive effects on a high CNS level and, therefore, *the TVOC measure cannot be expected to be useful for assessments of, e.g., acceptability of the effects.*

No documented or theoretical background exists for the use of the TVOC indicator in relation to other health effects (e.g. cancer, allergy, and neurological effects). *TVOC should only be used in the low-exposure range (below 25 mg/m^3, where such effects are less likely to appear).*

The consequences of varying exposure duration depend on the type of effect and type of compounds in the mixture. Some of these have an acute effect which may fade due to adaptation; others only show effects after accumulation of exposure. As the sampling inevitably will average exposures over, e.g., hours, the probability of immediate (acute) and delayed (subacute) responses cannot be separated. For these practical reasons the TVOC indicator is probably also less suitable for the evaluation of effects of fluctuating concentrations. Moreover, TVOC is expected to be less suitable for evaluating the likelihood of odor problems. However, if it is used for this purpose, the smallest possible sampling period (e.g. less than 15 min) and many repeated samples should be used. The use of direct-reading instruments may be considered as an alternative.

The EU-ECA group also found that *the indicator should refer to the effects appearing in a normal population (excluding allergic and otherwise hypersensitive persons) exposed up to 24 h per day in a normal environment, with humidity, temperatures and concentrations of other pollutants (e.g. non-organic volatiles or particulates) within normal ranges.* The TVOC should, therefore, be used with special caution in indoor environments which are outside normal ranges with respect to these factors.

The possibility cannot be excluded that specific VOCs appear indoors which may turn out in the future to be much more potent in causing effects on humans than the average VOCs. In this case, these (e.g. formaldehyde) should be evaluated individually, and a list of such compounds should be established. A draft list of special compounds for which low thresholds for effects or high detection limits cause the probability of sensory effects to be grossly underestimated if this TVOC-concept is used has not yet been officially established. This list should also include compounds which are likely to cause other types of health effects than irritation and which, therefore, must be considered in a total evaluation of the indoor air quality. Potential candidates for the list are mentioned above.

The WHO definition of VOCs refers to the behavior of the compounds in traditional analytical procedures and not to their ability to cause discomfort and health effects through environmental exposures. Also, some organic compounds outside the VOC range as defined by WHO may contribute to relevant sensory effects. Consequently, the organic compounds relevant to the TVOC concept from the toxicologist's point of view may not be defined strictly by the WHO definition. However, no exact definition of the relevant compounds can be made at present. The pragmatic short cut suggested by

ECA-WG13 includes a limited VOC range of the relevant compounds as shown with the list of compounds of interest. If many compounds are found outside the VOC range in the gas-chromatographic analysis, this must be reported together with the TVOC value.

4.2.4.3 Guidelines Based on TVOC

Based on theoretical considerations and experience from occupational hygiene, it can be argued that a sufficiently high total concentration of any complex mixture of VOCs is likely to evoke odor as well as sensory irritation among the majority of those exposed.

However, the controlled human exposure studies are few, and the results have not been confirmed. Also, the results of epidemiological studies are inconsistent. Therefore, is it not possible today to conclude whether or not sensory irritation is associated with the sum of mass concentrations of VOCs at the low exposure levels typically encountered in non-industrial indoor air. Thus, at present, *no precise guidance can be given on which levels of TVOC are of concern from a health and comfort point of view, and the magnitude of protection margins needed cannot be estimated.*

This is also the conclusion of a Nordic consensus group (Andersson et al., 1997), which found insufficient data for assigning exact guideline values in their review of a number of field and laboratory experiments dealing with the effects of VOCs. These studies were found to report exposures unsystematically and to give an imprecise estimate of the ranges in which effects can be expected.

The general need for improved source control to diminish the pollution load on indoor environments from health, comfort, energy efficiency and sustainability points of view leads the EU-ECA group to the recommendation that *VOC levels in indoor air should be kept as low as reasonably achievable (ALARA).* Such an ALARA principle will require that TVOC concentrations in indoor environments – when determined with the proposed procedure on representative samples of buildings and spaces – do not exceed the typical levels encountered in the building stock of today, unless there are very good and explicit reasons.

At present an old review (Mølhave, 1986, 1991) is the only available attempt to draw conclusions on dose-response relations. Mølhave clearly states the inadequacy of the data used for the estimates and that these values must be seen in the context that very little health data is available, that the most recent study referred to is from the mid 80's, and that the researchers are using different definitions of TVOC. It appears that at TVOC concentrations of ca. 3 mg/m^3 complaints are evident in all investigated buildings, occupants having symptoms and odors being perceived in experimental studies. Physiological effects are seen at 5 mg/m^3, and exposures up to 8 mg/m^3 lead to significant mucosal irritation in eyes, nose and throat. Whereas headaches are reported to occur at levels around 3 mg/m^3 in field studies, chamber studies indicate no such effect at levels below 25 mg/m^3. At TVOC levels between 0.2 and 3 mg/m^3, irritation and discomfort may occur if other types of exposures interact.

Acknowledgements

This Chapter is based on the work of a working group within the European commission [EU-ECA-WG 13 (1996)]. The members are Prof. Birgitta Berglund, Prof. Jacques De Ceaurriz, Dr. Lars Mølhave (chairman), Prof. Thomas Lindvall, Prof. Marco Maroni, Prof. A. Kettrup/Dr. Uwe Risse, Dr. Anthony C. Pickering, Dr. Heinz Rothweiler, Dr. Bernd Seifert, Dr. Maged Younes (associate member). The work on this paper received additional support through the Danish Healthy Building Study from the Danish Technical Research Council (STVF). The present publication is this author's condensation of the EU-ECA report and therefore does not necessarily reflect the opinions of ECA.

References

Andersson K., Bakke J.V., Bjørseth O., Bornehag C.G., Clausen G., Hongslo K., Kjellman M., Kjærgaard S.K., Levy F., Mølhave L., Skerfving S. and Sundell J. (1997): TVOC and health in non-industrial indoor environments. Reports from a Nordic scientific consensus meeting at Langholmen in Stockholm 1996. Indoor Air, *7*, 78–91.

Berglund B. and Olsson M. (1993): Perceptual and psychophysical models for odor intensity interaction. In: Garriga-Trillo A., Minon P.R., Garcia-Gallego C., Merino J.M. and Villarino A. (eds) Fechner Day '93. Palma de Mallorca, Spain. International Society for Psychophysics, 35–40.

Berglund B., Brunekreef B., Knöppel H., Lindvall T., Maroni M., Mølhave L. and Skov P. (1992): Effects of indoor air pollution on human health. Indoor Air, *2*, 2–25.

Brown S.K., Sim M.R., Abramson M.J. and Gray C.N. (1994): Concentrations of volatile organic compounds in indoor air–a review. Indoor Air, *4*, 123–134.

Clausen P.A., Wolkoff P., Holst E. and Nielsen P.A. (1991): Long term emission of volatile organic compounds from waterborne paints. Methods of comparison. Indoor Air, *1*, 562–576.

De Bortoli M., Knöppel H., Pecchio E., Peil A., Rogora L., Schauenburg H., Schlitt H. and Vissers H. (1986): Concentrations of selected organic pollutants in indoor and outdoor air in northern Italy. Environ. Int., *12*, 343–350.

Devos M., Patte F., Rouault J., Laffort P. and Van Gemert L.J. (1990): Standardized human olfactory thresholds. Oxford University Press.

EU-ECA-IAQ (European Concerted Action "Indoor Air Quality and its Impact on Man") (1991): Effects of indoor air pollution on human health. Report No. 10, EUR 14086 EN. Office for Official Publications of the European Communities, Luxembourg.

EU-ECA-WG13 (1996): The use of TVOC as an indicator in IAQ investigations. Report of working Group 13 of European Collaborative Action on Indoor Air Quality and its Impact on Man. JRC, Ispra, Italy.

Gammage R.B., White D.A., Higgins C.E., Buchanan M.V. and Guerin M.R. (1986): Total volatile organic compounds (VOCs) in the indoor air of East Tennessee homes. Proceedings of the 1986 EPA/APCA Symposium on Measurement of Toxic Air Pollutants. U.S. Environmental Protection Agency, Environmental Monitoring Systems Laboratory / Air Pollution Control Association, Raleigh, North Carolina, 104–115.

Hudnell H.K., Otto D.A., House D.E. and Mølhave L. (1992): Exposure of humans to a volatile organic mixture, II, Sensory assessment. Arch. Environ. Health, *47*, 31–38.

Kjærgaard S.K., Mølhave L. and Pedersen O.F. (1991): Human reactions to a mixture of indoor air volatile organic compounds. Atmos. Environ., *25a*, 1417–1426.

Koren H.S., Graham D.E. and Devlin R.B. (1992): Exposure of humans to a volatile organic mixture, III, Inflammatory response. Arch. Environ. Health, *47*, 39–44.

Krause C., Mailahn W., Nagel R., Schulz C., Seifert B. and Ullrich D. (1987): Occurrence of volatile organic compounds in the air of 500 homes in the Federal Republic of Germany. In: Seifert B., Esdorn H., Fischer M., Rüden H. and Wegner J.(eds) Indoor Air '87, Proceedings of the 4th International Conference on Indoor Air Quality and Climate. Berlin, 17–21 August 1987, Vol. 1, 102–106.

Mølhave L. (1986): Indoor air quality in relation to sensory irritation due to volatile organic compounds. ASHRAE Trans., paper 2954, 92(1), 1–12.
Mølhave L. (1991): Indoor climate, air pollution and human comfort. J. Expo. Anal. Care Environ. Epidemiol., *1*, 63–81.
Mølhave L. (1992): Controlled experiments for studies of the sick building syndrome. Ann. New York. Acad. Sci., *641*, 46–55.
Mølhave L. (1998): Principles for evaluation of health and comfort hazards caused by indoor air pollution. Accepted for Indoor Air.
Mølhave L. and Nielsen G.D. (1992): Interpretation and limitations of the concept "total volatile organic compounds" (TVOC) as an indicator of human responses to exposures of volatile organic compounds (VOCs) in indoor air. Indoor Air, *2*, 65–77.
Mølhave L., Bach B. and Pedersen O.F. (1986): Human reactions to low concentrations of volatile organic compounds. Environ. Internat., *12*, 167–175.
Mølhave L., Jensen J.G. and Larsen S. (1991): Subjective reactions to volatile organic compounds as air pollutants. Atmos. Environ., *25a*, 1283–1293.
Mølhave L., Liu Z., Hempel-Jørgensen A., Pedersen O.F. and Kjærgaard S.K. (1993): Sensory and physiological effects on humans of combined exposures to air temperatures and volatile organic compounds. Indoor Air, *3*, 155–169.
Mølhave L., Clausen G., Berglund B., DeCeaurriz J., Kettrup A., Lindvall T., Maroni M, Pickering A.C., Risse U., Rothweiler H., Seifert B. and Younes M. (1997): Total volatile organic compounds (TVOC) in indoor air quality investigations. Indoor Air, *7*, 225–240.
Otson R. and Fellin P. (1992): Volatile organics in indoor environment: sources and occurrence. In: Nriagu J.O. (ed) Gaseous pollutants, characterization and cycling. Wiley, New York, 335–421.
Otto D.A., Mølhave L., Rose G., Hudnell H.K., and House D. (1990a): Neurobehavioral and sensory irritant effects of controlled exposure to a complex mixture of volatile organic compounds. Neurotoxicol. Teratol., *12*, 649–652.
Otto D.A., Hudnell H.K., Goldstein G. and O'Neil J. (1990b): Indoor air–Health. Neurotoxic effects of controlled exposure to a complex mixture of volatile organic compounds. Report EPA/600/1–90/001. Research Triangle Park, NC 27711, HERL, U.S. Environmental Protection Agency.
Otto D.A., Hudnell H.K., House D.E., Mølhave L. and Counts W. (1992): Effects of exposure to a volatile organic mixture. I: Behavioral assessment. Arch. Environ. Health, *47*, 23–30.
Rothweiler H., Wäger P.A. and Schlatter C. (1992): Volatile organic compounds and some very volatile organic compounds in new and recently renovated buildings in Switzerland. Atmos. Environ., *26A*, 2219–2225.
Rylander R. and Jacobs R.R. (1994): Organic dusts, exposure, effects, and prevention. C.R.C. Press, London.
Seifert B. (1990): Regulating indoor air. In: Walkinshaw D.S. (ed) Indoor Air '90, Proceedings of the 5th International Conference on Indoor Air Quality and Climate, Toronto, Canada, July 29–August 3, Vol. 5, 35–49.
Wallace L., Pellizzari E.D. and Wendel C. (1991): Total volatile organic concentrations in 2700 personal, indoor and outdoor air samples collected in the US EPA Team Study. Indoor Air, *4*, 465–477.
WHO (1986): Indoor Air Quality Research: report on a WHO meeting. Stockholm 1984, EURO-reports and Studies 103. WHO, Copenhagen, Denmark, 1–64.
WHO (World Health Organisation) (1989): Indoor air quality: organic pollutants. Euro Reports and Studies No. 111. World Health Organisation, Regional Office for Europe, Copenhagen.

Subject Index